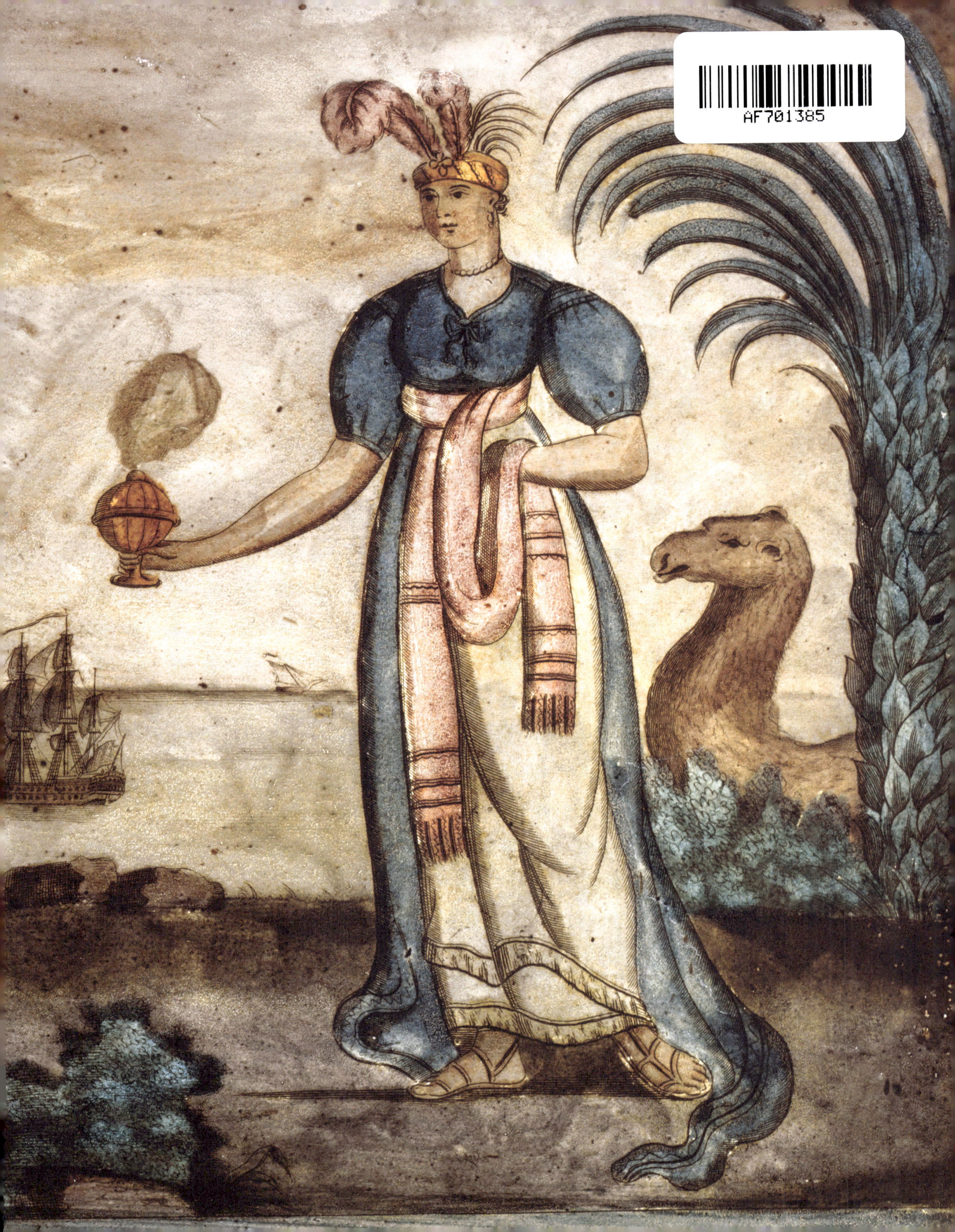

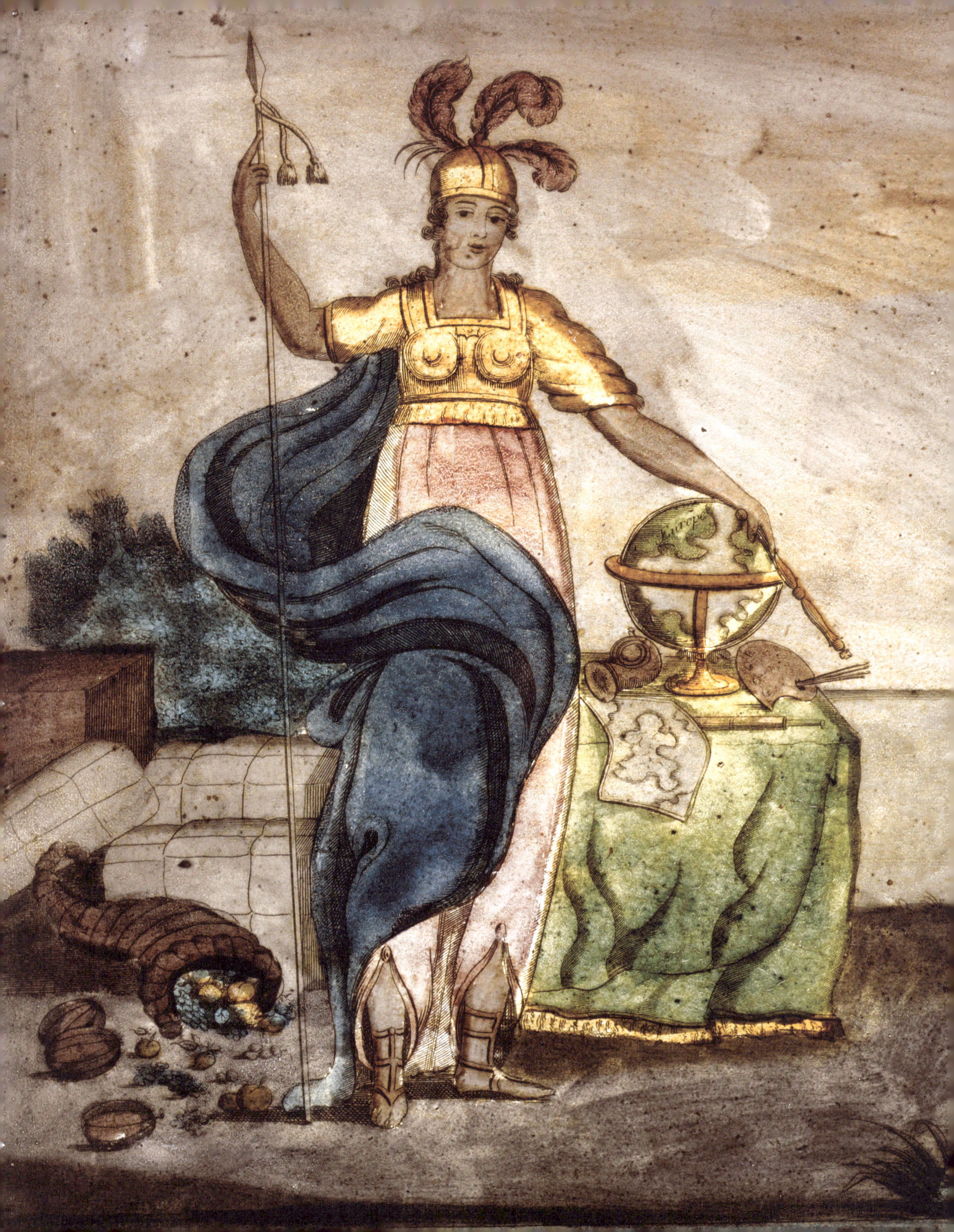
Europe

Ceramics in America

CERAMICS IN AMERICA 2016

Edited by Robert Hunter

Published by the CHIPSTONE FOUNDATION

Distributed by University Press of New England

Hanover and London

Cover Detail, Russell Aitken, *Wyoming Europa*, 1939. Glazed earthenware. H. 12". (Private collection; photo, Randl Bye.)

Endpapers D. Gerolina, (*front*) *Europe and Asia* and (*back*) *America and Africa*, England, ca. 1807. Reverse painting on glass. 14" x 10". (Private collection.)

Design and production: Wynne Patterson, VT
Copyediting: Mary Gladue, CT
Proofreading: Fronia Simpson, VT
Permissions: Angelika Kuettner, VA

Published by the Chipstone Foundation
Distributed by University Press of New England
1 Court Street
Lebanon, New Hampshire 03766

Printed in China
ISSN 1533-7154
ISBN 978-0-9827722-8-7

Contents

Editorial Statement

Ceramics in America is an interdisciplinary journal intended for collectors, historical archaeologists, curators, decorative arts students, social historians, and studio potters. We seek articles on the broad role of historical ceramics in the American context, including essays on ceramic history, archaeological research, technology, social history, studio pottery, and collecting. Further information about this process can be found at the Chipstone Foundation's website, www.chipstone.org. Authors are encouraged to contact the editor before submitting materials.

The Chipstone Foundation offers significant honoraria for full-length manuscripts accepted for publication, and reimburses authors for photography that is approved in writing by the editor.

If an article is accepted for publication, the author should submit photocopies of the title page and copyright page for all cited published sources. Authors are encouraged to provide digital files for all images, along with permissions to reprint. Low-resolution digital images or photographs of poor quality will not be accepted for publication.

Questions and comments from readers and authors are welcome.

Robert Hunter
Editor, Ceramics in America
P.O. Box 401
Yorktown, Va. 23690

Robert Hunter

Introduction

▼ EUREKA! The word was first shouted by the Greek scientist Archimedes when, while in his bath, he discovered the relationship between displaced water volume and weight. Although some of my best thinking occurs during shower time, ceramic discoveries are usually made in the less hygienic environments of grimy flea markets, dusty archives, or in the dirt and mud of archaeological digs.

Yelling "Eureka!" is most often associated with the unearthing of gold by nineteenth-century frontier miners working grubstake claims, but ceramic scholars and collectors are known to use it in the context of new insights or new treasures. The articles for the 2016 volume of *Ceramics in America* are, to varying degrees, the result of just such Eureka! moments. Historian Martha McCartney was scanning through yet another roll of grainy microfilm when the long-hidden inventory of George Thorpe came into view. Thorpe was one of the victims of the uprisings in Virginia in 1622, when the Powhatan Indians decided to rid themselves of the English trespassers. McCartney immediately recognized the importance of the document as the earliest Virginia household inventory yet revealed. She discusses the frontier context of Thorpe's belongings in her article "George Thorpe's Inventory of 1624: Virginia's Earliest Known Appraisal."

Accompanying McCartney's text is a beautifully illustrated photo essay by Bly Straube titled "Ceramics in Early Virginia." Straube has made a career out of curating exciting ceramic finds from this country's earliest English settlements. In this combined presentation, she shares that knowledge, bringing three-dimensional life to the George Thorpe inventory while also reminding us that inventory data are often biased, particularly when it comes to household breakables.

Researchers of all sorts have a fundamental understanding that historical documents on paper offer only a partial glimpse into our past. We sometimes forget that fired clay has long been a medium on which to record the written word. Richard Miller uses writings found on ceramics to explore the origins and meanings of a well-known but poorly understood type of American nineteenth-century slipware. In his broad-ranging article "Norwalk (Connecticut) Slip-script Pottery, the Potters, and Related Ware," Miller deciphers the literal and hidden meanings of the slip-decorated slipware to offer a provocative glimpse into the relationships between New England potters and consumers.

Private collections are among the best sources for finding overlooked nuggets of information in the ceramics field, but these closely guarded spaces often require special access, lacking that, we usually have to wait for estate sales or gifts to museum repositories. One such recent gift, to the Cleveland Museum of Art, led scholar Tom Folk to a rich vein of research. Ceramic artists throughout history have always found inspiration across time and space. In his article "The Allegory of Europa in Twentieth-Century American Sculpture," Folk contextualizes a group of enchanting ceramic and bronze objects that invokes this theme. Of particular note is ceramic artist Russell Aitken's Americanized version of the ancient allegory, which serves as this year's cover.

Sometimes new discoveries are the result of a change in perspective. Isaac Asimov said that "the most exciting phrase to hear in science, the one that heralds new discoveries, is not 'Eureka!' but 'That's funny. . . .'" Revisionist thinking is a big part of the literature on contemporary crafts, whereby the latest theory is adjudicated against the canon of makers and their products. Ezra Shales takes such an approach in his article "Throwing the Potter's Wheel (and Women) Back into Modernism: Reconsidering Edith Heath, Karen Karnes, and Toshiko Takaezu as Canonical Figures." He casts a fresh critical eye on three women's careers in art, craft, and design in the mid-twentieth century, connected by their use of one of the world's oldest ceramic tools—the potter's wheel.

Deciphering the physical history of repairs, alterations, and outright fakes is a basic element of ceramic connoisseurship. We are fortunate this year to have several interrelated submissions that explore these aspects. A fascinating tale of deceit and fakery is told by Ellen Archie, Ronald W. Fuchs II, Jennifer Mass, and Erich Uffelman in their study "'The most dangerous imitations': A Group of Spurious Chinese Export Porcelain Decorated with Fame and the American Eagle." Buyer beware!

Not all alterations to ceramics are made with an intent to deceive. In some cases, honest wear can alter the surface of an object, hiding its embellishments. In their article "A Chinese Export Porcelain Mystery Solved Using Intensive Surface Analysis," Shirley M. Mueller and Matthew Bunney utilized high-powered optical analysis to reveal the original armorial decoration on an important early-eighteenth-century Chinese porcelain service.

Archaeology and personal experience show us that ceramic objects are among the most fragile of human artifacts. Breakage is nearly inevitable and cuts across all social and economic boundaries. Even the rich and famous have experienced the tragedy of a dropped tray of porcelain teaware. Fortunately, such pieces are among the most easily repaired. Angelika R. Kuettner's "Simply Riveting: Broken and Mended Ceramics" provides some fascinating historical examples of the various methods used in America to repair household china.

A surprising archaeological case study of home-repaired ceramics relating to the Fredericksburg Ferry Farm of George Washington's mother is presented by Mara Kaktins, Melanie Marquis, Ruth Ann Armitage, and Daniel Fraser in "Mary Washington's Mended Ceramics: A Study of Eighteenth-Century Glues." After they detected residues on previously mended creamware fragments, the authors undertook an exploration of eighteenth-century glue recipes that might have been available to the widow Mary's household.

A fortuitous flea market find led George L. Miller and Emily Brown to a twentieth-century repair firm in Philadelphia that used the time-honored technique of riveting to fix cracked and broken ceramic objects. The authors' research, presented in "A Harry A. Eberhardt–Repaired Chinese Porcelain Saucer," shows that Henry Francis du Pont, one of America's greatest china collectors, sought out Eberhardt and his riveting technique to repair damaged ceramics in the Winterthur Museum collection.

The development and application of scientific techniques have the greatest potential for uncovering new information for students of historical ceramics. For the past decade, geologist J. Victor Owen has been applying methodologies, initially created for geologic research, to what he calls "artificial rocks"—the soft-paste porcelains of eighteenth-century British and American manufacturers. His work has yielded some excellent physical markers for identifying specific porcelain bodies and their firing characteristics along with a mountain of data. In "Statistical Evaluation of Analytical Data for Eighteenth-Century American and British Sulphurous Phosphatic Porcelains," Owen and coauthors John D. Greenough and Nick Panes began to look at data using multidimensional scaling techniques that offer further refinement of Owen's porcelain-divining capabilities.

During my active days as an archaeologist, I knew that the next turn of the shovel or trowel might uncover a once-in-a-lifetime discovery. Since taking on the editorship of *Ceramics in America* in 2001, I have been sorting through the boxes and cabinets of artifact collections with the same anticipation, that the next drawer might contain "the one." During a recent visit to the laboratory of the Commonwealth Heritage Group to look at eighteenth-century slipware being excavated in Philadelphia's Old City, I had one of those lifetime encounters with the American ceramic equivalent of the Holy Grail.

The search for physical evidence of American hard-paste or true porcelain manufacture has eluded researchers since the twentieth-century publication of circa 1743 claims by Andrew Duché. Duché boasted that he had found "the true material and manner of making porcelain," although no example of this feat had ever been found—until now. In "An Eighteenth-Century American True-Porcelain Punch Bowl," my colleague Juliette Gerhardt and I discuss this recently recovered object from the site of the newly constructed Museum of the American Revolution in Philadelphia.

When I first examined the small, white punch bowl sitting among the thousands of other Philadelphia ceramic artifacts, I immediately thought to myself, "Could this be it?" To be certain, I had to send a sample of the body to Victor Owen in his lab at St. Mary's University. After a few nearly sleepless nights, Victor's email arrived four days before Christmas and was filled with his palpable excitement: "I analysed your sherd this morning. It has a silicious-aluminous composition. Two types of ware can have this: (1) true (hard paste) porcelain, and (2) fine earthenware." I am thrilled to share this Eureka! moment in the pages of this journal. For the rest of the story, the results of the analysis are fully presented by Victor, Joe Petrus, and Xiang Yang in "Bonnin and Morris Revisited: The Geochemistry of a True Porcelain Punch Bowl Excavated in Philadelphia."

A materials science approach to sorting out relatively anonymous American ceramics is becoming increasingly common. An example that makes clear the utility of clay analysis is presented in "A Comparative Scientific Study of James Morgan and the Kemple Family Stoneware" by authors Johanna R. Bernstein, Arthur F. Goldberg, and Jennifer Mass. It is important to acknowledge that their groundbreaking stoneware study was submitted in 2009 and is finally coming to publication here.

Publishing new discoveries is an important aspect of ceramics scholarship and the advancement of knowledge in the field, although the shelf life of articles, books, and blogs can vary immensely depending on subsequent finds. In the 2014 issue of *Ceramics in America*, Noël Hume published his analysis of an English pearlware puzzle jug ostensibly made for John Bloome, owner of the ship *Hopewell*. We thought we had the final word on its interpretation, but in just two years another commemorative ceramic, a pearlware pitcher celebrating Lord Admiral Duncan's victory over a Dutch fleet in 1797, led Noël to revise his thinking. When encountering evidence that conflicts with their theories, some scholars react with "Oh damn" rather than "Eureka!", but in his article "A New Bloome," Noël turns it into a triumph of wit, research, and substance for ceramic historians.

At first glance, the remarkable pair of mid-nineteenth-century porcelain water pitchers with painted ships and gilt-inscribed names seemed destined to reside among the innumerable memorialized ceramics that never fully tell their story. Much to my surprise, when Oliver Mueller-Heubach began his document search for Baltimore steamship captain George Russell, up popped a local newspaper account of the pitchers' presentation, down to the day, hour, location, and names of the presenting parties. I wasn't around to record Oliver's "Eureka!" or perhaps "thar she blows" moment, but the excitement of his discovery is still being celebrated. Even more of the story was to surface and is recounted in "The Captain George W. Russell Presentation Pitchers," a remarkable tale of mid-nineteenth-century dealers.

Apparently print-based scholarship is alive and well in this digital age. Amy Earls has compiled and edited six reviews of recently published ceramics scholarship, which conclude this issue of the journal. The reviews suggest that a high standard for ceramic-based publications will continue for the foreseeable future in the study of ceramics in the American context.

Online digital databases for archaeological collections, still appearing at a frustratingly slow pace, will one day make comparative research between various American sites fruitful and commonplace. In the meanwhile, for the 2017 volume of *Ceramics in America*, I asked archaeologists from fifteen American historical cities or towns—from Boston to St. Augustine to New Orleans—to select and discuss their "top ten" ceramic artifacts from their localities. It is my hope that this exercise will be an enlightening and entertaining way to expand our knowledge about the history of ceramics use in America and to pave the way for future comparative studies.

Ceramics in America

Figure 1 Henricus Hondius's *Nova Virginiae Tabula*, map of the Virginia colony and the Chesapeake Bay, 1630. 15½" x 19½". (Geographicus Rare Antique Maps Cooperation Project.)

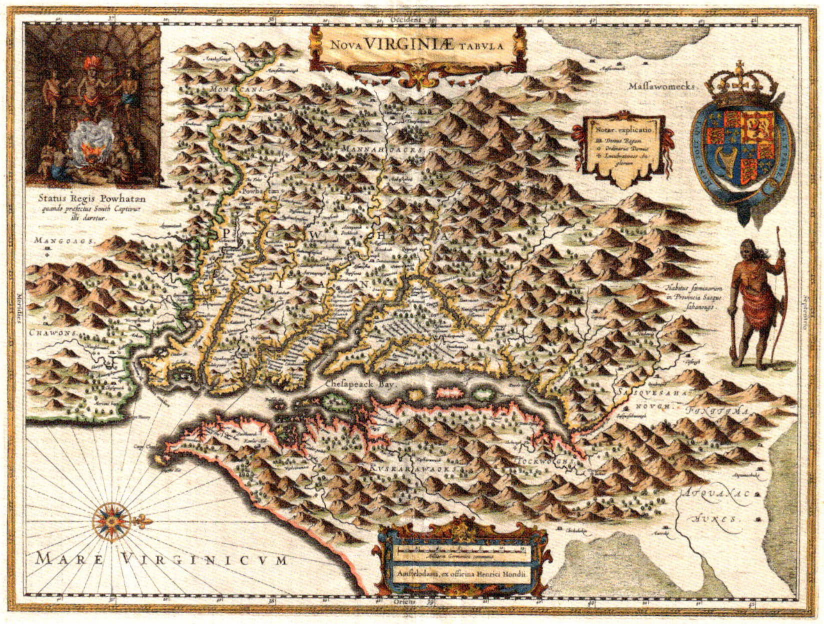

Figure 2 Detail of the map illustrated in fig. 1 showing the location of Berkeley Hundred. On March 22, 1622, Opechancanough, the paramount chief of the Powhatan Chiefdom, commenced a series of deadly surprise attacks on English settlements and plantations along the James River. More than three hundred colonists were slain, nearly a third of the young colony's population. George Thorpe, an investor in the Virginia Company of London and a resident at Berkeley Hundred, was among the dead.

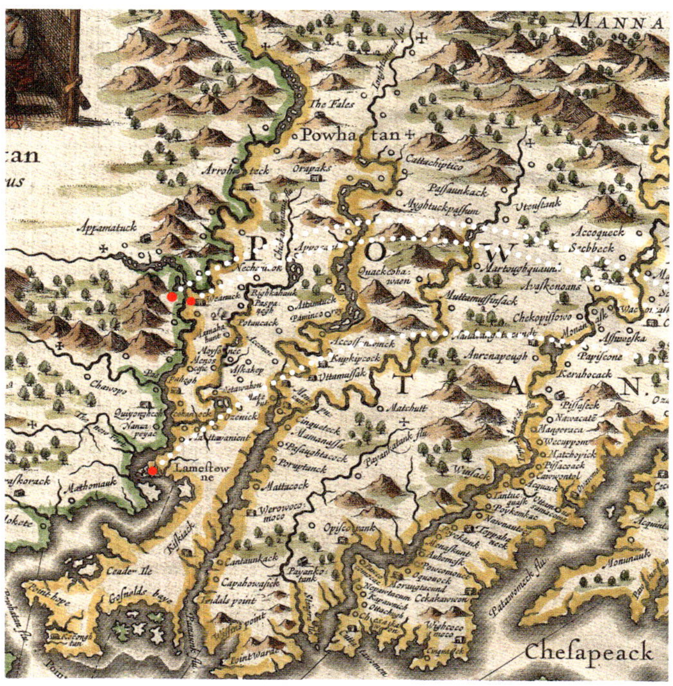

Jordan's Journey
Berkeley Hundred
Jamestown

Martha W. McCartney

George Thorpe's Inventory of 1624: Virginia's Earliest Known Appraisal

Beverly Straube

Ceramics in Early Virginia (photo essay)

EDITOR'S NOTE: In this article, Martha McCartney presents the historical context for the earliest household inventory yet discovered for Virginia. Peter Wilson Coldham, a British scholar whose knowledge of Virginia records in overseas archives was unparalleled, surmised that inventories of early colonists' estates were discarded after the probate process had been completed. This surviving document is thus rare indeed, and promises to provide many exciting insights into the material world of Virginia's early colonists. Historians, curators, and archaeologists have long used inventories to better understand a given household in terms of economic and social status. However, those in these same disciplines are also aware of the limitations and biases inherent in inventory data. In particular, archaeologists have learned that ceramic listings, if present at all, are often ambiguous and usually incomplete. In an attempt to offer a corrective, Beverly "Bly" Straube has contributed the illustrations for this article, with captions that explain some types of early-seventeenth-century ceramics that have been archaeologically recovered from other Virginia household sites similar in composition and economic standing. It is hoped that this combined presentation will underscore the rich material culture of what is often considered a frontier outpost in the New World.

▼ GEORGE THORPE, an investor in the Virginia Company of London and a member of the Society of Berkeley Hundred, died at Berkeley Hundred on March 22, 1622, at the hands of attacking Indians. An inventory of his possessions, compiled two years after his death, is the earliest known account of a Virginia colonist's estate. While few ceramics are listed, the inventory reflects the material world of an English gentleman in Virginia.

George Thorpe was the son of Nicholas Thorpe, Esquire, of Wanswell Court in Gloucestershire, and Mary Wilkes, his wife. Both families were prominent in England's West Country. On February 20, 1598, when George was admitted to the Middle Temple, one of the Inns of Court, he was described as a gentleman "Late of Staple Inn." When Nicholas Thorpe made his will in early 1600, he named twenty-four-year-old George as executor and residual heir but made generous bequests to numerous members of his extended family. These legacies depleted his assets but George was able to retain Wanswell Court and a portion of his patrimony.[1]

On July 11, 1600, George Thorpe married Sir Thomas Porter's daughter, Margaret, in the Church of St Martin-in-the-Fields, London. His in-laws

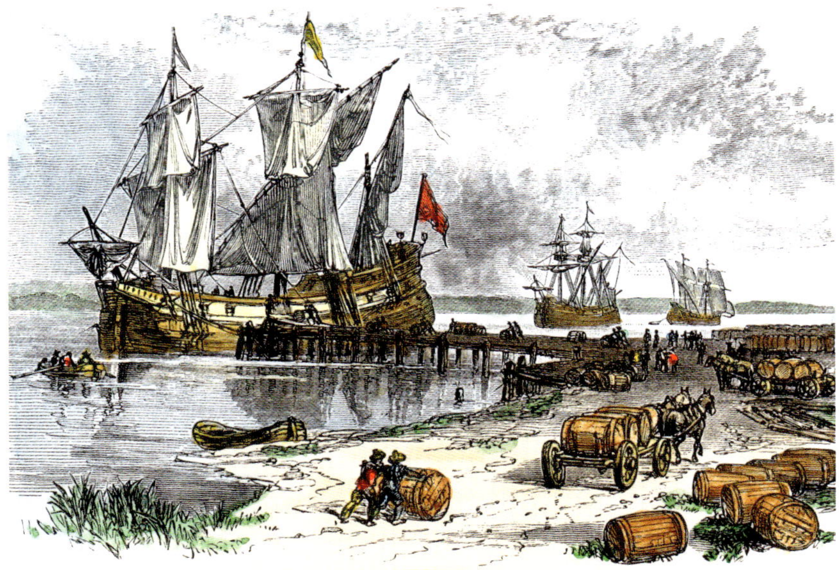

Figure 3 Anonymous, *Sailing Ships Loading Tobacco at Jamestown in Virginia Colony, 1600s.* Hand-colored woodcut of a nineteenth-century illustration, ca. 1880. (© North Wind Picture Archives / Alamy Stock Photo.) Archaeological research of the English settlements in Virginia—from the 1607 James Fort to the dispersed settlements that characterized the later seventeenth century—has revealed that individuals with the means to do so imported high-quality European and Asian ceramics to mediate the practical, social, and cultural aspects of their lives. The assemblages, largely transshipped from London and Plymouth, England, include wares that are uncommon in English households. Even the English coarse earthenwares, drawn chiefly from the potteries of London and Surrey-Hampshire, comprise many specialized shapes that are rare in contexts of the mother country.

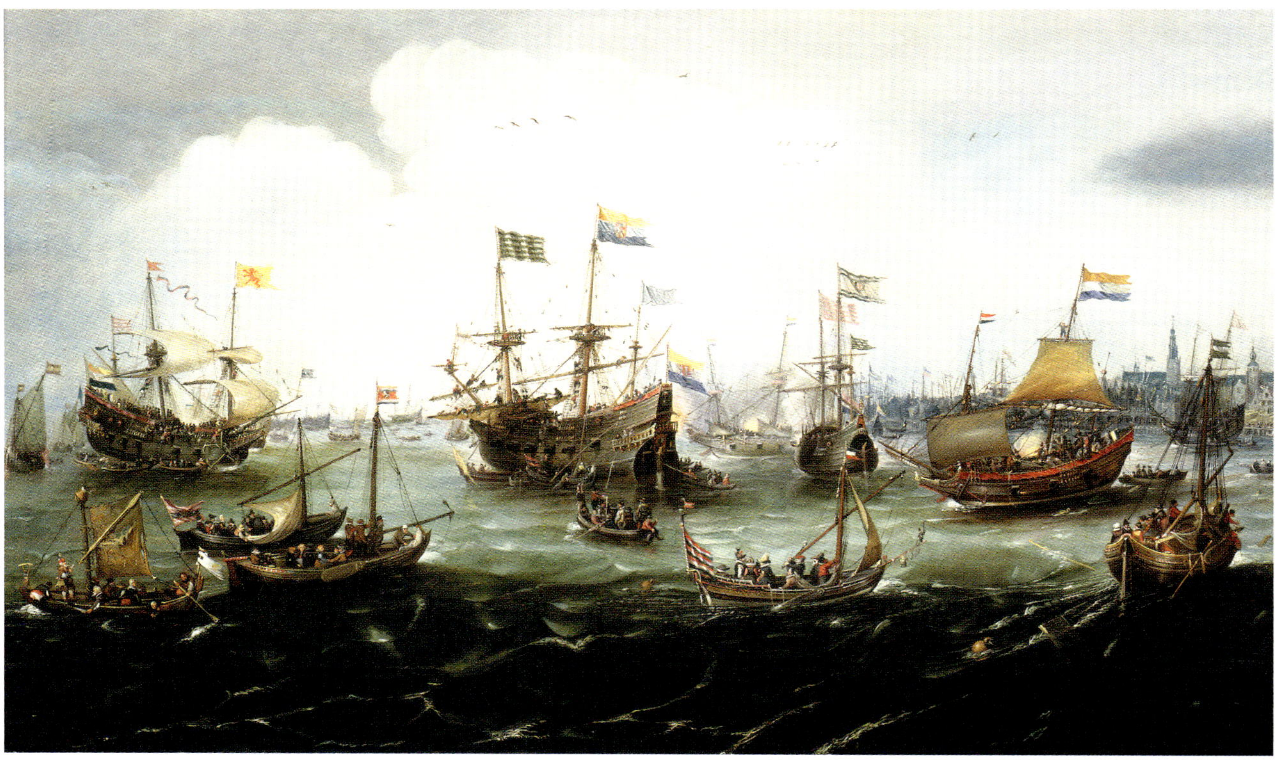

Figure 4 Andries van Eertvelt, *The Return to Amsterdam of the Second Expedition to the East Indies on 19 July 1599*, 1610–1620. Oil on panel, 22" x 36". (© National Maritime Museum, Greenwich, London.) Dutch traders made significant inroads to the Chesapeake area in the first quarter of the seventeenth century by establishing close mercantile relationships with the English settlers in Virginia and Maryland. Colonists encouraged this trade, which was opposed by London merchants and the English government, because they could exchange tobacco for household and luxury goods at a much better rate than that provided by the mother country. This was the golden age of Dutch maritime, and commercial activities and merchants (some of them English factors based in the Netherlands) were engaged in the extensive movement of goods around the world. Many of the Chinese, Dutch, German, Italian, and Portuguese wares found on early Virginia sites can probably be attributed to Dutch traders, some of whom established trading posts in the colony.

included at least four titled noblemen who went on to become major investors in the Virginia Company, chartered by James I in 1606. These influential connections paid off in time, and George received his first court appointment, that of Gentleman Pensioner, a post he held from 1606 through 1614. As a Gentleman of the King's Privy Chamber, his official duties placed him in London, where he had access to the monarch and other members of the royal household. He also had to be available for military duty, and in 1608 he and seven of his servants were among the "able and sufficient" Gloucestershire men considered "fit for His Majesty's Service in the Wars." In 1609 George Thorpe obtained a fifty-one-year lease for some fishing grounds along the Severn River, where his forebears had held fishing rights for generations. Arnold Oldisworth, a fellow leaseholder and one of George's kinsmen, joined him in a costly and futile lawsuit stemming from the meandering river's impact on riparian rights.[2]

Margaret Thorpe died in March 1610.[3] When George remarried, in February 1611, he wed another Margaret, a well-connected Londoner who was the daughter of David Harris of Bristol and the niece of Sir John Bennett. Together, the Thorpes produced five children, at least three of whom survived infancy. George was elected to the House of Commons in 1614, representing Portsmouth, and took part in the so-called Addled Parliament, whose members debated the king's right to levy taxes on commodities without their consent. On June 29, 1615, George was among the adventurers King James I authorized to establish a plantation in the Somers Isles, also known as the Bermudas. Five years later George invested in the Virginia Company of London and was named to the company's council.[4]

The Society of Berkeley Hundred

In 1618 George Thorpe and his cousin Sir William Throgmorton joined Richard Berkeley and John Smyth of North Nibley, two longtime investors in the Virginia Company, in forming a partnership they called the Society of Berkeley Hundred.[5] Later, Sir George Yeardley joined in. The Society's purpose was to establish a private plantation in Virginia, a settlement they planned to call Berkeley Hundred. They received a charter from the Virginia Company on February 3, 1619, and two weeks later, when Thorpe sent a letter to Yeardley who was en route to Virginia, he enclosed a copy of that document, noting that it had been written by his own "Virginian boy," that is, an Indian youth.[6] In April 1619 Virginia Company officials informed newly arrived Governor George Yeardley that they were planning to send George Thorpe to the colony to serve as deputy for the so-called College land, a settlement that encompassed ten thousand acres near the head of the James River. Thorpe's contract outlined his administrative duties and stated that he was to be compensated in land and supplied with servants or tenants.[7]

Planting a Settlement

During the spring and summer of 1619, the Society of Berkeley Hundred's investors discussed plans to procure the fifty or so servants and tenants

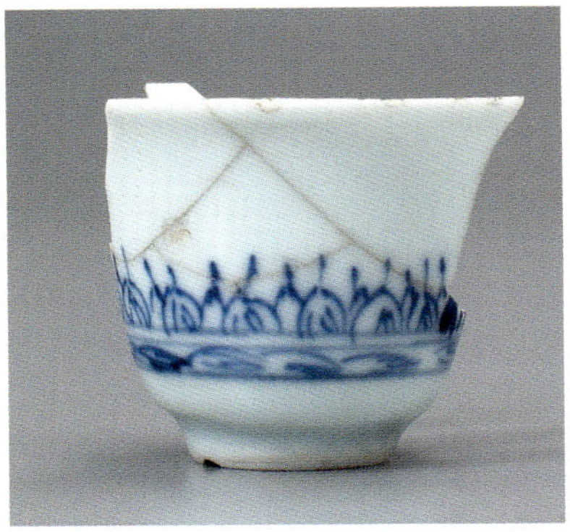

Figure 5 Wine cup, Jingdezhen, China, 1600–1640. Hard-paste porcelain. H. 1½". (Courtesy, Virginia Department of Historic Resources; photo, Robert Hunter.) More than a dozen of these diminutive cups with flame-frieze decoration have been reported from early-seventeenth-century Virginia archaeological sites. These fine cups were once thought to be Imperial ware, reserved for use by the Chinese elite. Now they are recognized as status vessels made for export. They have been found on a number of Dutch shipwrecks (such as the *Witte Leeuw*, 1613, and the *Hatcher*, 1643) and terrestrial sites dating to the first half of the seventeenth century. This example was recovered from the Walter Aston Site, Charles City County, Virginia, in a 1630–1650 context.

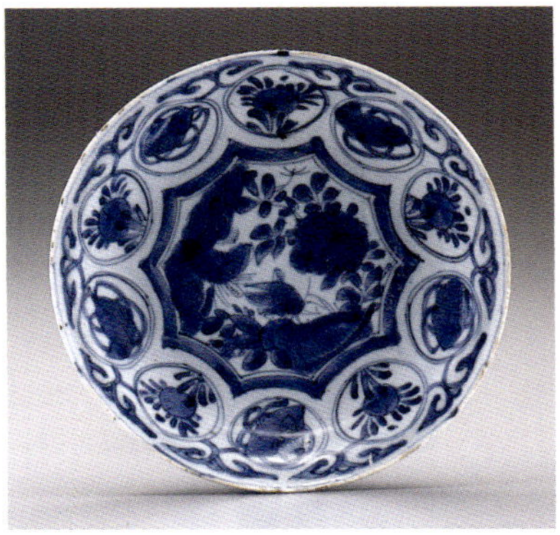

Figure 6 Dish, China, 1613. Hard-paste porcelain. D. 8⅛". (Private collection; photo, Gavin Ashworth.) This Chinese porcelain dish was recovered from the site of the *Witte Leeuw* shipwreck of 1613. Cargoes of porcelain wares were being received in Europe on Portuguese ships from the turn of the fifteenth century, and then by the Dutch trade in the late sixteenth century. Although not common and found only in households with high social status, examples have been found archaeologically on a number of early Virginia sites and are noted in probate inventories such as the "6 litle pursline [porcelain] dishes" in Thorpe's inventory; see Appendix, p. 31.

they needed to establish their plantation. They also began accumulating the provisions and other supplies their settlers would require once they reached Virginia. In July 1619 George Thorpe informed his fellow investors that he had received the weaponry that would be sent to Berkeley Hundred. As the summer drew to a close, so did preparations for the Society's Virginia adventure. Edward Williams, a Bristol merchant, signed a contract on August 18, 1619, agreeing to allow the Society of Berkeley Hundred to use the *Margaret*, a ship of 45-tons burthen. He promised to see that his vessel was in good repair and sufficiently victualed and he agreed to provide a crew of nine: a master plus seven other men and a boy. The Society's investors, on the other hand, stipulated that Henry Penry was to be the ship's master and that Toby Felgate was to serve as pilot. The *Margaret* was scheduled to set sail on September 15, 1619, weather permitting. Captain John Woodlief, the Buckingham merchant commissioned to serve as governor of the Berkeley Hundred settlement, agreed that the ship would stay in Virginia no more than fifty days. At the end of August, the Society of Berkeley Hundred's investors provided Captain Woodlief with some very specific instructions on how the settlement was to be organized.[8]

A list of the cargo the Society's investors put aboard the *Margaret* reveals that their venture was carefully planned and costly. George Thorpe supplied some items that would be critical to the settlers' defense: twenty-four muskets and two lighter firearms known as calivers; three barrels of gunpowder and matches; lead for forming into bullets; pike heads; three corselets or torso-covering body armor; and sixteen swords, sword belts, and bandoleers (belts strung with containers for gunpowder). He sent along literally thousands of large and small glass beads and some copper, in anticipation of trading with the Indians, and he furnished substantial quantities of seeds that could be used to plant gardens for sustenance and medicinal purposes. He also had a large quantity of buttons packed into a barrel and readied for shipment. Thorpe, reputedly a religious man, made provisions for the colonists' spiritual well-being by supplying two church Bibles, two copies of the *Book of Common Prayer*, and volumes called *The Practice of Piety* and *The Playne Man's Path Way*.

Staple foods, such as cheese, oatmeal, peas, meal, spices, and butter, were carted to the *Margaret*. Also loaded aboard was an abundance of treenware—wooden platters, dishes, trenchers, bowls, and spoons for dining. Culinary equipment included a dozen skimming dishes and saucers, a ladle, spits for roasting meat, kettles, pots and pothooks, skimmers, and a deep wooden bowl in which mustard could be prepared. Horn cups and metal cans (cylindrical drinking vessels that came in various sizes) were provided for the consumption of beverages. Common household items such as lanterns, candlesticks, andirons, pot-hangers, shovels and tongs, soap, dowlas sheets and bolsters, locks, and padlocks were put aboard. Most of the items loaded onto the *Margaret* were packed in chests and barrels. Proper decorum was to be maintained in the settlement whose flag was to be mounted on a tasseled pike. A drum was sent along for drummer George Hale, who would use it to summon the settlers to meals and worship or to sound an alarm.

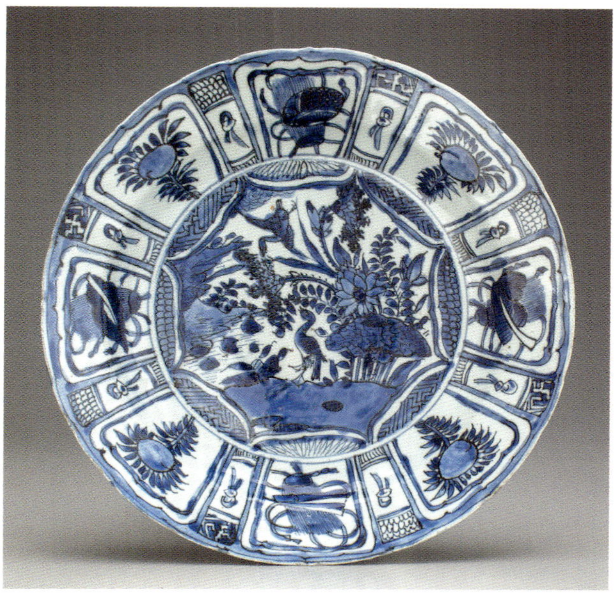

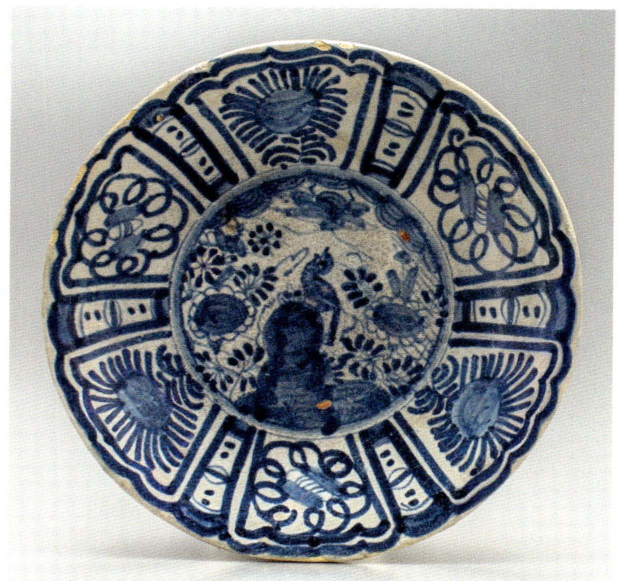

Figure 7 Dish, China, 1600–1620. Hard-paste porcelain. D. 12½". (Private collection; photo, Robert Hunter.) *Kraak* porcelain was China's first porcelain mass-produced for export to Europe. It was considered too coarse for the upper-class Chinese and too fragile for everyday use by the poor. The hand-painted designs, in underglaze blue only, are of landscapes and animals. Human figures are rarely depicted, making this porcelain suitable for Islamic markets as well. Buddhist and Daoist good-luck symbols make up the paneled border decorations.

Figure 8 Dish, Pickleherring potworks, Southwark, London, England, 1625–1630. Tin-glazed earthenware. D. 7⅞". (Private collection; photo, Robert Hunter.) In decorating their wares, early London and Netherlandish tin-glazed potters adopted the popular style from late Ming porcelain of China's Wan Li period (1573–1619). One of the most popular motifs, "bird-on-rock," is shown on the central design of this dish and appears on English tin-glazed wares. Evidence of the early London tin-glazed dishes is relatively scarce in Virginia archaeological contexts of the first half of the seventeenth century.

The Berkeley Hundred colonists were furnished with substantial quantities of readymade clothing, articles such as suits of apparel, shirts, falling bands (large, flat, loosely fitting collars), yarn stockings, caps, handkerchiefs, garters, and two hundred pairs of shoes. They also were supplied with

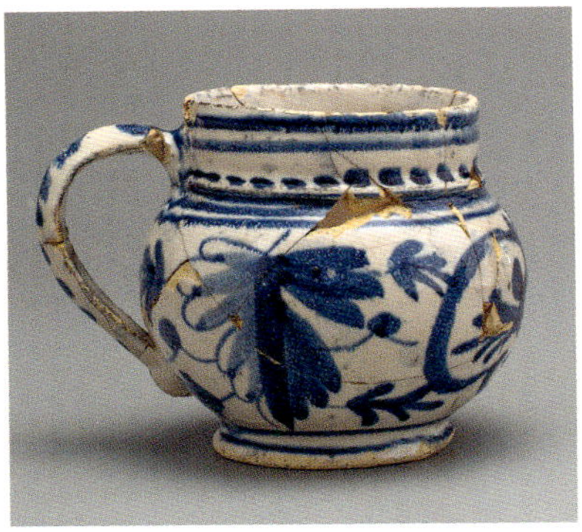

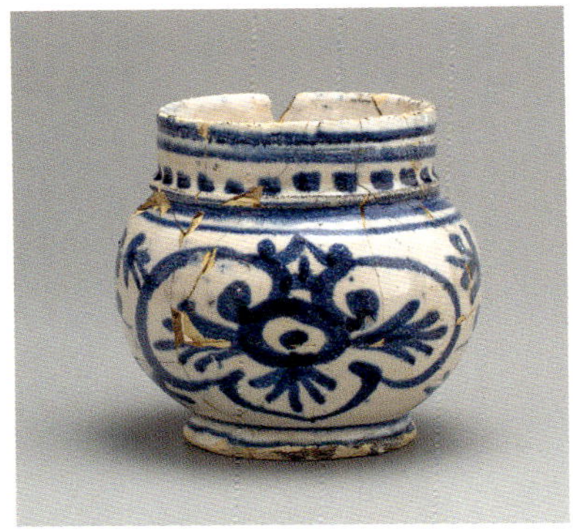

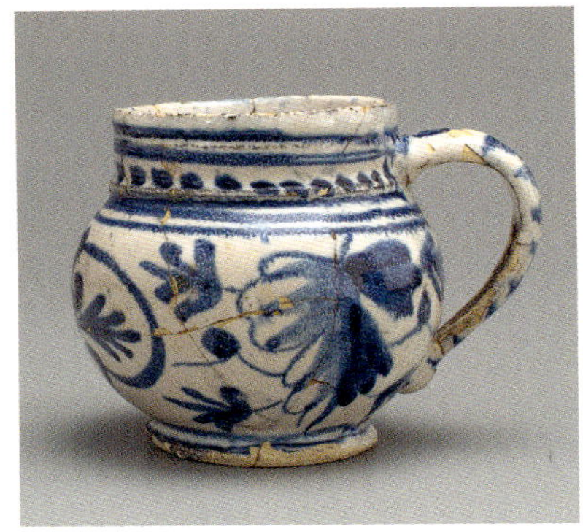

Figure 9 Mug, Lisbon, Portugal, 1635–1660. Tin-glazed earthenware. H 3⅜". (Courtesy, Virginia Department of Historic Resources; photo, Robert Hunter.) Portuguese tin-glazed wares are frequently found in early Virginia archaeological contexts. A parallel vessel to the mug shown here was found in York County, Virginia, and is associated with a high-status household that had dealings with merchants in the Netherlands. The Dutch trade may account for the presence of these mugs in Virginia, although parallels have also been found in Devon, England, an area of increasing commercial activity in the colony after the Virginia Company of London lost its charters for settlement in 1624. This example was excavated at Jordan's Journey, Prince George County, Virginia, in a 1630–1650 context.

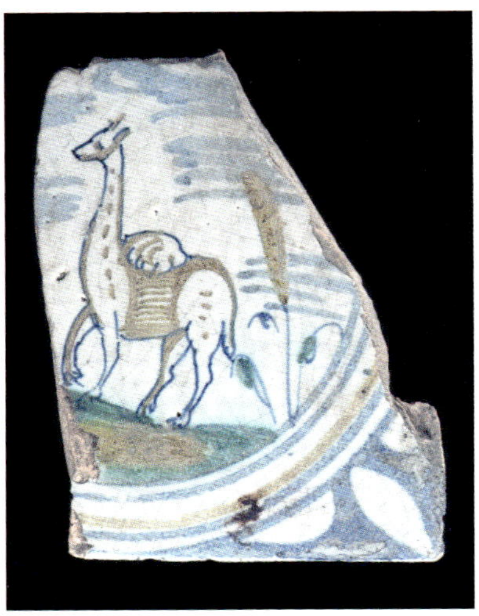

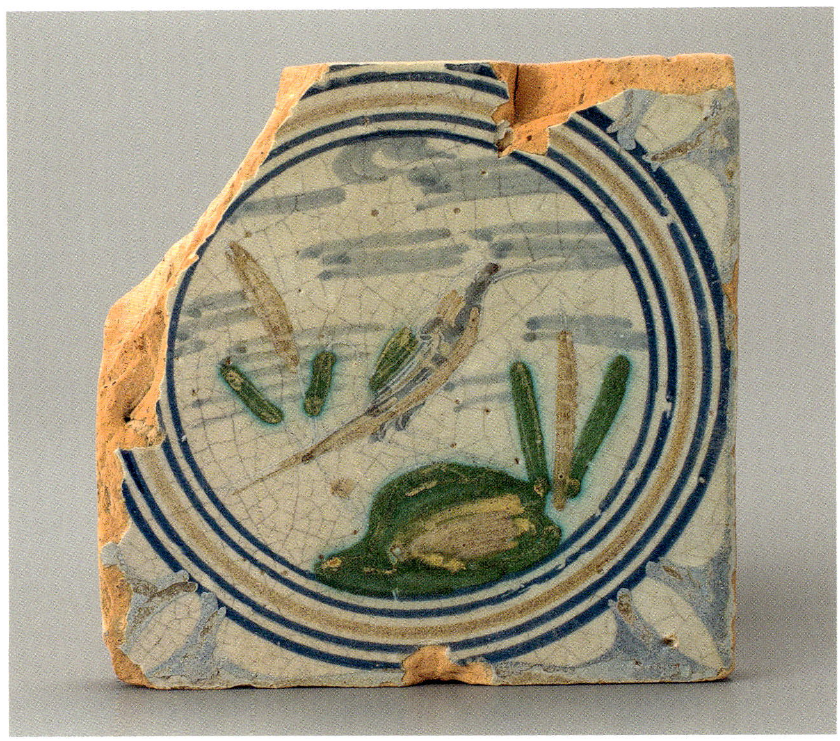

Figure 10 Floor tiles, Pickleherring potworks, Southwark, London, England, 1618–1650. Tin-glazed earthenware. Both are 5¼" square. (Courtesy, Virginia Department of Historic Resources; photo on right, Robert Hunter.) Excavated at Jordan's Journey, Prince George County, Virginia, 1620–1635 context. These medallion tiles, one depicting a camel and the other a bird, are copies of earlier tiles produced in the Netherlands and reflect the influence of Christian Wilhelm, an immigrant potter from the Low Countries who established the Pickleherring pothouse in London. Decorative floor tiles such as these indicate high-style accommodations.

points (the metal-tipped laces they used to fasten their clothing) and with thread, needles, buttons, and shoe thread, presumably so that they could keep their clothing and footwear in usable condition. Also present were substantial quantities of dowlas for making sheets and bolsters, plus frieze (a coarse woolen), silk taffeta, probably for use in making clothing for the settlement's leaders. Buckram, a coarse fabric used as a stiffener in clothing and for wrapping merchandise prior to storage or travel, was also included. Tools for clearing land, planting crops, building houses, and making furniture were provided to the settlers, along with literally thousands of nails and nets and hooks for catching fish. Specialized implements used in surveying, blacksmithing, coopering, carpentry, tailoring, and joinery were furnished to the skilled workers the Society was sending to the colony. Quires of writing paper were dispatched to Berkeley Hundred's sergeant, to kitchen clerk Roland Painter and his sons, and presumably to the settlement's leaders. Also present were bread and wine, elements intended for use in Holy Communion. Some fifteen gallons of aqua vitae, five-and-a-half tuns of beer, and cider were loaded aboard the ship, along with water, fish, bacon, suet, bread, and biscuits. Naval supplies that might be needed at sea, such as canvas for repairing sails, caulking material, and a supply of rope, were loaded aboard. All of these items were enumerated in the *Margaret*'s manifest.[9]

The *Margaret* left Kingsrode in Bristol on September 16, 1619, with the Society of Berkeley Hundred's first group of settlers and in accord with George Thorpe's instructions. Ferdinando Yates documented the voyage. After a very difficult crossing, the ship arrived at the mouth of the James River on the evening of November 30 and landed at Old Point Comfort. Among the thirty-six Berkeley Hundred colonists under Captain John Woodlief's command were a number of artisans and skilled workers, some of whom were multitalented. Thomas Coopy, for example, was a carpenter, smith, turner, and fowler, whereas Humphrey Plant was a sawyer and carpenter. Thomas Davis was a cooper and shingle-maker; gun-maker and smith Charles Coyse was a fisherman but also knew how to make pitch and tar. Also included were shoemaker Christopher Nelmes, joiner Richard Godfrey, and others capable of serving as cooks, gardeners, coopers, and smiths. The expertise of surgeon John Singer would have been invaluable while the *Margaret* was at sea and also after the settlers touched land.

The Society of Berkeley Hundred made contractual agreements with their tenants, all of whom were considered free men, and promised an allotment of land in exchange for a certain number of years' work. For instance, husbandman Robert Coopy was to receive thirty acres of land in exchange for three years of work and a modest annual rent. On December 4, Governor George Yeardley had Secretary of the Colony John Pory record the names of the men aboard the *Margaret* when it reached Jamestown. Shortly thereafter, he sent them to the land he had chosen for them, a location on the upper side of the James River, in the immediate vicinity of the plantation still known as Berkeley. According to the Society of Berkeley Hundred's instructions, when Captain John Woodlief and the first contingent

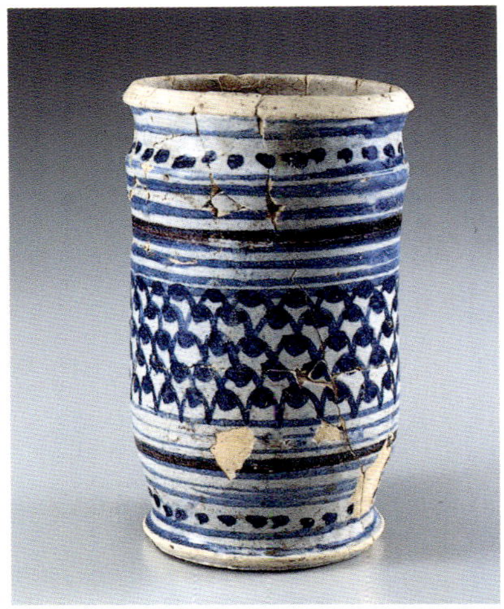

Figure 11 Apothecary jar, Aldgate, London or Netherlands, 1590–1610. Tin-glazed earthenware. H. 7¼". (Courtesy, Preservation Virginia; photo, Gavin Ashworth.) Jars like this were produced in the Netherlands and were among the first tin-glazed earthenwares produced in England. The form developed in the Middle East as a container for medicinal substances, which also included spices and herbs that we may consider homeopathic remedies today. Apothecary jars are among the most common ceramic forms found on early-seventeenth-century archaeological sites in Virginia. This example was excavated from the James Fort in Jamestown, Virginia, 1607–1610.

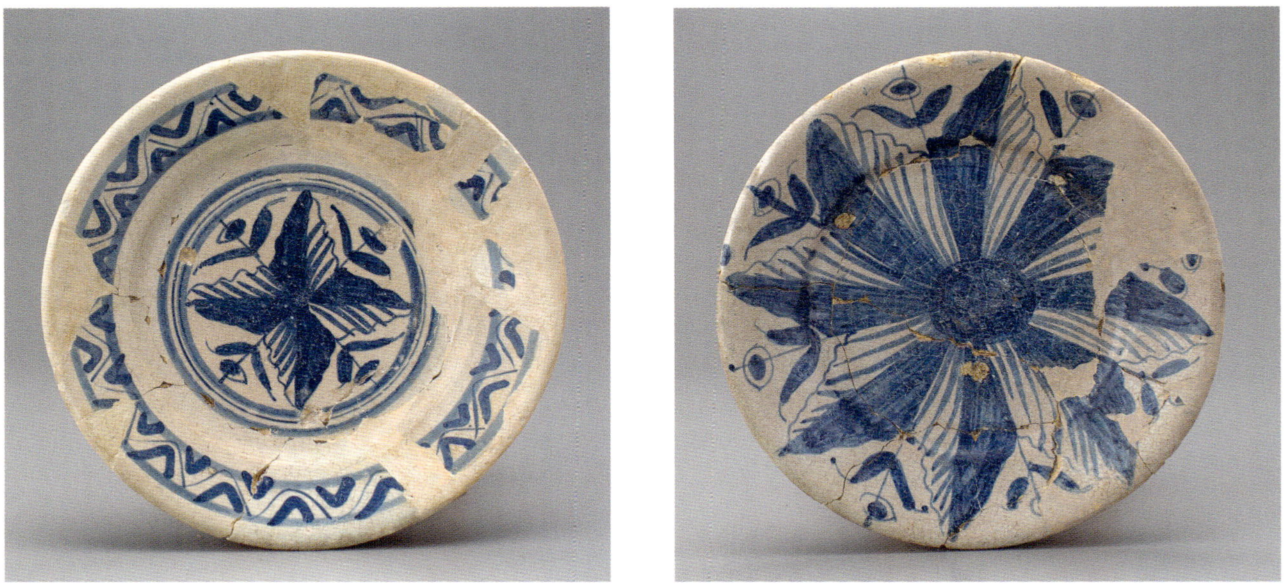

Figure 12 Dishes, Northern Netherlands, 1620–1640. Tin-glazed earthenware with lead glaze on back. D. (*left*) 7¾"; D. (*right*) 7⁹⁄₁₆". (Courtesy, Virginia Department of Historic Resources; photo, Robert Hunter). In the mid-sixteenth century, Italian potters introduced tin-glazed earthenware to the Northern Netherlands and early designs were borrowed from Italian maiolica. By the beginning of the seventeenth century, a distinctive Dutch style had developed in the region that relied heavily on geometric designs. Drug jars and dishes are the most common forms found in Virginia. These dishes were found at the Chesopean Site, Virginia Beach, Virginia, in a context roughly dating to the middle decades of the seventeenth century.

of settlers reached their assigned acreage, they were to kneel and give thanks for their safe arrival. That day of thanksgiving, to be observed annually, was America's first.[10]

The Society's investors ordered Captain Woodlief to see that substantial, "homelike" houses were built, dwellings that were roofed over with boards. Two of the settlement's buildings were to be very sturdily constructed. One was to serve as a storehouse in which arms, ammunition, armor, victuals, and tools were to be kept. The other was a substantial hall where the colonists, who were expected to lead a communal lifestyle, could assemble for meals and gather for meetings and daily worship. Until Berkeley Hundred's storehouse was built, the community's weapons, staple foods, and tools were to be kept in the common warehouse at Jamestown or in the granary at nearby Bermuda Hundred. The settlers were to erect a palisade that was seven-and-a-half feet high, enclosing four hundred acres of land in which crops could be grown and livestock could graze. Captain Woodlief, as com-

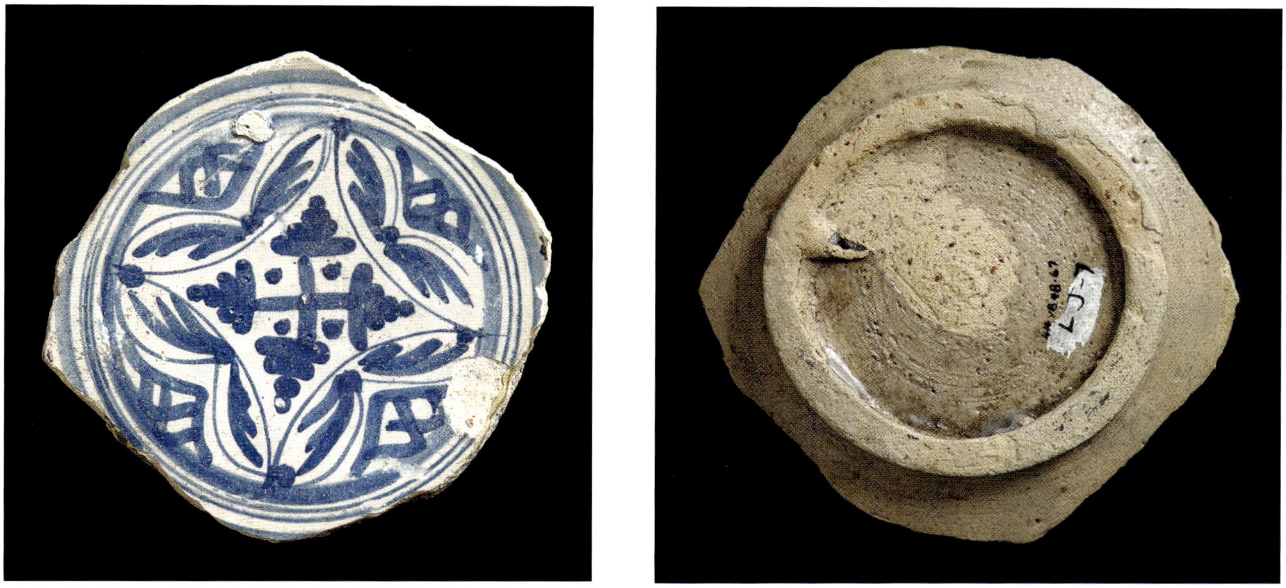

Figure 13 Dish base (front and back), Northern Netherlands, 1620–1640. Tin-glazed earthenware with lead glaze on back. (Courtesy, Virginia Department of Historic Resources; photo, Robert Hunter.) Contacts with China that developed after 1600 led to a movement away from the polychrome decorations of the Mediterranean to the blue-and-white motifs of the Far East. The undersides of these early dishes were lead-glazed in order to save the more expensive tin enamel for application to the upper surface. Many of the dish footrings are pierced to allow hanging. This and the lack of wear on the surface of some vessels suggest that dishes were not necessarily used for the consumption of food but to display status. This piece was excavated at the Chesopean Site, Virginia Beach, Virginia.

munity commander, was to be assisted by ensign Ferdinando Yates, who was placed in charge of the settlement's armor and agricultural tools. John Blanchard was to be steward of the household and clerk of the store, and Roland Painter was designated clerk of the kitchen. Henry Perce, a gentleman, was to be usher of the hall and Thomas Partridge, if willing, was to serve as bailiff of husbandry. Yates, Blanchard, Painter, Richard Godfrey, and Thomas Coopy, considered men of discretion, were to take their meals at the captain's table. Each night, several male settlers were to take turns standing watch, forming a *corps de garde*. Captain Woodlief, as Berkeley Hundred's chief merchant, was authorized to trade with the Indians and to represent his settlement's best interests when trading with other colonists and merchants.[11]

George Thorpe of Berkeley Hundred

George Thorpe sailed from England on the *London Merchant* on March 27, 1620, less than four months after the *Margaret* landed at Berkeley Hundred. He quickly demonstrated that he was a capable leader and on June 28, 1620, the Virginia Company's council named him to the colony's Council of State, whose members served as advisers to the governor and functioned as a judiciary body. Thorpe's activities in Virginia, recorded in his correspondence and in records generated by the Virginia Company, reveal that he was knowledgeable about the legal system and the laws by which the Virginia colony was governed. In fact, Secretary John Pory, who became acquainted with Thorpe after he reached Virginia, described him as "an Angell from Heaven" and said that he was of inestimable benefit to the colony. Thorpe was held in such high esteem that the Virginia Company's court ordered him to assist Governor George Yeardley in investigating Lady Cisley Delaware's allegations against Sir Samuel Argall's mishandling of her late husband's estate. During the latter part of 1620, Yeardley heaped praise upon Captain George Thorpe and told company officials that he

Figure 14 Storage jar, attributed to Thomas Ward, James City County, Virginia, 1620–1635. Lead-glazed earthenware. H. 11⅛". (Courtesy, Virginia Department of Historic Resources; photo, Robert Hunter.) Thomas Ward arrived at Jamestown in 1619 as the indentured servant to a bricklayer and is the first documented potter in the English colonies. The forms and styles of his vessels reflect a combination of the London-area redwares (like this thumb-impressed jar) and the products of the potteries on the border of Surrey and Hampshire counties. This example was excavated at the Nomini Plantation Site in Westmoreland County, Virginia.

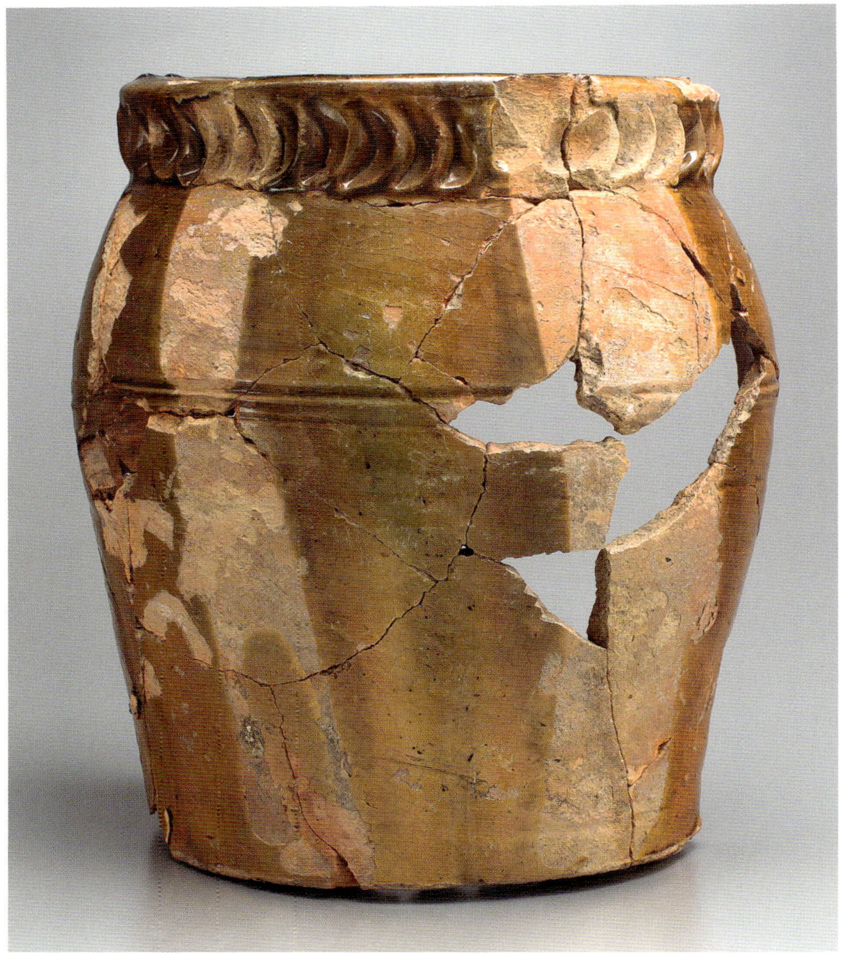

Figure 15 Detail of the thumb-impressed rim on the storage jar illustrated in fig. 14. When the pot was thrown on the wheel, an extra coil was applied to reinforce the rim. The thumb impressions are both functional and decorative. This technique is also found on London redwares of the period.

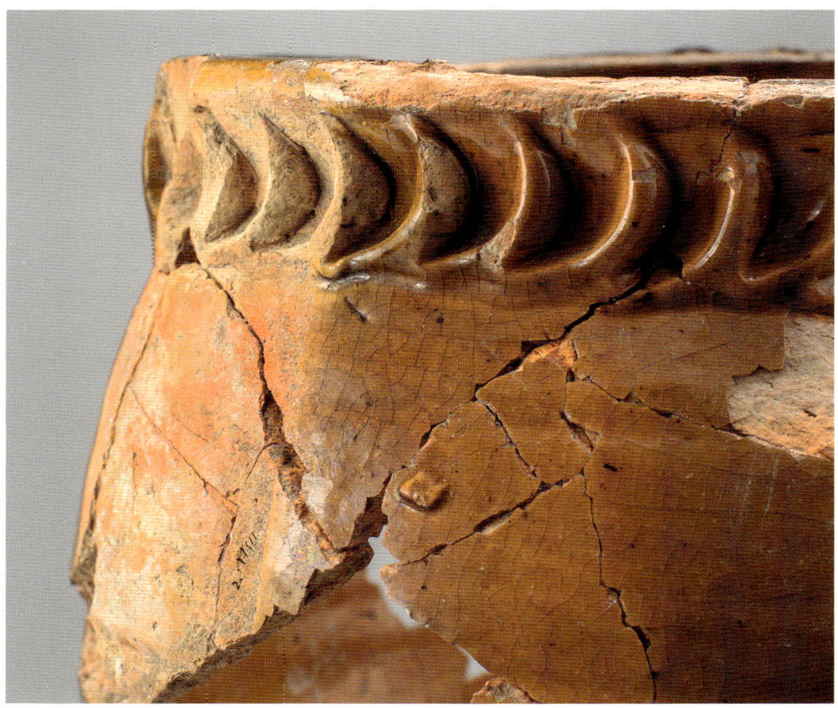

would "be the most ffitt man" to take over his governorship and would have no problem following their orders.[12]

In a letter to Berkeley Hundred investor John Smyth dated December 19, 1620, George Thorpe acknowledged that some of the people the Society had sent to the colony were dead. He attributed some deaths to "disease of theire minde"[13] but he declared that his own health had never been better. He said that the victuals sent from England were overpriced and pointed out that the newly arrived colonists had not expected to drink water instead of beer. He added that the settlers had found a way to make a good beverage from Indian corn and that sometimes he preferred it to good, strong English beer. In closing, Thorpe said that he wanted his wife and children to join him in Virginia and asked Smyth to make sure that they were all right. Virginia Company officials seem to have been optimistic about Berkeley Hundred's future and in 1621 they sent George Thorpe a barrel of seeds.[14]

During late spring 1620 there was a change in the Society of Berkeley Hundred's leadership, for Sir William Throgmorton transferred his interest in the Society to William Tracy, a Virginia Company investor and resident of Hayles in Gloucestershire. An undated amendment to the Society's charter stipulated that the Berkeley Hundred colonists were to produce salable staple commodities, such as corn, wine, oil, flax, pitch and tar, soap ashes, iron, and clapboard, and were not to rely "wholly or chiefly upon Tobacco." Tracy told John Smyth that he needed to supply the plantation with cattle and horses if the colonists were to have a bloomery. He said that he should be able to borrow some goats and silkworms from Lady Elizabeth Dale, whose land in Shirley Hundred was nearby. By July 1620 William Tracy had been named to Virginia's Council of State. He began making plans to go to Virginia aboard the ship *Supply*, taking along his wife, Mary, son Thomas, daughter Joyce, and some servants. Near the end of August, the Society of Berkeley Hundred gave Tracy and George Thorpe a commission to govern Berkeley Hundred and Captain John Woodlief's authority was set aside.[15]

The Second Group of Settlers

When the *Supply* was readied for departure in September 1620, the same types of foodstuffs, clothing, tools, and household goods were loaded aboard as had been sent on the *Margaret* the previous year. This time some medicinal drugs were included, probably for the healing work of Robert Pawlett, a surgeon and ordained clergyman, who was among the fifty passengers accompanying William Tracy to Berkeley Hundred. Numerous pairs of "women's scissors," shears, and tailors' shears were sent along, as well as sleeping mats, rugs, two grindstones, hundreds of yards of fabric of various types, shoes, and readymade clothing. The second group of Berkeley Hundred colonists took along a much larger supply of defensive weaponry than had been loaded aboard the *Margaret* in 1619, probably at the recommendation of the settlers already living in Virginia. Also aboard the *Supply* were seeds for the colonists' gardens and fruit trees, scion wood for grafting, spices, and large quantities of blue and white trade beads.

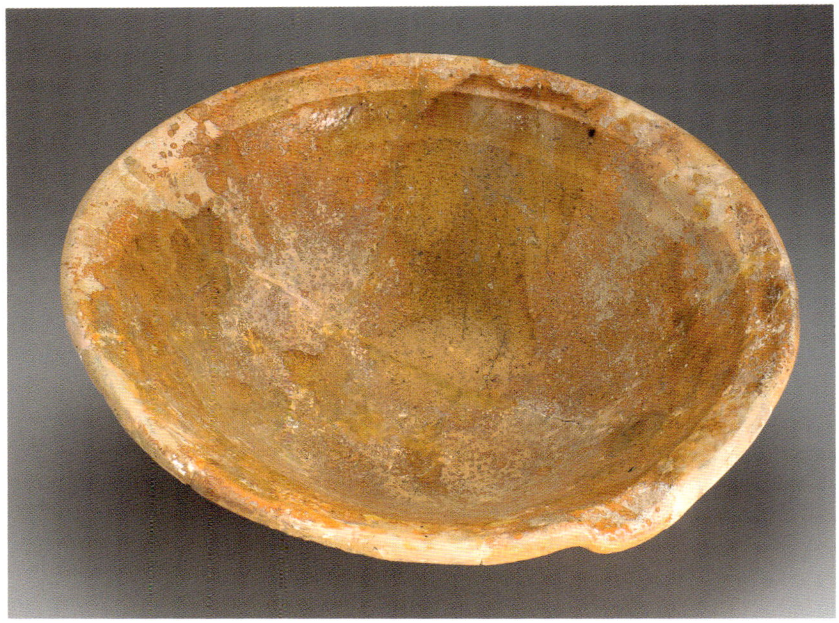

Figure 16 Milk pan, Virginia, attributed to Thomas Ward, James City County, Virginia, 1620–1635. Lead-glazed earthenware. D. 15". (Courtesy, Virginia Department of Historic Resources; photo, Robert Hunter.) Since the late sixteenth century, milk pans were used as settling vessels to separate cream from milk. The form was not common in Virginia before the large-scale importation of cows in the second quarter of the seventeenth century.

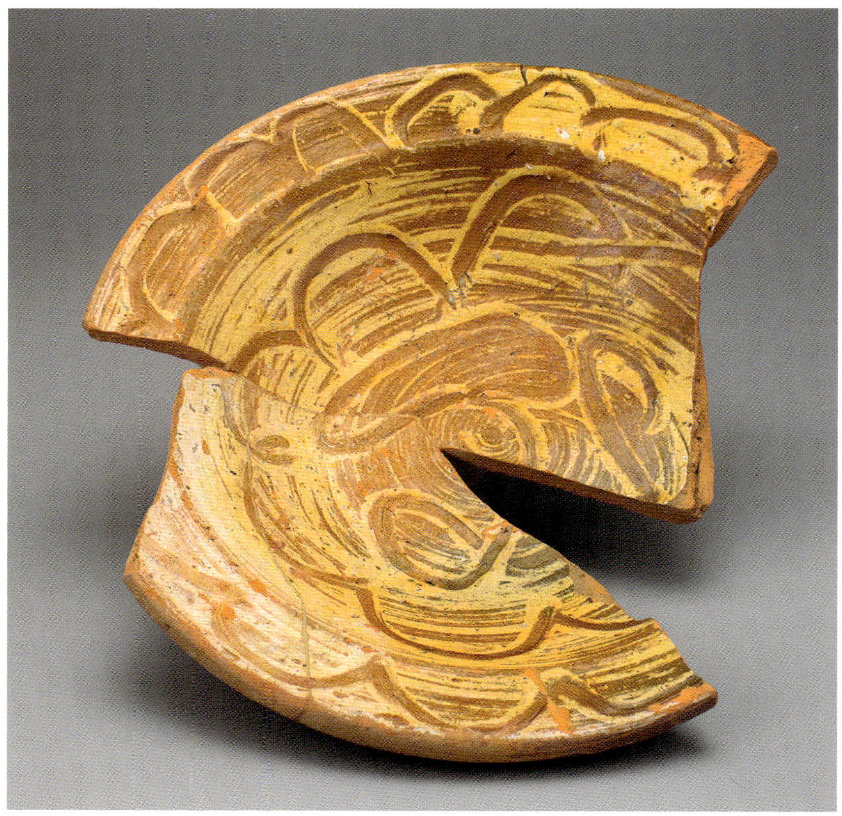

Figure 17 Dish, South Somerset, England, 1620–1630. Lead-glazed earthenware. D. 13¾". (Courtesy, Virginia Department of Historic Resources; photo, Robert Hunter.) This dish with wet slip-sgraffito decoration has visual attributes that are characteristic of wares produced in the village of Donyatt in South Somerset. It is therefore considered a Donyatt-type, until chemical analysis of the clays can link it to a specific kiln in the well-known potting area of southwest England. Pottery from this region was arriving in Virginia before 1610, as ships embarking for the New World stopped in at Plymouth to avail themselves of goods they could acquire more cheaply than in London. This example was excavated at Jordan's Journey, Prince George County, Virginia, in a 1630–1650 context.

Barrels of beef and pork were brought on board, along with cod and other fish procured from Newfoundland. Also on hand were more than sixty gallons of aqua vitae, eighteen tuns of beer, and sixty gallons of sack. A hogshead of new cider was sent to George Thorpe along with a copy of the commission he received from the Society of Berkeley Hundred and the

Virginia Company's instructions, encouraging the production of artificial wine and discouraging tobacco production.

The *Supply*, with fifty men, women, and children, arrived at Berkeley Hundred on January 29, 1621. Among the passengers were eight gentlemen, two of whom decided to return to England at once.[16] The second group of Berkeley Hundred colonists, like the first, was composed of servants and tenants, most of whom were skilled workers and experienced husbandmen. The Society's tenants were promised food and shelter until they could be given a furnished house. As sharecroppers, they were to be allocated acreage on which they could grow food crops and keep livestock. Society servants who completed their three- or four-year terms of indenture also stood to receive land.[17]

Figure 18 Jug, South Somerset, England, 1600–1650. Sgraffito slipware. H. 8⅜". (Courtesy, Virginia Department of Historic Resources; photo, Meg Eastman, Virginia Historical Society.) English potters began producing sgraffito slipwares by 1600. The earliest examples are decorated with simple patterns of incised lines. The decoration became more elaborate over time, as demonstrated on a 1666–1690 bottle from St. Mary's City, Maryland, decorated with birds and letters. This example was found at Jordan's Journey, Prince George County, in a 1620–1635 context.

The Rev. Robert Pawlett's contract with the Society of Berkeley Hundred stipulated that he was to commence serving as the community's minister, surgeon, and physician as soon as he reached Berkeley Hundred. He was supposed to share a dwelling with George Thorpe and William Tracy, who were obliged to pay him twenty pounds apiece. Pawlett's stipend was to be

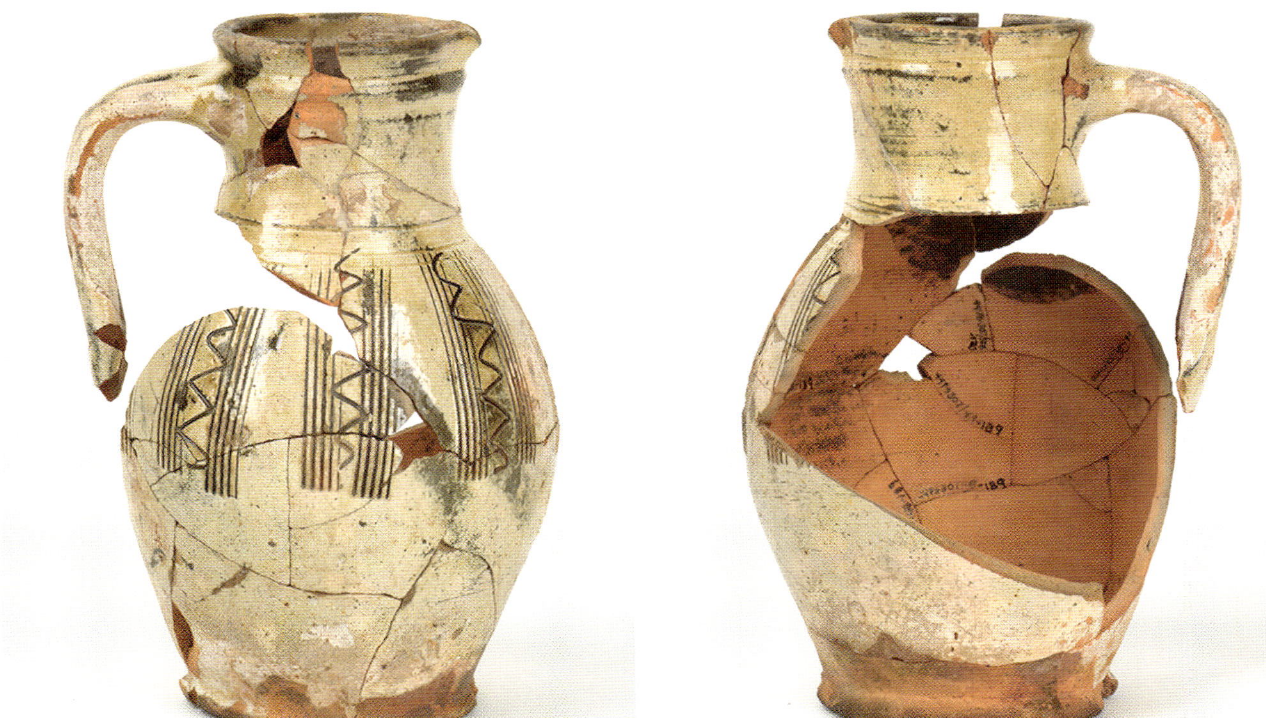

supplemented by Society investors Richard Berkeley and John Smyth. At the end of a year, Pawlett had the option of returning to England or staying on in the colony in a salaried position. On July 16, 1621, the Virginia Company designated Pawlett a provisional councilor and eight days later he was named to the Council of State.[18]

Christianizing the Indians

William Tracy died within two months of his arrival at Berkeley Hundred, leaving George Thorpe in sole command. Thorpe continued to be responsible for the servants at the College and he tried to advance the Falling Creek ironworks and other Virginia Company projects. His correspondence reveals that he was optimistic about the colony's economic future, but as a deeply religious man he believed that it was important to convert the Indians to Christianity. In mid-May 1621 he informed Virginia Company officials that most colonists treated the natives disdainfully rather than charitably. He recommended that the company send over some apparel and household goods that could be given to Indian leaders to curry their affection. In the wake of

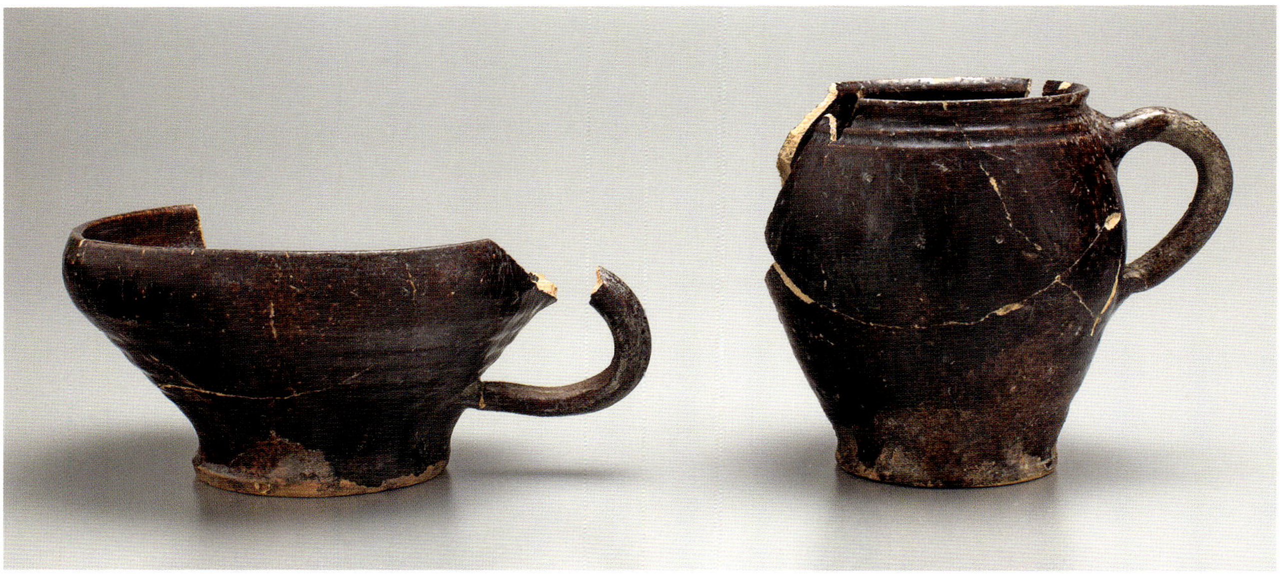

Figure 19 Cup and mug, Surrey-Hampshire border ware, 1600–1650. Lead-glazed earthenware. H. of cup 3"; H. of mug 4½". (Courtesy, Virginia Department of Historic Resources; photo, Robert Hunter.) These two drinking vessels were made in the potteries located on the Surrey-Hampshire border in England. For three hundred years beginning in the fourteenth century this area was a major source of quality utilitarian wares for the London market. Border wares in both everyday household forms and those with a specialized function—such as alembics, schweinetopfs, and double dishes—were part of the ceramic assemblage of the first Jamestown colonists in 1607. The earthenware has been found on seventeenth-century sites on the eastern North American continent from North Carolina to Newfoundland. These examples were found at Wills Cove, Suffolk, Virginia, in a 1630–1650 context.

Thorpe's letter, the Virginia Company urged the colony's governing officials to promote religious conversion and to treat the Indians justly so that the peace would be maintained and old quarrels would not be resurrected.[19]

On June 22, 1621, George Thorpe informed company officials that he was going to visit the Indians' principal leader, Opechancanough, whom he hoped to convert to Christianity. He said that the Indians had expressed their concern that the colony's incoming governor, Sir Francis Wyatt, would not uphold the peace agreements that had been forged during Governor George Yeardley's administration. Even before Thorpe's letter reached England, Virginia Company officials ordered Wyatt to see that "the best meanes bee used to draw the better disposed of the natives to converse with o^{r} people" and expressed their hope that the Indians "may growe to a likeing and love of Civility, and finallie bee brought to the knowledge and love of god and true religion."[20]

When Governor Wyatt arrived in October 1621, he sent Thorpe to see Opechancanough, with assurances that he would continue his predecessor's policies. Virginia Company officials also offered encouragement to the natives and in January 1622 sent gifts to Opechancanough and his brother,

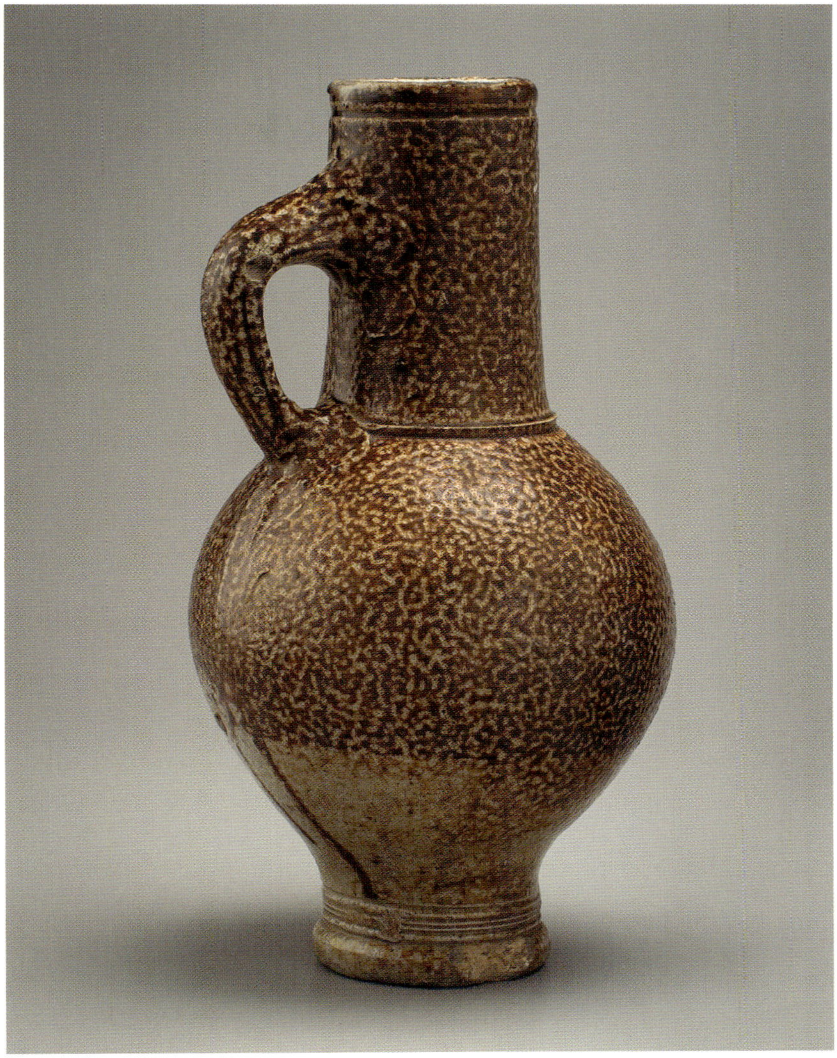

Figure 20 Jug, Frechen, Germany, 1600–1620. Salt-glazed stoneware. H. 8¾". (Private collection; photo, Robert Hunter.) Vessels from Frechen, a small village near Cologne, were the most widely traded German stonewares of the seventeenth century. While this wide-mouthed jug is undecorated, it is a form that was highly prized by English consumers, who mounted the necks and bases with silver. Fragments of these German vessels have been found on many early archaeological sites in the Tidewater region of Virginia.

along with a message of friendship. Contemporary accounts reveal that Thorpe went to great lengths to win the Indians' trust and affection and that he focused his attention on Opechancanough. However, between January and March 1622 relations between the Indians and colonists rapidly deteriorated after one of the paramount chief's favorites, a war captain named Nemattanow, also known as Jack of the Feather, killed a settler named Morgan. Two or three days later, when two of Morgan's male servants encountered Jack, who was wearing their master's cap, they assumed the worst and killed him. This enraged Opechancanough, who, according to Captain John Smith, orchestrated a devastating attack, which occurred only two weeks later.[21]

George Thorpe's Martyrdom

On Friday, March 22, 1622, the Powhatan Indians attacked the numerous plantations that were sparsely scattered along both sides of the James River. Especially hard hit were the settlements in the river's upper reaches. When

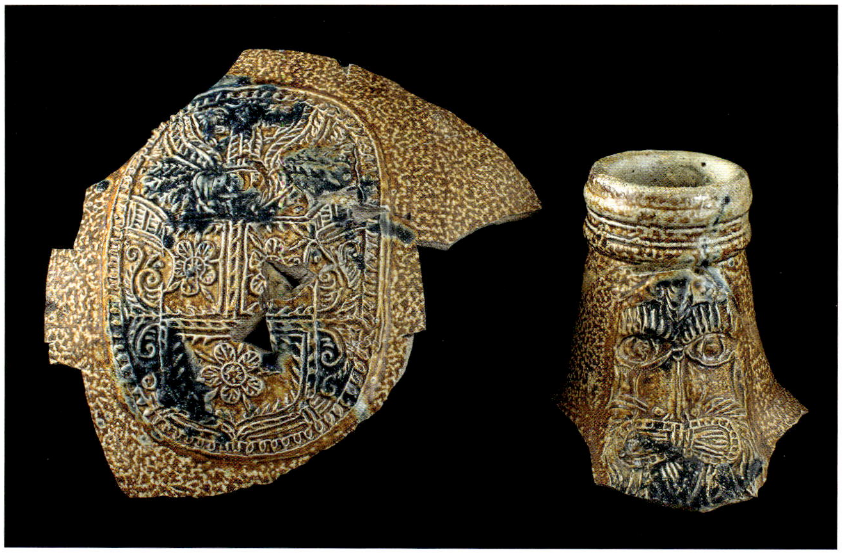

Figure 21 Jug neck and body, Frechen, Germany, 1620–1630. Salt-glazed stoneware. (Courtesy, Virginia Department of Historic Resources; photo, Robert Hunter.) These fragments are from a large Bartmann ("bearded man" in German) jug, also known since the seventeenth century as a "Bellarmine." Shipped to Jamestown as containers for alcoholic beverages, these jugs are among the most common ware types found in early colonial contexts. The random accents of cobalt coloring on this vessel made it more costly than undecorated examples. These fragments were found at the Chesopean Site in Virginia Beach, Virginia.

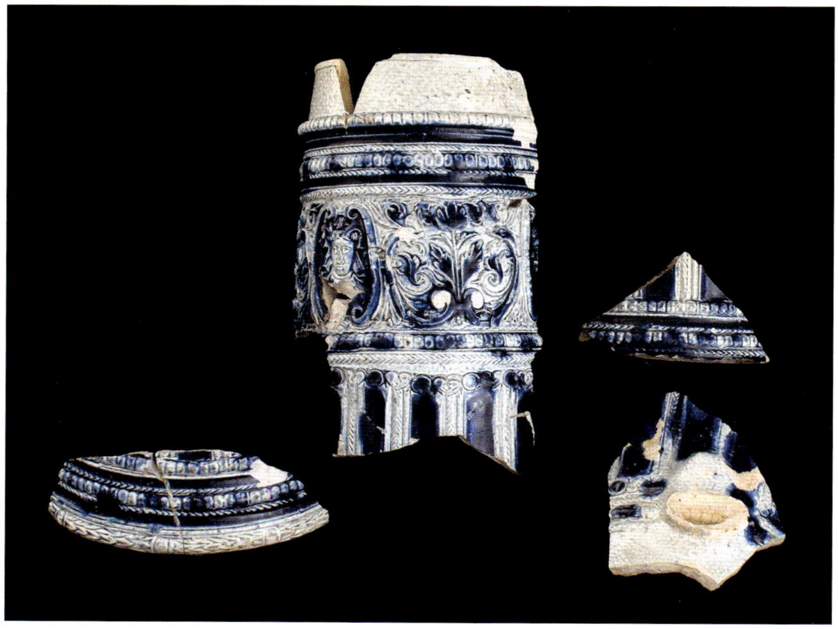

Figure 22 Schnelle fragments, Westerwald, Germany, 1600–1620. Salt-glazed stoneware. (Courtesy, Virginia Department of Historic Resources; photo, Robert Hunter.) Westerwald stoneware is commonly found in North American contexts dating 1620–1775, but it is rarely found in England prior to the second half of the seventeenth century. This suggests that the source of earlier Westerwald vessels in England's colonies was a result of the extensive Dutch trade. The *schnelle*, or elongated tankard, form was first used in the sixteenth-century German potting industries in Siegburg and Raeren. It is seldom documented in colonial Virginia contexts. These fragments were excavated at the Chesopean Site in Virginia Beach, Virginia.

the Indians descended on Berkeley Hundred, George Thorpe and ten others were killed. Edward Waterhouse, who was in England, recounted the Indian assault in highly inflammatory terms, basing it on what he learned from returning colonists and letters he had received. He described Thorpe as a religious martyr who was so intent on converting the Indians to Christianity that he punished anyone who antagonized them. He said that Thorpe frequently conferred with Opechancanough, gave him numerous presents, and had an English-style house built for him, a structure whose functional door lock and key fascinated the Indian chief. In fact, Captain John Smith said that the native emperor assured the settlers that "he held the peace so firme, the sky should fall or he dissolved it." According to Waterhouse, Opechancanough professed interest in the god the English worshiped,

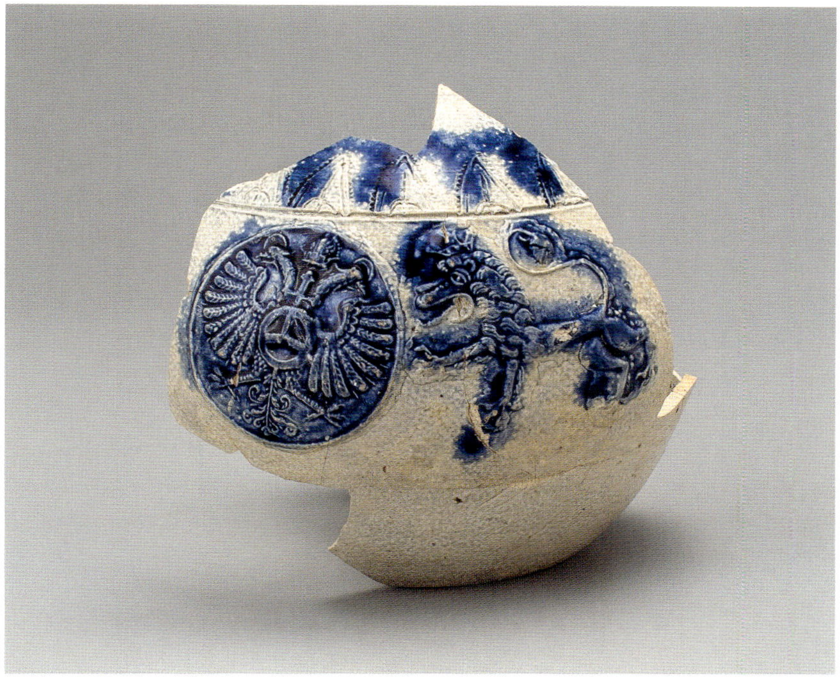

Figure 23 Jug fragment, Westerwald, Germany, 1630–1650. Salt-glazed stoneware. (Courtesy, Virginia Department of Historic Resources; photo, Robert Hunter.) Stoneware products of the Westerwald continued to find their way into the Virginia colony throughout the seventeenth century. Mugs, chamberpots, and examples of these globular, big-bellied jugs are frequently seen in archaeological contexts. Relief decoration in the form of sprigged applied rampant lions and medallions is typical on these decorative but utilitarian wares. This example was found in Gloucester Point, Virginia.

which he said was "much better than theirs," and led Thorpe to believe that he was in the process of religious conversion. Too late, Thorpe and the other colonists learned that it was a hoax. Waterhouse said that Thorpe's manservant, who sensed that some treachery was afoot, tried to warn Thorpe, but Thorpe "was so void of all suspition, and so full of confidence, that they had sooner killed him, than hee could or would believe they meant any ill against him."[22] His naïveté proved fatal.[23] Waterhouse went on to describe the Indians as a "Viperous brood" who not only killed George Thorpe, but also mutilated his corpse.[24]

Aftermath

The colonists living in the upper part of the James River who managed to survive the native assault were evacuated to positions of greater safety. Although a substantial number were taken to Jamestown Island, most were transported to Flowerdew Hundred, Shirley Hundred, and Samuel Jordan's plantation, Jordan's Journey, all of which were strengthened and held. At Berkeley Hundred, the Indian attack undoubtedly caused momentary chaos, since the settlement's leader, George Thorpe, had been killed. Nonetheless, the survivors seem to have acted quickly in an attempt to reestablish order. Sixteen men from Berkeley Hundred were sent to West and Shirley Hundred Island, where some of the Society of Berkeley Hundred's cattle were taken, but most of the Berkeley survivors went to Jordan's Journey.[25] Archaeological research suggests that after moving to Jordan's Journey they resumed the communal lifestyle to which they had grown accustomed at Berkeley Hundred. This mode of living contrasted sharply with the other planter households at Jordan's Journey, which enjoyed a more independent existence.

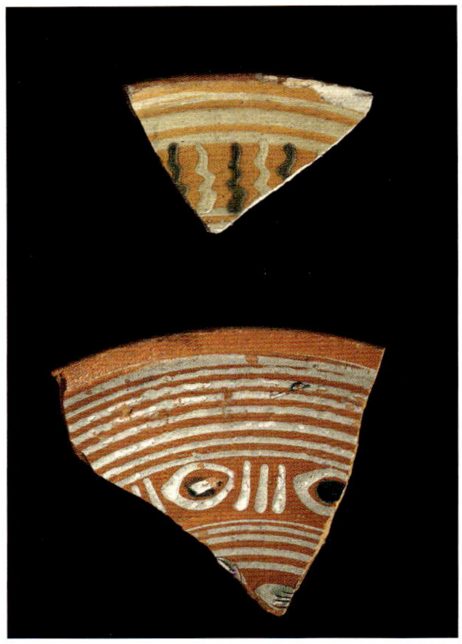

Figure 24 Dish rim fragments, Germany, 1570–1630. Lead-glazed earthenware, with slip decoration. *Top*: Weser slipware; *bottom*: Werra ware. (Courtesy, Virginia Department of Historic Resources; photo, Robert Hunter.) Both of these highly decorative German slipware types were exported in great numbers from Bremen to the Low Countries and from there to England in the late sixteenth and early seventeenth centuries. The wares are found in small numbers on Virginia's early-seventeenth-century archaeological sites, with Werra ware being much more common than Weser. These fragments were found at the Chesopean Site in Virginia Beach, Virginia.

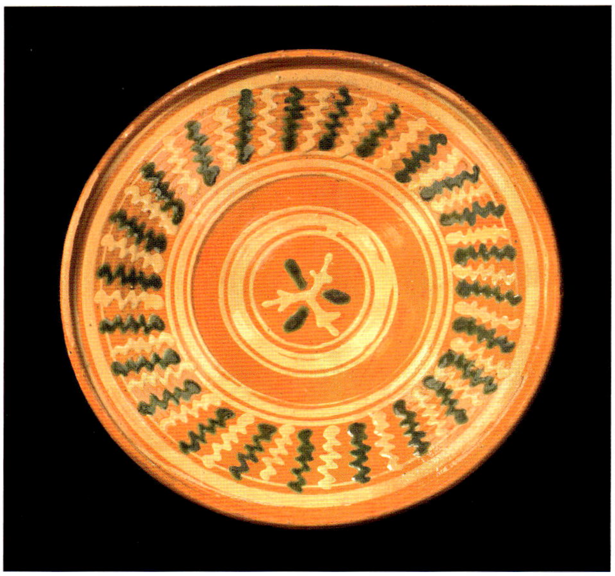

Figure 25 Weser slipware dish, Germany, 1570–1630. Lead-glazed earthenware, with slip decoration. D. 12¼". (Private collection; photo, Mark Heffron.) Weser ware was made in a number of production sites between the Weser and Leine rivers. Unlike Werra ware, Weser is a hard white fabric that was fired once. Designs are simple, consisting of geometric dots, lines, and squiggles in red-brown and green on a yellow ground. Weser ware is present but uncommon on early-seventeenth-century Virginia sites. Dishes comprised a major part of the industry's output but hollow-ware forms have also been excavated: a small jug was found at Jamestown, Virginia, and a double-handled pot with rouletted decoration was located in Hampton, Virginia.

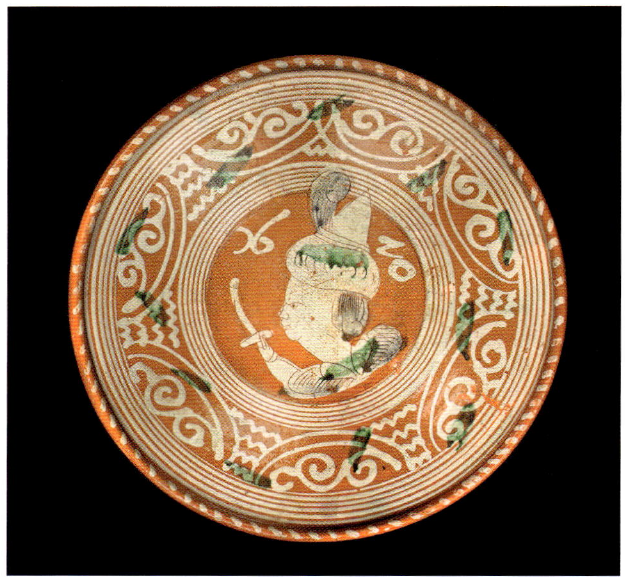

Figure 26 Werra slipware dish, Germany, dated 1620. Lead-glazed earthenware, with slip decoration. D. 13⅛". (Private collection; photo, Mark Heffron.) This ware used to be known as Wanfried, for the village on the Werra River where large numbers of wasters were found in the late nineteenth century. Now kiln sites for the ware have been located more broadly along the river, with many biscuit wasters indicating that the vessels were fired twice. Dishes and bowls are most commonly found in Virginia on sites dating to the first quarter of the seventeenth century. The low-fired red fabric is decorated with white, green, dark brown, and yellow colors, with the central motif typically outlined in sgraffito. Common designs include flora, fauna, soldiers, and individuals holding wineglasses. Often the dishes are dated.

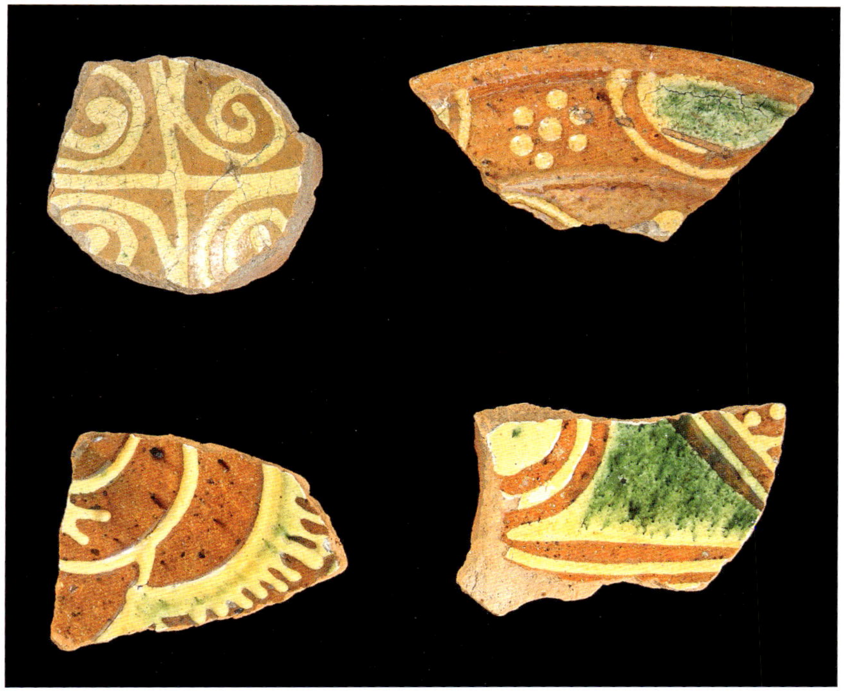

Figure 27 Dish fragments, North Holland, 1620–1650. Lead-glazed earthenware, with slip decoration. (Courtesy, Virginia Department of Historic Resources; photo, Robert Hunter.) German and Dutch potters were using a variety of slip decorating techniques in the early 1600s. The distinctive Dutch slip-trailed wares are sometimes found in American contexts of the 1630s–1660s. Two-handled drinking-bowl forms are known, in addition to broad dishes represented by these fragments found at the Chesopean Site in Virginia Beach, Virginia.

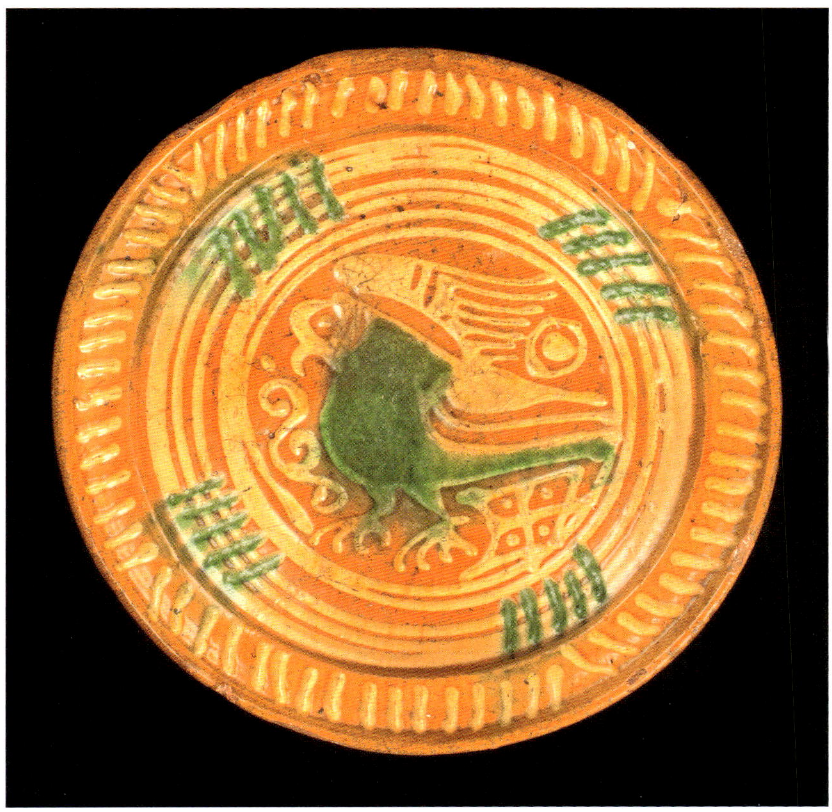

Figure 28 Dish, North Holland, 1620–1650. Lead-glazed earthenware, with slip decoration. D. 10⅞". (Private collection; photo, Mark Heffron.) The highly decorated Dutch slipwares usually have exuberant animal, bird, floral, pomegranate, and geometric designs. White slip, trailed directly onto the earthenware bodies, was often highlighted with splashes of green oxide beneath a clear lead glaze.

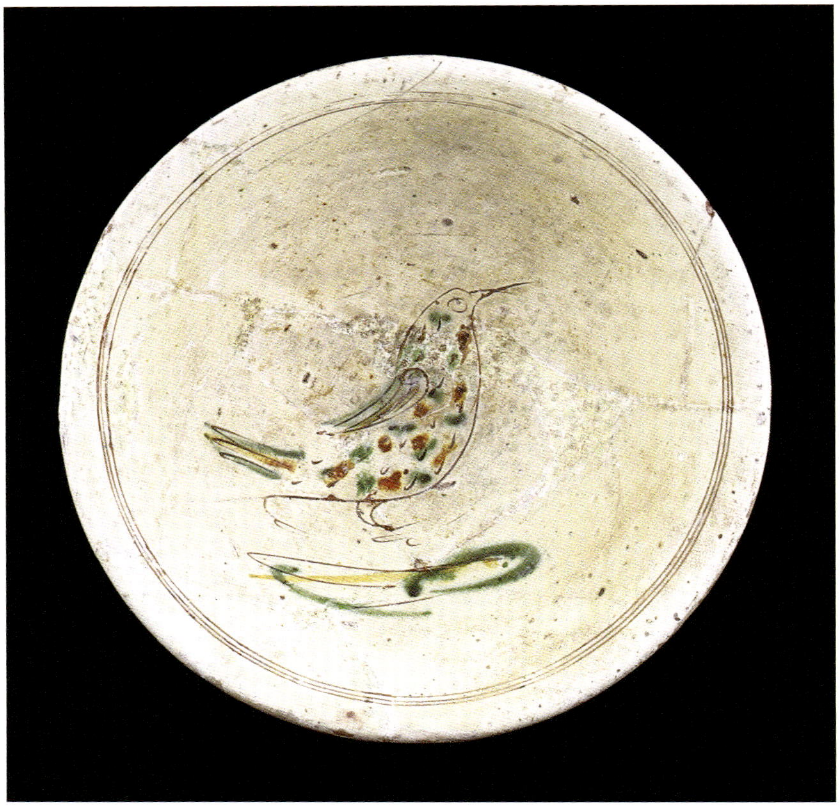

Figure 29 Bowl, northern Italy, 1575–1625. Sgraffito slipware. D. 9¾". (Courtesy, The National Park Service, Colonial Historical National Park; photo, Gavin Ashworth.) Sgraffito slipware bowls and flanged dishes produced in Pisa, Italy, are found in Virginia contexts dating to the second and third quarters of the seventeenth century. The red alluvial clay is covered on the interiors of all forms with a thick white slip through which motifs are incised. Ocher and green copper oxide are used to highlight the design, which in Virginia is usually a flower or a bird, as shown here. This bowl was recovered from Jamestown Island, Virginia, in a second-quarter-of-the-seventeenth-century context.

In August 1622 John Smyth, one of the Society of Berkeley Hundred's main investors, compiled a list of the Berkeley Hundred settlers he believed had survived the Indian attack. As it turns out, his list was highly inaccurate, for he omitted at least sixteen people's names. Demographic records reveal that at least three prominent gentlemen—Richard Milton, Thomas Palmer, and John Gibbs—went to Jordan's Journey and stayed on, as did Margaret Finch, Edith Halliers, William Popleton, and Richard Sheriffe Jr., a cooper. Sheriffe's father, a carpenter in the employ of the Society of Berkeley Hundred, died at Jordan's Journey between April 1623 and February 1624. At least four other Berkeley Hundred refugees were still residing at Jordan's Journey in 1625. Ten people, including two nuclear families, who were living at Jordan's Journey in 1624 were gone by early 1625 and may have died there.[26]

In July 1623, when Virginia Company investors resolved to send supplies to the colony for the relief of surviving settlers, John Smyth indicated that he was dispatching aid to his "servants now living in Virginia in Berckeley [*sic*] Hundred and such others as this next August I sende over to encrease them."[27] However, after he learned that the surviving settlers had abandoned their plantation, he changed his mind. On May 24, 1624, when the Virginia Company's charter was revoked and Virginia became a Crown colony, many of those who had invested in the colonization effort were left to ponder what the future might hold. In 1627 William Farrar, the community commander at Jordan's Journey, was ordered to make an

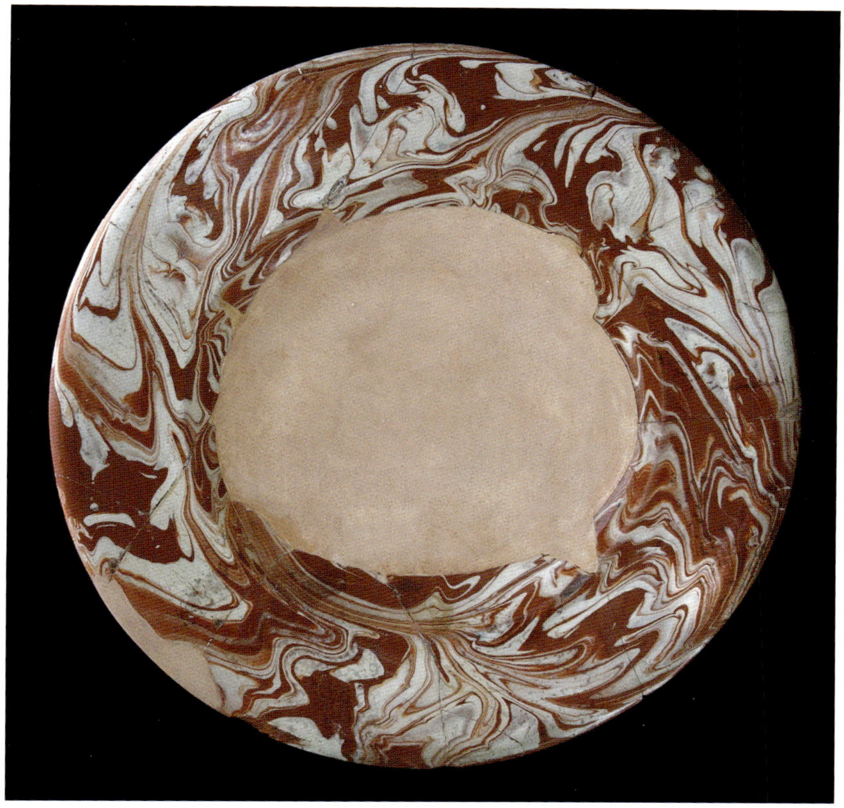

Figure 30 Dish, Pisa, Italy, 1600–1650. Marbled slip-decorated earthenware. D. 10⅝". (Courtesy, Virginia Department of Historic Resources; photo, Robert Hunter.) This ware, made in Pisa, is uncommon in England until close to the mid-seventeenth century. It frequently appears on Virginia sites earlier, which—like the other Italian, Netherlandish, and Portuguese wares—may be a result of the robust Dutch trade with the English colonies. Dishes and bowls are the most common forms. This dish, the center portion of which has been reconstructed, was excavated at the Chesopean Site in Virginia Beach, Virginia.

accounting of the Society of Berkeley Hundred's cattle, and several years later, when he was in England, he conferred with Society investors about their plantation's future. He would have had firsthand knowledge, obtained from the Berkeley Hundred settlers living at Jordan's Journey. In 1632 the Society of Berkeley Hundred's remaining investors discussed reviving their settlement, but financial support was so limited that the project never got underway. Finally, in 1637 Governor John Harvey issued to a syndicate of London merchants and mariners a patent for eight thousand acres called Berkeley Hundred, noting that they had purchased the property.[28]

The Settlement of Thorpe's Estate

On June 20, 1622, George Thorpe's wife, Margaret, who was still unaware of her 46-year-old husband's death, asked John Smyth of Nibley for financial assistance, in accord with an agreement the two men had made. Several months later, after word of George's death spread through Gloucestershire, an inquisition was held at Gloucester Castle. A jury then noted that the widowed Margaret Thorpe had life rights in the Manor of Wanswell, in accord with a prenuptial agreement she and George made on February 20, 1611, the day before they wed, and that the couple's eldest son and primary heir, William, was only ten years old. The Court of Exchequer acknowledged that a month before George Thorpe left for Virginia he had succeeded in having the entail removed from some of his inherited property, a strategy that would free his executors to sell his acreage in order to pay his debts.[29]

As it turned out, nearly two more years would pass before Captain George Thorpe's estate was presented for probate.[30] On March 7, 1624, when Virginia's General Court convened as a judicial body, Sir George Yeardley was ordered to appoint men to appraise Thorpe's personal estate. The justices expressed their uncertainty about which debts were attributable to the decedent personally and which to the Society of Berkeley Hundred. Thorpe's inventory, presented at court on April 10, 1624, is the earliest known tabulation of a Virginia colonist's personal effects and includes items attributed to him at the time of his death on March 22, 1622.[31]

Thorpe's inventory was compiled by three respected men and near-neighbors: Richard Biggs of West and Shirley Hundred, Samuell Sharpe of Flowerdew Hundred, and Thomas Palmer, who had lived at Berkeley Hundred but settled at Jordan's Journey after the Indian attack.[32] A sentence at the conclusion of the inventory stated that Yeardley and Thorpe's appraisers expected compensation for their work. Yeardley also wanted to be paid for "preservinge and keeping the goods." It is unclear whether he had Thorpe's goods taken to his Flowerdew or Weyanoke plantations, which

Figure 31 Costrels, northern Italy, 1580–1650. Marbled slip-decorated earthenware. H. (*left*) 10¼", H. (*right*) 10½". (Private collection; photo, Mark Heffron.) Fragments of these distinctive-looking wine flasks have been found in early Virginia contexts. The four molded lion's-head lugs were meant to accommodate a strap or cord for suspension. The bichrome decoration of yellow and dark brown without the addition of green is considered to date to the later years of production. The narrow form of the example on the right is also a later shape.

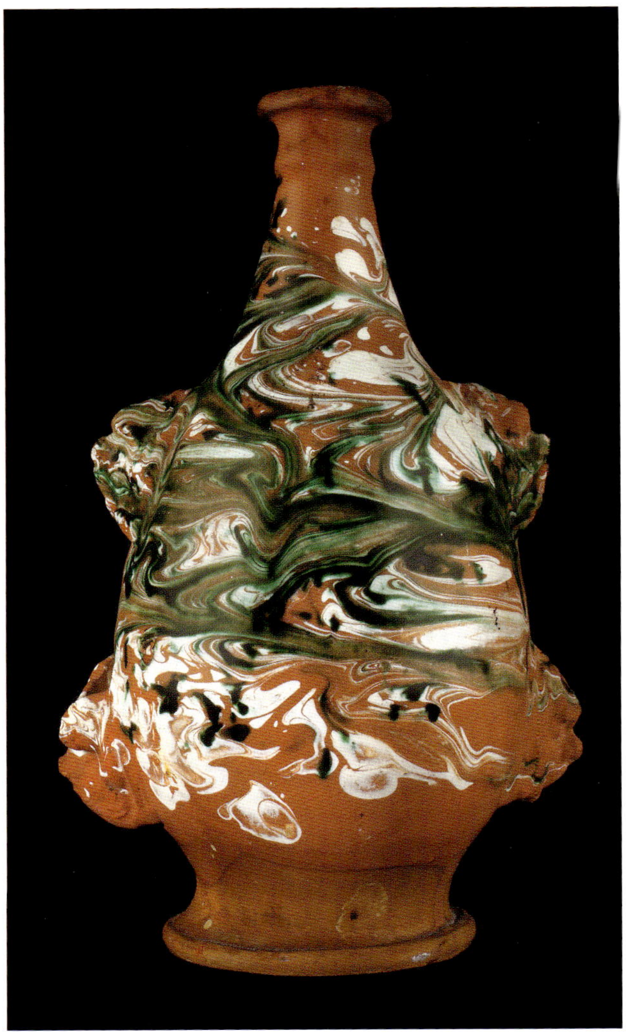

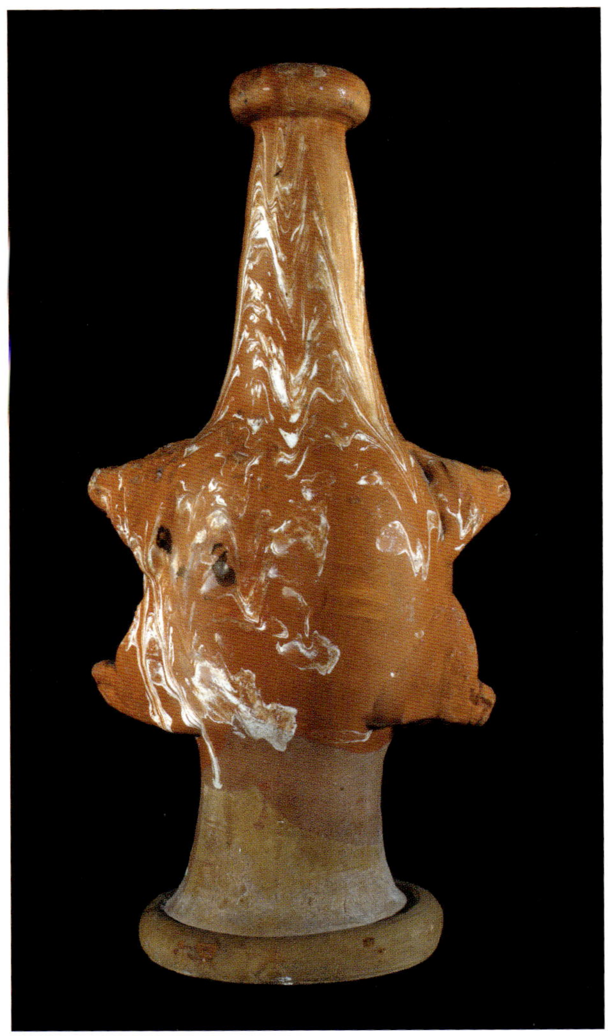

were nearby, or had them brought to Jamestown. In any case, Thorpe's wine was "drunke out amonge the people that fetcht downe his goods."[33] While in court on March 7, 1624, Yeardley reported that either Thorpe or the Society of Berkeley Hundred's investors owed eight barrels of corn to the estate of Captain Nathaniel West, whose family owned Westover. In July 1624 the General Court ordered Thorpe's administrators to repay West's widow, Frances, and Sir George Yeardley was told to settle Thorpe's debt with Alice Davison, the widow of Secretary of State Christopher Davison. In the same session of court, John Gibbs and Richard Milton testified that Thorpe had borrowed some corn from a Mr. Dade, probably John Dade who came to Virginia on behalf of Cecily West, Lady Delaware.[34]

In February 1625 several other people or their legal representatives made claims against the late George Thorpe's estate. A partial summary of those debts reveals that when Thorpe provided for the people under his command, he sometimes procured staple foods and livestock from other colonists at his own expense. Thorpe's creditors included his cousin, Lady Elizabeth Dale, from whom he had borrowed cattle and goats, and the Rev. David Sandys and the Rev. Richard Buck, two clergymen who probably performed religious services at Berkeley Hundred or the College. Thorpe also owed Sir George Yeardley for hiring the men who built an English-style house for Opechancanough. Court records reveal that Yeardley had employed Robert Fisher, a carpenter and Jordan's Journey resident, who charged for five weeks' work. The decedent's greatest debts, calculated in tobacco, were owed to Yeardley, which probably explains why he was ordered to see that Thorpe's estate was appraised.[35]

George Thorpe's Personal Effects

On August 14, 1634, a purportedly accurate copy of George Thorpe's inventory was presented to officials in England, in response to legal action undertaken by his eldest son, William, a resident of Bristol.[36] George Thorpe's inventory not only reflects his status as a gentleman and community leader, it provides insight into the manner in which he was living at Berkeley Hundred. His possession of three featherbeds, three pairs of old sheets, and three pillows raises the possibility that he and his housemates, the Rev. Robert Pawlett and William Tracy, lived in the large and sturdily built structure in which the Berkeley Hundred colonists were supposed to take their meals and hold worship services, as they had been instructed to share accommodations. Thorpe was credited with two pewter porringers, six small porcelain dishes, a gilded bowl, a silver cup, five silver spoons, a knife in a case, a pair of knives with gilded handles, and a gilded spoon. Some or all of these items most likely were his personal possessions, although some objects may have belonged to his housemates. However, Thorpe also was credited with a brass baking pan, a large brass pot, several pie plates, platters and saucers, and twenty-three wooden trenchers, utilitarian objects that probably belonged to the Society of Berkeley Hundred. As the settlement's leader, they may have been in his possession at the time of his death. The trenchers' presence, along with four tablecloths and more than five-

and-a-half dozen napkins (roughly half of which were made of diaper, a cotton or linen fabric that was woven in a simple geometric pattern), suggests that many of the residents dined communally, in accord with the Society of Berkeley Hundred's instructions.

The late George Thorpe was credited with a substantial quantity of men's clothing, some of which may have belonged to some of the settlement's other gentlemen, perhaps his housemates. On hand were five black suits, two of which were made of satin, and one "very old" suit that was ash-colored or gray. Thorpe also had an assortment of black cloaks, most of which were old. One was made of shallot (a soft silk) and lined with plush. Two others were velvet-lined and a third one was interfaced with "stuffe," a coarse woolen. One of Thorpe's cloaks was made of taffeta, a luxury fabric.[37] He also had a mohair gown (a silky-soft garment made from the hair of Angora goats) and a waistcoat made of a wool-silk blend known as fustian. The men who appraised Thorpe's estate noted that he was in possession of a velvet jerkin, a type of sleeveless jacket whose laces would have allowed it to fit snugly. The appraisers proffered that the jerkin was "Mr. Owldsworths," a reference to Arnold Oldisworth, who arrived at Berkeley Hundred on January 29, 1621, was appointed to the Council of State, and died sometime before July 16, 1621.[38]

Thorpe's inventory included seven old doublets or close-fitting jackets. Two were made of shamways or chamois and the remainder were fabricated of grosgrain silk or satin. He also had five pair of silk stockings, two pair of linen stockings, and one pair made of worsted, along with a pair of black silk garters and an ample supply of points, metal-tipped laces that would have enabled Thorpe to fasten his clothing.[39] Thorpe's wardrobe included four pairs of hose plus a pair of old trunk hose (puffed breeches or "pumpkin pants") that the appraisers disdainfully described as "quite out of fashion." He also had some enameled spurs. He apparently liked to keep his head covered while he slept, for he owned a decorated nightcap and an old quilted cap. Thorpe also had five Holland caps, headgear made from linen. His attire included a neck cloth and two kerchiefs. The presence of thirty-four falling bands suggests that he had been storing some of the five dozen brought to Virginia as part of the Berkeley Hundred settlers' supplies.

Thorpe's household furnishings included a pair of curtains, a bolster, an ample supply of sheets and pillow bears, and cloths for three cupboards. He had custody of two large chests and five trunks, one of which was iron-bound and contained documents; two books; Berkeley Hundred's colors; and bow strings, probably to a longbow or crossbow, both weapons documented archaeologically in early Virginia. Thorpe was credited with utilitarian items such as a copper still, a box of spices, a barrel of tools, fish hooks, sea compasses, and a quadrant. He also had a case of bottles to which Mr. Mathews (probably Samuel Mathews) laid claim and three runlets of a beverage purportedly produced in Virginia, perhaps the artificial wine that the Berkeley settlers were supposed to make or the tasty beer that Thorpe claimed to have made from Indian corn. Also on hand was a runlet of Rosa Solis, or Rosolio, a bright yellow cordial best known as a restorative.[40]

The Berkeley Hundred Settlers at Jordan's Journey

During the early 1990s, when archaeologists from the James River Institute for Archaeology undertook excavations at Jordan's Journey, they surmised that the remains of a sizable, substantially built early-seventeenth-century dwelling, designated 44PG300 in the inventory of the Virginia Department of Historic Resources, was occupied by a relatively large group of people.

Evidence of nearly 150 early-seventeenth-century ceramic vessels was found at 44PG300. According to curator Beverly Straube, most were English and continental earthenwares. Stoneware from England and Germany made up more than one fifth of the total number of vessels, and there were numerous containers used for food service and beverage consumption. Fragments of Dutch, English, Portuguese, and Spanish coarseware storage jars, milk pans, butter pots, and cooking vessels were found in abundance and constituted nearly half of the ceramic vessels that were recovered. Iberian olive jars and Italian earthenware costrels were present, along with German blue-and-gray stoneware jugs from the Westerald region and brown stoneware from Frechen. Archaeologists also found pottery that had been produced at Jamestown, perhaps by Martin's Hundred potter Thomas Ward or one of his apprentices, who temporarily withdrew to Jamestown Island after the 1622 Indian attack.[41]

A nearly complete Portuguese tin-glazed cup was found at 44PG300 along with fragments of several delftware drug jars, chargers, plates, and dishes (see fig. 9). A Chinese export porcelain wine cup was unearthed that dates to the first quarter of the seventeenth century. A rarity in its day, it would have been owned by someone of elevated social status. Surprisingly few of the ceramic vessel fragments found at 44PG300 are associated with food consumption, perhaps because most of the site's occupants used pewter plates or treen, evidence of which rarely is recovered during archaeological excavations.[42]

Because slipware from Somerset County, England, was present in abundance at 44PG300, archaeologist Taft Kiser has surmised that the site's occupants were recent arrivals from the Bristol area of England. Most of the Society of Berkeley Hundred's settlers disembarked from Bristol, in southwestern England, and the Society's principal investors were based in nearby Gloucestershire.[43] Thus, archaeological and documentary evidence suggests that when the settlers took refuge at Jordan's Journey in the wake of the 1622 Indian attack, they lived communally at 44PG300.

Conclusion

The manifests of the ships *Margaret* and *Supply* reveal that the Society of Berkeley Hundred tried to anticipate their servants' and tenants' needs in Virginia. However, conspicuously absent from the manifests were any references to the personal belongings that Berkeley Hundred's gentlemen and their families brought to Virginia. George Thorpe's inventory reveals that his lifestyle was commensurate with his social status, for he had possession of an abundance of clothing, curtains, bedding, and case furniture, along with some porcelain dishes, a gilded bowl, some pewter porringers, a silver cup, and several silver spoons.

The few mentions of ceramics in Thorpe's household inventory, though vague, offer an important snapshot for ceramic historians. The listing of "6 litle pursline [porcelain] dishes" is probably the earliest documentary confirmation of Chinese porcelain in Virginia and possibly all of the North American colonies. The "6 small platters & 2 sawcers" are most likely English earthenwares, as are the "2 pie plates, 2 deepe dishes." When examined in the context of the inventory document, ceramics appear to occupy a meager position in the Thorpe household. However, the fuller picture of ceramic usage in early Virginia is clearly dependent on archaeology for understanding the universe of fragile ceramics which rarely survive the passage of time. Curators of seventeenth-century "period rooms" should take particular note of the culturally diverse profusion of color, pattern, and form represented in the cosmopolitan ceramics assemblages of early Virginia. When both historical and archaeological evidence are combined, it is clear that early Virginia colonists had a rich and varied material culture in what is sometimes perceived as a monochromatic frontier existence.

ACKNOWLEDGMENTS

The authors wish to acknowledge the kind assistance of Taft Kiser, who drew on his considerable knowledge of early Virginia archaeology. We also thank Dee DeRoche, chief curator, Virginia Division of Historic Landmarks, and Carol Marks Bowman, executive director of the Prince George Historical Society. We are indebted to Joseph P. Gromacki for arranging photography of objects from his collection.

1. Lothrop Withington, *Virginia Gleanings in England* (Baltimore, Md.: Genealogical Publishing Company, 1980), p. 73; Eric Gethyn-Jones, *George Thorpe and the Berkeley Company* (Gloucester. U.K.: Alan Sutton Publishing, 1982), pp. 33–36; Alexander Brown, *The Genesis of the United States*, 2 vols. (Boston and New York: Houghton, Mifflin, 1890), 2:1031.

2. Gethyn-Jones, *George Thorpe*, pp. 36–38, 42–43, 48–49; British Archives, London, Public Record Office (hereafter PRO), Class: Court of Exchequer 407, fols. 37, 41; John Smyth of Nibley Papers, New York Public Library, Manuscript Nos. 14–16 (microfilm, Rockefeller Library, Colonial Williamsburg Foundation, Williamsburg, Virginia); Susan M. Kingsbury, *Records of the Virginia Company of London*, 4 vols. (Washington, D.C.: Government Printing Office, 1906–35), 1:332.

3. She may have produced two short-lived sons, John and Richard, who were baptized at St Martin-in-the-Fields, her home parish, on January 28, 1602, and October 6, 1603, respectively. Gethyn-Jones, *George Thorpe*, p. 43.

4. Ibid., pp. 43–45, 50 n. 31, 56; Philip Barbour, ed., *Travels and Works of Captain John Smith, President of Virginia and Admiral of New England, 1580–1631*, 3 vols. (Chapel Hill: University of North Carolina Press, 1986), 2:282–83, 370.

5. Smyth was steward of the "Hundred and Liberty of Berkeley," whose focal point was Berkeley Castle in Gloucestershire (Gethyn-Jones, *George Thorpe*, p. 18). All four Society members were Gloucestershire men.

6. The boy may have been one of the natives Sir Thomas Dale brought from Virginia in 1616 aboard the ship that also carried Pocahontas and John Rolfe. In 1904 Lothrop Withington noted that on September 19, 1619, a Georgius Thorpe was baptized at St Martin-in-the-Fields. When he was buried on September 27, he was identified as a "Homo Virginiae," or man of Virginia (Withington, *Virginia Gleanings*, p. 75).

7. Brown, *Genesis of the United States*, 2:1005; Kingsbury, *Records of the Virginia Company*, 3:130–34, 136.

8. Kingsbury, *Records of the Virginia Company*, 1:332, 336, 347, 371, 375, 379; 3:138–39, 148, 151, 193–96, 199–201, 207–10, 374, 240; 4:185.

9. Kingsbury, *Records of the Virginia Company*, 3:178–86. While at sea, the food for the ship's passengers would have included limited amounts of dried fish, ship biscuits (hardtack), butter, cheese, peas, oatmeal, oil, vinegar, and perhaps pork. In addition to water, they would have consumed beer and cider (Ferrar Papers Manuscript No. 210 [microfilm, Rockefeller Library, Colonial Williamsburg Foundation, Williamsburg, Virginia]).

10. Kingsbury, *Records of the Virginia Company*, 3:109–14, 197–99, 210–11, 213, 230.

11. Ibid., 3:130–34, 196, 199–201, 207–10, 375, 403.

12. Ibid., 3:122–29, 305, 377, 448–49; Ferrar Papers, Manuscript No. 194; Gethyn-Jones, *George Thorpe*, pp. 33–34.

13. He probably was referring to despondency, a malady exacerbated by malnutrition.

14. Kingsbury, *Records of the Virginia Company*, 3:417–18; Ferrar Papers Manuscript No. 322.

15. Kingsbury, *Records of the Virginia Company*, 1:382; 3:134, 271–74, 289–91, 293 367–68, 374–81, 393–96.

16. Arnold Oldisworth, a gentleman and kinsman of George Thorpe, disembarked at Berkeley Hundred and stayed on. Martha W. McCartney, *Virginia Immigrants and Adventurers, 1607–1635: A Biographical Dictionary* (Baltimore, Md.: Genealogical Publishing Co., 2007), p. 522.

17. Kingsbury, *Records of the Virginia Company*, 3:385–94, 396–406, 410–11, 426; Barbour, ed., *Travels and Works of Captain John Smith*, 2:20–21.

18. McCartney, *Virginia Immigrants*, p. 540.

19. Kingsbury, *Records of the Virginia Company*, 3:446–49, 469.

20. Ibid., 3:470.

21. Ibid., 3:462, 470, 583–84; Barbour, *Travels and Works of Captain John Smith*, 2:287, 293; Smyth of Nibley Manuscript No. 43.

22. Kingsbury, *Records of the Virginia Company*, 3:553.

23. Colonist William Caps, a hardliner who tended to be bombastic, blamed George Thorpe's soft-heartedness for the Indian attack. Kingsbury, *Records of the Virginia Company*, 4:76.

24. Barbour, *Travels and Works of Captain John Smith*, 2:294; Kingsbury, *Records of the Virginia Company*, 3:552–53, 555, 567. Thomas Thorpe, a Society of Berkeley Hundred tenant killed at Berkeley Hundred, may have been one of George Thorpe's kinsmen.

25. Barbour, *Travels and Works of Captain John Smith*, 2:301; Kingsbury, *Records of the Virginia Company*, 3:612, 652.

26. Gethyn-Jones, *George Thorpe*, p. 277; Smyth of Nibley Manuscript No. 3; McCartney, *Virginia Immigrants*, pp. 152–53, 211, 298, 325–26, 357, 494, 531, 568, 636.

27. Kingsbury, *Records of the Virginia Company*, 4:246.

28. Gethyn-Jones, *George Thorpe*, pp. 246–49; Virginia Land Office Patent Book 1, Pt. 1, p. 410; Smyth of Nibley MS 41, 42.

29. PRO Class: Chancery 142/395 f 107; Exchequer 133/155; Gethwyn-Jones, *George Thorpe*, pp. 266–68.

30. Assembly minutes and court records for 1623 and 1624 are fragmentary. However, they reveal that after the devastating Indian attack in 1622, Virginia officials focused on the colonists' safety and security and tried to weaken the natives by undertaking systematic retaliatory raids. In addition to the turmoil that resulted from the native assault, another destabilizing change occurred: revocation of the Virginia Company's charter.

31. When Eric Gethwyn-Jones examined Thorpe's inventory in 1982, he questioned the April 10, 1634, date assigned by its transcriptionist, Council Secretary Benjamin Harrison or his clerk. More recent research establishes April 10, 1624, as the correct date. On March 10, 1624, Sir George Yeardley was ordered to have the estate appraised. The General Court convened in early April 1624 and had adjourned by April 17. Afterward, Thorpe's estate was settled. Moreover, by 1634 Sir George Yeardley and all three of the three men who had appraised Thorpe's estate and signed the inventory were dead. Gethwyn-Jones, *George Thorpe*, p. 218; William W. Hening, ed., *The Statutes At Large: Being a Collection of All the Laws of Virginia*, 13 vols. (Richmond, Va.: Samuel Pleasants, 1809–23), 1:120–21; McCartney, *Virginia Immigrants*, pp. 135, 531, 631, 774.

32. Biggs died in 1626, Sharpe and Yeardley in 1627, and Palmer died in 1629. McCartney, *Virginia Immigrants*, pp. 135, 531, 631, 774.

33. Smyth of Nibley Manuscript No. 43.

34. H. R. McIlwaine, ed., *Minutes of Council and General Court of Colonial Virginia* (Richmond, Va.: The Library Board. 1924; Library of Virginia, 1979), pp. 11, 17–18; Hening, *The Statutes At Large*, 1:120–21; Smyth of Nibley Manuscript No. 43.

35. McIlwaine, *Minutes of Council*, pp. 36, 40, 47–48, 73–74; Peter Wilson Coldham, *The Complete Book of Emigrants, 1607–1660* (Baltimore, Md.: Genealogical Publishing Co., 1987), p. 118. Traditionally, a decedent's greatest creditor was named his administrator.

36. William Thorpe was baptized in his infancy at Berkeley Church on September 2, 1612, and in 1634 would have just come of age. His mother, Margaret Harris Thorpe, made her will on June 29, 1629, and died sometime before February 1630. She made bequests to her three living children: sons William and John and daughter Margaret. Gethyn-Jones, *George Thorpe*, pp. 56, 269.

37. Ten ells (or roughly 10½ yards) of taffeta sarcenet, a soft silk, were included in the manifest of the *Margaret*, raising the possibility that Thorpe's cloak was made in Virginia, perhaps by Berkeley Hundred tailor Christopher Bourton.

38. Smyth of Nibley Manuscript No. 43; McCartney, *Virginia Immigrants*, p. 522.

39. Laces fastened a man's breeches through holes in his doublet that permitted the shiny aglet tips to dangle on the exterior.

40. Smyth of Nibley Manuscript No. 43. Rosa Solis was made from a mixture of sundew (an insectivorous bog plant), aqua vitae, and hot, provocative spices (like cubebs, grains of paradise, and galingale) that had been blended together and then distilled. When cordial waters first arrived in England during the late 1400s, they were considered medicinal and were supposed to invigorate the heart and revitalize the spirits. Rosa Solis gained popularity in social situations and, according to seventeenth-century writer William Salmon, "stirs up lust." In fact, in Salmon's day it was used at banquets as an accompaniment to lascivious food items (www.historicfood.com/rosolio.htm).

41. Tim Morgan et al., *Archaeological Excavations at Jordan's Point: Sites 44PG151, 44PG300, 44PG302, 44PG303, 44PG315, 44PG333*, 2 vols. (Williamsburg, Va.: Virginia Company Foundation, 1995), pp. 193–234, 236; McCartney, *Virginia Immigrants*, p. 719.

42. Beverly A. Straube, *Finds from Jordan's Journey, A 1620–1635 Settlement in Prince George County, Virginia* (Richmond. Va.: Virginia Department of Historic Resources, 1996), p. 52; Morgan et al.. *Archaeological Excavations*, p. 246.

43. Robert Taft Kiser, "The Ceramics Missed by the Muster: Jordan/Farrar 44PG302, 1620–1635," manuscript on file, Virginia Department of Historic Resources, Richmond, 1994, p. 39; McCartney, *Virginia Immigrants*, p. 63.

Appendix

A Copy of the Inventory of George Thorpe, Deceased March 22, 1622 (Smyth of Nibley Manuscript No. 43)

An Inventorie of all and singular the goods and effectes of Captayne George Thorpe Esquire deceased, valued and prised the 10th day of April Anno 1634 [1624] by Samuell Sharpe, Richard Biggs and Thomas Palmer as followeth, vizt.

	lbs. tobacco
Imprimus a wrought challot cloak lined wth plush	0100
Item an old bl: silke grogran suite	0050
Item an old bl: cloth cloake lined wth velvett	0050
Item an old bl: satten suite	0030
Item a ffust. Wastecoate	0008
Item 4 pair of old silk stockings	0020
Item one paire of old enameld spurs	0000½
Item one pair of bl silke garters	0003
Item 3 dozen of old silke points	0001
Item a knott of Bowstrings	0000½
Item 2 paires of old lynen stockings	0000½
Item 2 pa^r of gloves	0004
Item one Truncke	0005
Item a walnutt chest wth locke & key	0008
Item a bl: Satten Suite	0040
Item an old bl: silke grograne dublett & old satten hose	0010
Item a very old ash col: Satten suite	0020
Item 4 paire of sheetes	0026
Item 5 very old shirtes	0005
Item 2 cupboard clothes	0005
Item 2 pillow beares	0002
Item sixe handerchirs	0001
Item an old wrought night capp	0003
Item an old quilt capp & 2 old holland capps	0001
Item 2 Towells	0001
Item 2 books & a small remnant of calico	0016
Item a pa^r of old worne shooes, galloshes & pantaples	0005
Item 33 fallinge bands	0004
Item 8½ yards of coarse holland	0016
Item a paire of gold waights	0001
Item a paire of fine old sheets & a pillowbeare	0008
Item 33 coarse napkins	0012
Item 3 coarse table cloathes	0012
Item 3 old holland capps, 2 kerchers, a neck cloath & two napkins	002
Item 16 pa^r of diaper napkins, a table cloath, a cupboard cloath & a towell	040
Item a fallinge band & a candlesticke	000¼
Item 2 pie plates, 2 deepe dishes & a voyder	008
Item 6 small platters & 2 sawcers	006
Item 2 pa^r of swasse [Swiss] compasses, 23 wooden trenchers, a small lookinge glasse, 6 litle pursline [porcelain] dishes, a pa^r of sisers, 2 pewter porringers, with a few white fish hookes and lynes and a torne trunke past use	005
Item one trunke marked AC	003
Item 2 small remnants of broad cloath	028
Item one old bl: cloath cloake garded [girded]	020
Item an old bl: stuffe cloake	015
Item an old bl: taffatie cloake	012
Item an old broad cloth cloake lyned with stuff	016
Item a velvett jerkin laced, said to be Mr. Owldsworths [Oldisworth's]	008
Item a pa^r of old satten hose	006
Item a pa^r of old grogran hose	020

Item a suite of very old bl: cloath 008
Item 2 very old shamways dubletts 008
Item a trunke marked with T very old [no value listed]
Item a pa^{r} of trunke hose quite out of fashion 010
Item a pa^{r} of blew silke stockins 006
Item 4 very old satten dubletts 030
Item 2 pa^{r} of very old satten hose 020
Item 2 old grogran clokes not ded worne 040
Item a taffatie skairfe [scarf] 005
Item an old bl: cloath cloake turnd & lyned with velvet 025
Item a cloth cloake lyned with plush 060
Item a pa^{r} of worsted stockins & a pa^{r} of wrought hangers & 3 pa^{r} of white gloves 004
Item an old case of bottles 004
Item a great chest 006
Item a pa^{r} of russet bootes 004
Item 2 litle old sea compasses, 3 peeces of copper & a box of beads 002
Item a greene carpett 012
Item a peece of foxe furr 001
Item a bla: feather, a hammer & an anvil 002
Item 2 swords 010
Item 3 sheets & a blankett 012
Item 3 old feather beds, 2 boulsters, a quilt & 2 ruggs 150
Item 3 pillowes & 2 pillowbears 002
Item 3 pa^{r} of old sheets & 2 old tapestry hangings 023
Item five coverlitts 020
Item a coverlitt of fox skinn & one of raccoons 005
Item a pa^{r} of old curtaines 003
Item a copper still, old 003
Item 4000 of 5d nayles 006
Item 2 sugger loufs [sugar loaves] 005
Item a brasse bakeinge pan with a hole in the side 002
Item great brasse pott 015
Item a litle iron pott 002
Item a mohaire gowne 020
Item all the plate, rings & seales, the plate wayinge 43 oz. 1/8 valued & prised 150
Sum totalis 1323¾
Other goods not valued nor prised
Item the collers [colors] of Barkley to the Governor
Item an iron bound trunke full only of wrightings and papers not to be prised
Item a case of bottles cont. 14 botles, the which Capt. Mathewes doth challenge to be his
Item a small runlet of Rosasolus and 3 runns. Of Virginia, which were drunke out amonge the people that fetcht downe his goods.
Item a quadrant not prised
Item a box with some spices spent upon Barkeley people
Item a guilt boule, a silver cupp, 5 silver spoones, a case with a knife & a guilt spoone, a paire of knives with guilt handles
Sir George Yeardlie and the priasers doe desire to be recompenced for theire paines, he in preservinge and keeping the goods, they in comminge and spending of theire tyme in praisinge the goods. Moreover Sir George doth require 400 lb. of tobacco which he paid for Captain Thorpe to those that built Opachankenses house.
Samuell Sharpe
The marke of Richard Biggs R.B.
The marke of Thomas Palmer T. P.
Vera copia Ben Harryson Clic. Con. [Clerk of the Council]

Richard Miller

Norwalk (Connecticut) Slip-script Pottery, the Potters, and Related Ware

▼ IN SEPTEMBER 1953 auctioneer King Brown of Wilton, Connecticut, sold the collection of antiques formed by Andrew L. Winton (1864–1946) and his wife, Kate Barber Winton (1882–1974), at their home in Wilton. The Wintons were both respected food scientists—Andrew once headed the Chicago Food and Drug Inspection Laboratory, and Kate had been a microscopist at the Connecticut Agricultural Experiment Station—and they had collaborated on books and articles on food science beginning in 1907. Seven years after Andrew died, Kate sold their collection "Consisting mostly of furniture etc. of New England origin, the majority of which is from Fairfield County"; included in the sale was a "collection of Norwalk slipware and pottery."[1]

The Wintons had researched Norwalk's pottery industry and summarized their findings in an article that was published in 1934.[2] Bringing their scientific training to the project, they established the extent of Norwalk pottery manufacturing beginning in the eighteenth century, uncovered the names of several of the town's early potters, and located descendants who retained pottery that was attributed to their forebears. The couple also investigated the sites of early potteries that had disappeared long before but evidence of whose existence remained underground. On one outing, "equipped with a potato hook and weeder," they excavated the site of a pottery they identified as having belonged to Asa Hoyt Sr. (1744–1806) and found "hundreds of fragments of pottery, furnishing indubitable evidence as to the wares of the pottery."[3]

Among the examples of Norwalk pottery illustrated in the Wintons' article were three dishes having slip inscriptions reading, respectively, "Susan," "Adaline," and "pony up the Cash."[4] No fragments of this slip-script pottery, whose decoration was very different from the concentric-circle-and-line decoration that they had uncovered elsewhere in Norwalk, were found. In the eighty years since the Wintons' article was published, Norwalk has emerged as an important center for the production of earthenware and stoneware pottery, and many examples with slip inscriptions have been located. Evidence suggests that slip-script earthenware dishes began as souvenir ware made at the pottery operated by Absalom Day about 1824. Production soon expanded under the direction of Day's nephew, Asa E. Smith (1798–1880), and a variety of personal, secular, and religious inscriptions were written on dishes over the next thirty years. This pottery offers important information about the town's residents and their everyday life that is often omitted from the written record.

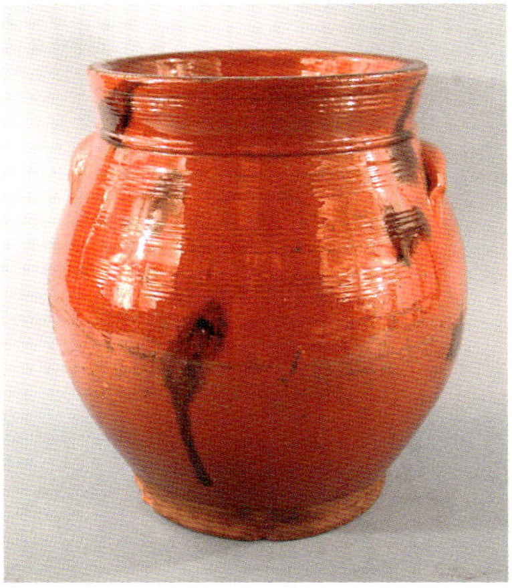

Figure 1 Jar, attributed to Day's Pottery, Norwalk, Connecticut, 1800–1825. Lead-glazed earthenware. H. 11¼". (Courtesy, Connecticut Historical Society.)

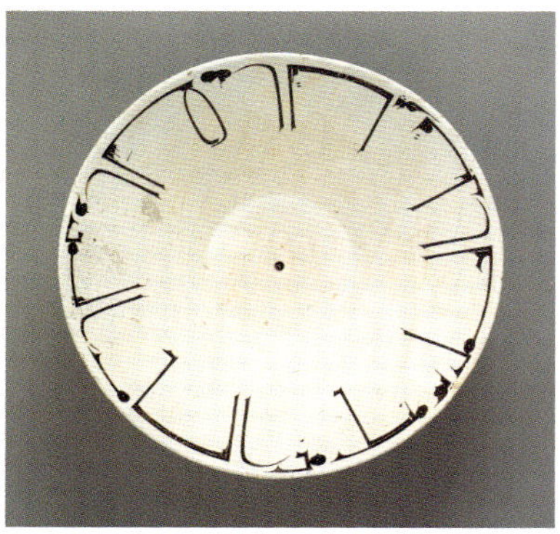

Figure 2 Bowl, Iran, Nishapur, tenth century. Slip-decorated earthenware. D. 18". Inscribed on obverse, in Arabic: "Planning before work protects you from regret; prosperity and peace." (Courtesy, The Metropolitan Museum of Art, 65.106.2.)

Pottery forms can yield information about production technologies, both local and regional preferences, foodways, the economics of the pottery industry, and cultural traditions. Those factors, along with function, determined form, and the form was paramount in successfully satisfying the purpose for which an object was intended. Decoration contributes nothing to the usefulness of pottery; identical forms, one decorated and the other plain, perform equally well. For Norwalk slip-script earthenware, however, functionality was secondary to decoration. Some dishes were used for preparing and serving food, but many show little evidence of use, indicating that the ornamental and commemorative value of the ware was primary. Many of the inscriptions document the names of original owners of the pieces, the foods they ate, the religious figures they revered, their heroes, and the biblical texts and everyday expressions that were meaningful to them.

In 1918 Albert Hastings Pitkin wrote: "Hartford seems to have been the center for the manufacture of Hollow Ware such as jugs, crocks, pitchers, etc.—So, from Norwalk came the heavy pie dishes, decorated with wavy lines of cream-colored slip, or with the owner's name."[5] Hollow ware was, in fact, an important component of Norwalk's pottery production, in addition to the dishes Pitkin identified. Settled in the seventeenth century, Norwalk developed along the Norwalk River, where it empties into Long Island Sound. The town's location made it well suited as a potting center. The sound was a major transportation thoroughfare linking Norwalk with other coastal Connecticut towns, Long Island, and New York City—all markets for the town's pottery. An "old potter's shop" probably owned by Asa Hoyt Sr. was in operation when British general William Tryon burned 130 Norwalk houses and numerous businesses in 1779, during the Revolutionary War.[6] After the war, potteries were reestablished: five were in operation at different times between 1825 and 1835, producing earthenware and stoneware, ceramic doorknobs, drawer pulls, and buttons. Several hollow ware forms splotched with manganese glaze are associated with both Norwalk and Huntington, Long Island, showing the close ties between potters in these towns (fig. 1). Norwalk's reputation throughout southern New England for quality earthenware led retailers as early as 1812 and as far away as Boston to refer to the town's products as "Norwalk ware." The market for slip-script ware, however, was local, centered primarily in Norwalk and nearby Fairfield County communities.[7]

How this distinctive pottery came to be made in Norwalk is unclear. Historical precedents exist, such as epigraphic pottery made during the Samanid Dynasty in the ninth and tenth centuries in what are now Iran and Uzbekistan (fig. 2). Norwalk's potters presumably had no firsthand knowledge of those dishes, bowls, and hollow wares, which feature written religious and secular sentiments that contrast with the dense, interlaced decorative patterns common on other Middle Eastern ceramics. First coated with a white engobe, they were then inscribed in blackish or purplish-brown slip in a stylized Kufic script that was also used to express the word of God in the Qur'an and on tiles decorating the façades and interiors of mosques and madrasahs. These stark inscriptions against a white ground (or white script on a dark ground) seem modern in appearance, and one

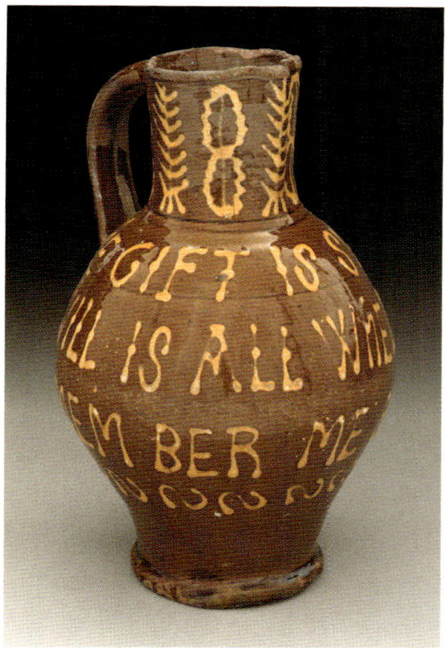

Figure 3 Jug, maker unidentified, Harlow, Essex, England, 1650–1700. Slip-decorated earthenware. H. 6⅞". Inscribed along sides: "THE GIFT IS SMALL GOO / D WILL IS ALL WHEN THES YOU SE / E REMEMBER ME BE MERY AND / WIES" (Courtesy, The Colonial Williamsburg Foundation.)

does not need to understand Arabic to sense the significance of what is written. Moreover, the inscriptions are often arranged in a circle and thus endlessly repeat, further emphasizing their meaning. Epigrams offering good wishes to the owner, practical advice for daily living, or celebrations of generosity are not uncommon. On these wares the inscription is the decoration, and meaning is communicated not through symbols but directly, with words.

Metropolitan slipware made at Harlow in Essex County, England, during the seventeenth century bears slip-script aphorisms and references to the Bible and the king (fig. 3). No European precedent for this pottery is known, and no example having a history of ownership in Norwalk has been found. The lack of evidence for the direct transfer of this style from the Middle East to Harlow and then to Connecticut suggests that ornamenting pottery with writing developed independently in places widely separated by distance and time. Norwalk's slip-script pottery may be viewed as the continuation of a practice that was used in the Middle East and found expression in England before it appeared in this New England town.[8]

Slip-script pottery was made elsewhere in Connecticut, New Jersey, and Pennsylvania both contemporaneously with Norwalk production and after it ended. Hervey Brooks (1779–1873), a potter in Goshen, Connecticut, produced a few initialed and dated pieces in the 1850s and 1860s (fig. 4), but his inscriptions are rudimentary and show no familiarity with Norwalk's products. Closer parallels are found in the work of an unidentified potter who may have worked in Pennsylvania. This potter made dishes with slip inscriptions that express sympathy for the temperance and antislavery movements that had gained widespread support in the first quarter of the nineteenth

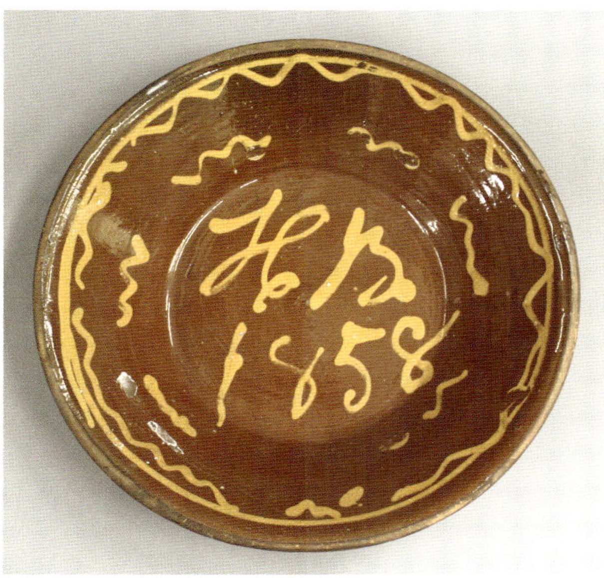

Figure 4 Milk pan, Hervey Brooks (1779–1873), Goshen, Connecticut, 1858. Slip-decorated earthenware. D. 13". Inscribed on obverse: "HB / 1858" (Courtesy, Litchfield Historical Society.)

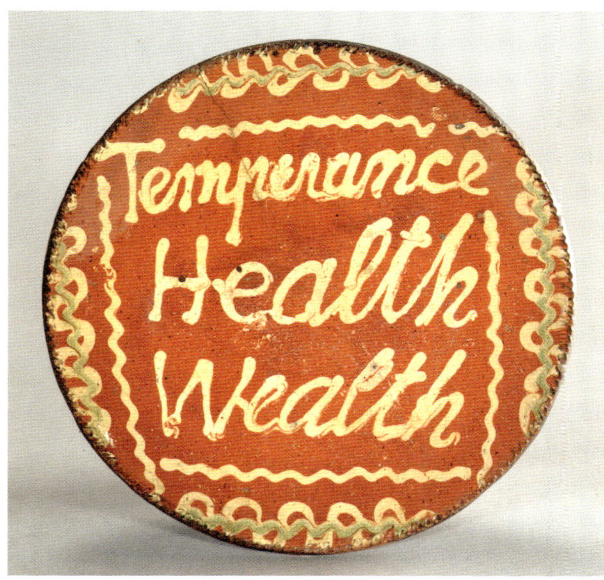

Figure 5 Dish, maker unidentified, probably Pennsylvania, ca 1835. Slip-decorated earthenware. D. 13½". Inscribed on obverse: "Temperance / Health / Wealth" (Courtesy, Historic New England.)

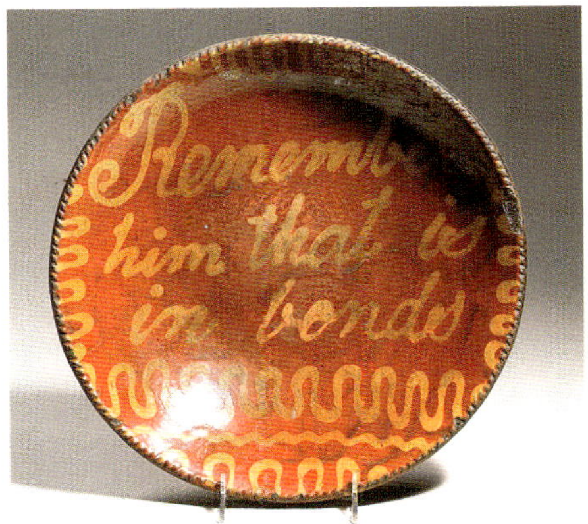

Figure 6 Dish, maker unidentified, probably Pennsylvania, ca. 1835. Slip-decorated earthenware. D. 14". Inscribed on obverse: "Remember / him that is / in bonds" (Courtesy, Sotheby's.)

century. Two dishes inscribed "Temperance / Health / Wealth" (fig. 5) indicate the benefits to be derived from abstinence. Another, counseling "Remember / him that is / in bonds" (fig. 6), relates to Hebrews 13:3 (KJV): "Remember them that are in bonds, as bound with them; and them which suffer adversity, as being yourselves also in the body." The inscription is a reminder that the bondage and adversity suffered by one are shared by all.[9]

It was in 1830 that Franklin County, Pennsylvania, potter George Wolfkiel (1805–1867) arrived in Bergen County, New Jersey, and acquired the pottery established in April 1827 by Henry Van Saun (d. 1829) in New Bridge (now River Edge).[10] Several dishes attributed to Wolfkiel are inscribed with the first name of women—including Mandy, Sally, Molly, Ginny, and Lucy—that ends with the descending letter "y," which is rendered with a long, coiled tail (fig. 7). A dish inscribed "Hard times / in Jersey" (fig. 8), referring to the financial panic of 1837, may be an early example of Wolfkiel's work.

Slip-script earthenware in Norwalk and elsewhere might not have been so appealing without the availability of public education. Newspaper and book publishing flourished during this time, spreading news of contempo-

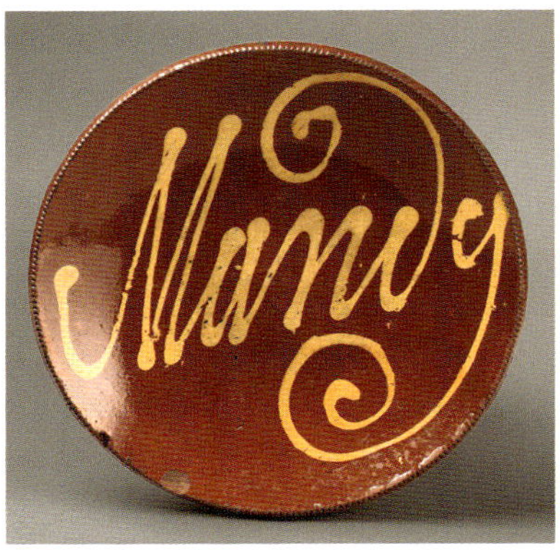

Figure 7 Dish, attributed to George Wolfkiel (1805–1867), Bergen County, New Jersey, 1847–1867. Slip-decorated earthenware. D. 12⅞". Inscribed on obverse: "Mandy" (Courtesy, Winterthur Museum, Garden and Library.)

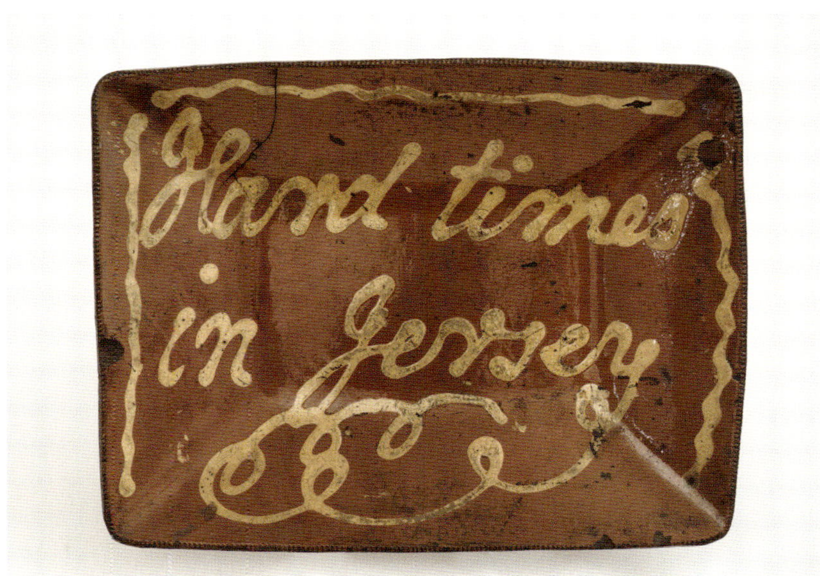

Figure 8 Dish, possibly George Wolfkiel (1805–1867), Bergen County, New Jersey, ca. 1837. Slip-decorated earthenware. 12¼" x 15⅝". Inscribed on obverse: "Hard times / in Jersey" (Courtesy, Wadsworth Atheneum Museum of Art, Gift of Mrs. Albert Hastings Pitkin in memory of her husband, 1918.1198.)

rary events and social movements. The result was that printers, publishers, and potters were confident that the literate public was large enough to make their enterprises worthwhile.[11]

A 1910 article titled "The Homely Pottery of Old New England" illustrated a "platter of dull red-ware with slip decoration. Made in Connecticut seventy-odd years ago."[12] The inscription on the dish appears to read "Angeline Cleveland," which the article states is inscribed on one example. Although the author did not say where in Connecticut this and the other inscribed pieces mentioned in the article were made, we know that Angeline Cleveland was the wife of William Cleveland, a Norwalk clerk. The couple was married in 1844, and the dish may have been made for Angeline about the time of her marriage.[13]

The 1850 Norwalk census lists nineteen men as potters, several of them related by blood or marriage. In 1847, when the Reverend William C. Hoyt, minister of the South Norwalk Methodist Church, and his wife, Betsey, were preparing to leave Norwalk for the reverend's next assignment, church members made a quilt as a going-away gift for the couple (fig. 9).[14] The quilt was composed of forty-eight individually sewn blocks of patterned, multicolored stars; in the center of each block is inscribed in ink the name of a church member. Several of the names belong to men whose occupations represent a sampling of the trades that were practiced in Norwalk, such as stone cutter Nathan D. Beers, hatters Nathan Kidney and Stephen Wilcox, wheelwright Samuel Waters, and shoemaker Hanford Wilcox.

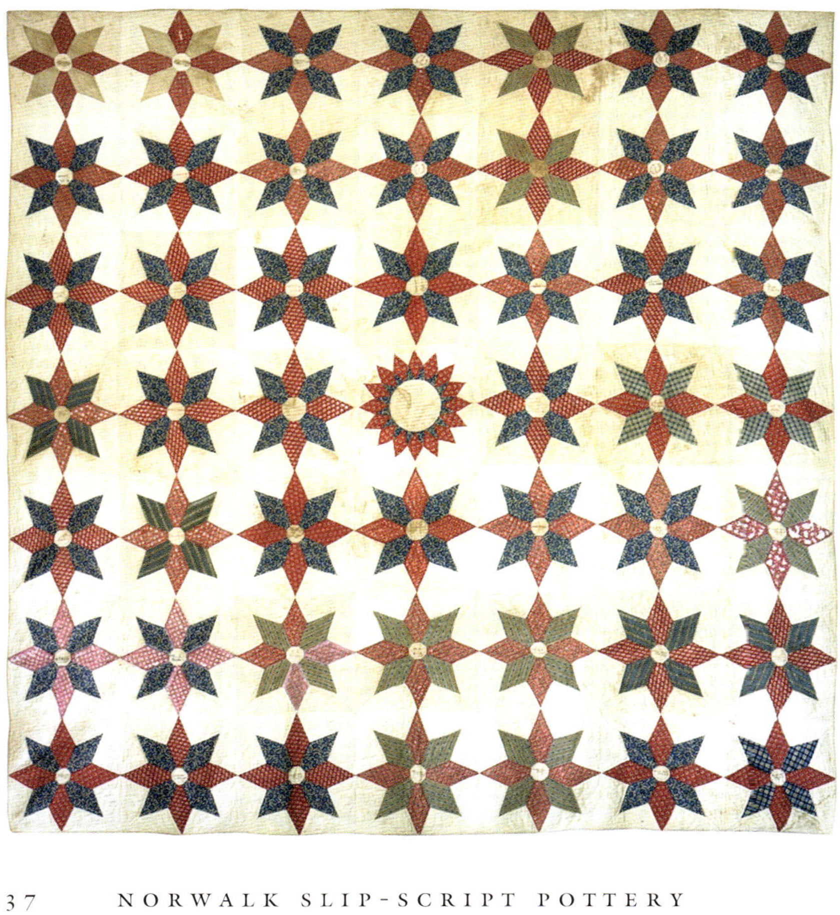

Figure 9 "Friendship" quilt, several makers, Norwalk, Connecticut, ca. 1847. Various printed cottons. 80" x 83". (Courtesy, Stamford Historical Society.)

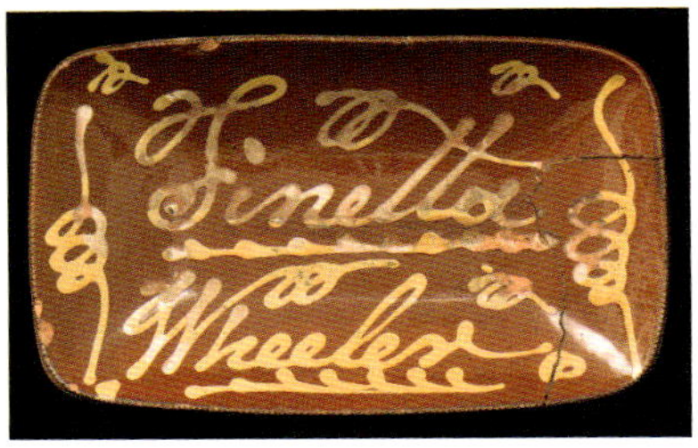

Figure 10 Dish, attributed to Smith Pottery, Norwalk, Connecticut, ca. 1844. Slip-decorated earthenware. L. 16". Inscribed on obverse: "Finetta Wheeler" (Courtesy, Norwalk–Village Green Chapter, Daughters of the American Revolution.)

Other blocks on the quilt bear the names of members of Norwalk's potting families, including Knapp, Merrill, Quintard, Selleck, Smith, and Waters. The names of the New Hampshire–born potter Lorenzo Dow Wheeler (1800–1864) and his wife, Finetta Lewis Wheeler (1805–1892), are inscribed on two of the quilt's blocks. The Wheelers moved from New York City to Norwalk between the births of their son Edward in 1828 and daughter Finetta in 1835. A rectangular dish inscribed "Finetta Wheeler" (fig. 10) may have been made for Mrs. Wheeler rather than the couple's daughter because of the resemblance of the form and the exuberant full-name inscription and infill designs to the dish made for another adult woman, Angeline Cleveland. In 1853 Wheeler was making "mineral knobs for doors, furniture, and shutters . . . composed of red, black, and white clays, mixed together, burned, and covered with ordinary Rockingham glaze."[15]

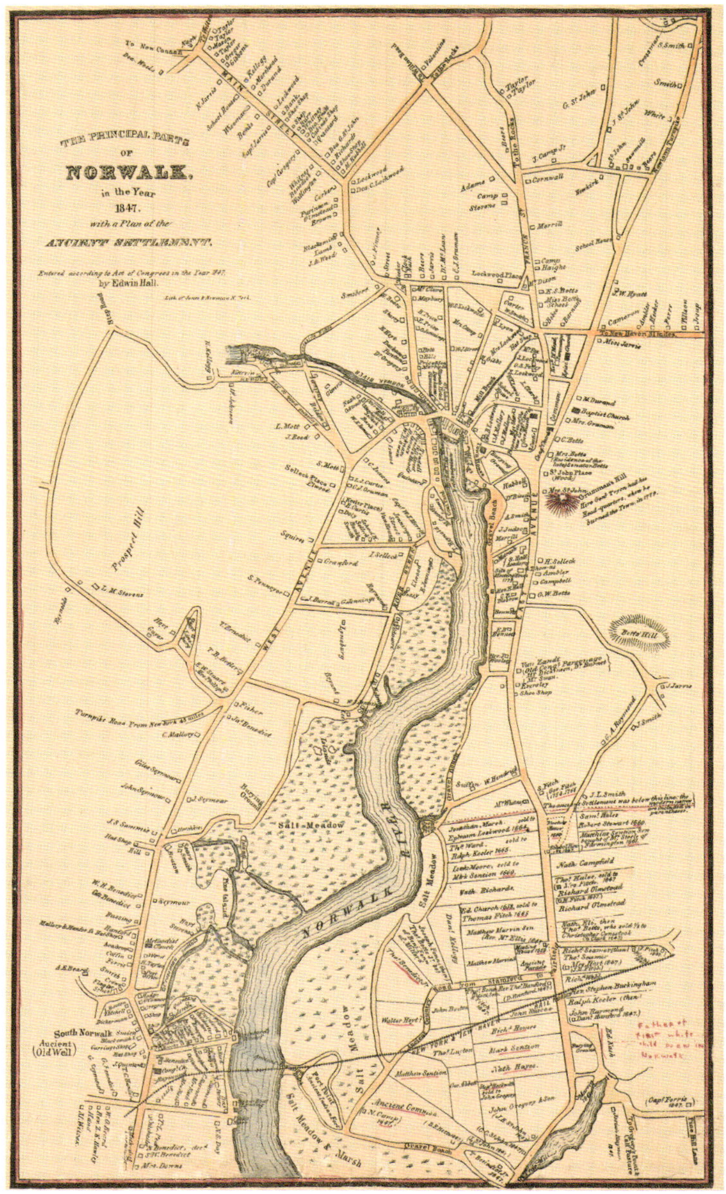

Figure 11 Map detail, *The Principal Parts of Norwalk in the Year 1847*, by Edwin Hall, printed by Jones & Newman, New York, New York, 1847–1849. Lithograph. Overall 17⅞" x 11". (Courtesy, Connecticut Historical Society.)

It has been suggested that the slip-script tradition was brought to Connecticut by New Jersey potters, although proof of this has not been found. The career of one Norwalk potter, however, provides circumstantial evidence of this transference. Absalom Day (1770–1843) was born in Day's Bridge (later Chatham), a settlement in Morris County, New Jersey, where the Day family had figured prominently since its founding.[16] Day was a potter in New Jersey before he arrived in Norwalk by 1792, the year he married Betsey Hoyt Smith (1777–1861). Day worked for Asa Hoyt Sr. at his pottery located at Washington and Water streets in Old Well, which later became part of South Norwalk. In 1796 Day acquired Hoyt's pottery, and it was still standing in 1847 when it appeared on a map of Norwalk (fig. 11) at the time it was operated by Day's nephew through marriage, Asa E. Smith, and his son Noah Smith Day (1798–1871).[17]

In 1797 Absalom Day advertised that he "continues the business of making & selling Earthen Ware" made of "clay imported from New-Jersey" (fig. 12).[18] Norwalk grew as a pottery production center despite not having

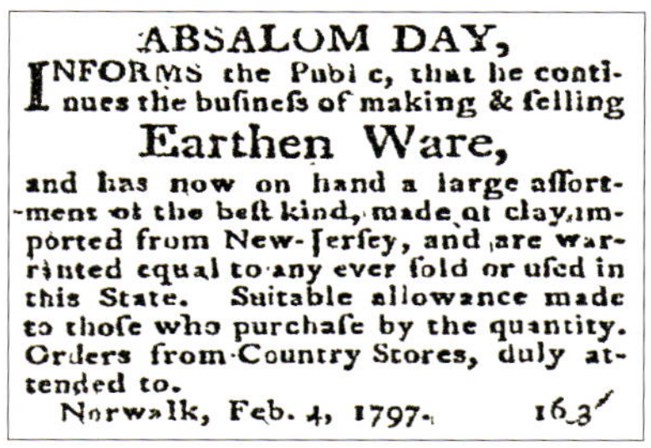
ABSALOM DAY,
INFORMS the Pubi c, that he continues the buſineſs of making & ſelling
Earthen Ware,
and has now on hand a large aſſortment of the beſt kind, made of clay imported from New-Jerſey, and are warranted equal to any ever ſold or uſed in this State. Suitable allowance made to thoſe who purchaſe by the quantity. Orders from Country Stores, duly attended to.
Norwalk, Feb. 4, 1797. 163

Figure 12 Announcement by Absalom Day dated February 4, 1797, published in the *Republican Journal* (Danbury, Conn.), February 4, 1797.

a potter's field nearby to supply clay of the quality and in the quantity that the potters required. Some potters got clay that had been dug in Huntington, Long Island, then shipped to Norwalk by sloop, but Day got his clay from New Jersey, which he knew from experience was plentiful and of high quality. Moreover, by fashioning New Jersey clay into Connecticut pottery, Day linked his birthplace and the town where he lived for the remainder of his life.

Absalom Day was also an influential Methodist preacher who is credited with establishing the First Methodist Church in South Norwalk. Prior to the church's construction in 1816, traveling Methodist ministers preached at Day's home. In 1802 Bishop Francis Asbury (1745–1816) "preached at the Old Well, at Absalom Day's, near Norwalk."[19] While on a preaching trip in 1818 the Reverend George Coles (1792–1858) visited Day's shop and observed an association between Day's potting activities and scripture:

> [After leaving New York City, I] passed to Norwalk, and put up with brother Absalom Day. . . . Brother Absalom Day followed the business of a potter. I looked into his workshop, and saw with interest what "power the potter hath over the clay, to make one vessel unto honor and another unto dishonor," as for example, of the same lump he would make two jugs—one for the purpose of holding milk, and the other rum.[20]

Coles saw this passage from Romans 9:21 manifested in Day's turning jugs for different purposes. Day knew the Bible and its references to potters,

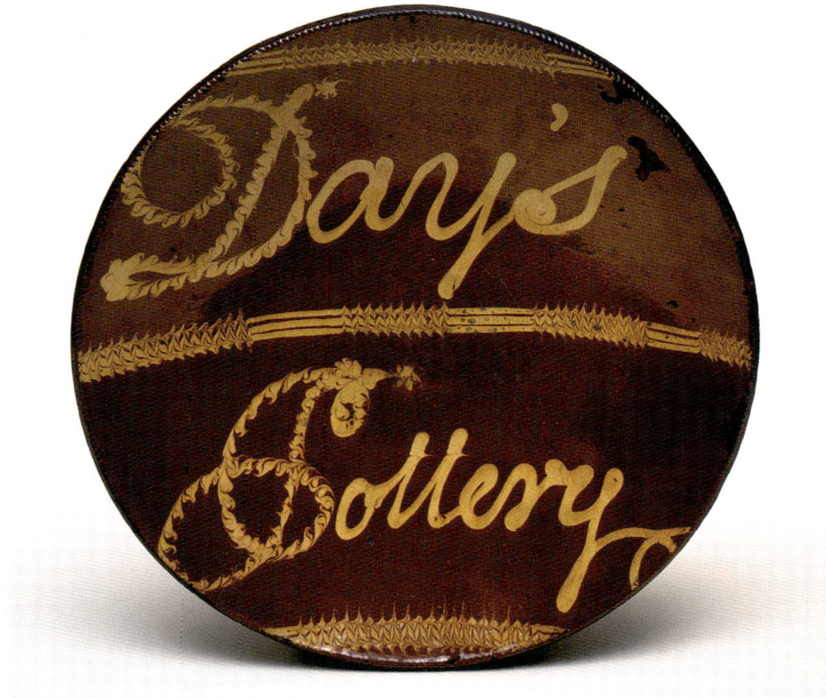

Figure 13 Dish, attributed to Absalom Day (1770–1843), Norwalk, Connecticut, ca. 1796. Slip-decorated earthenware. D. 15½". Inscribed on obverse: "Day's / Pottery" (Courtesy, High Museum of Art.)

and he would have been familiar with the verse. It probably had occurred to him that some of his jugs were utilized for purposes that were at odds with biblical teaching, but how they were used was out of his control. Day may have viewed his potting as an extension of his ministry, with each activity mirroring and reinforcing the other.

Two slip-script earthenware dishes attributed to Absalom Day are the earliest Norwalk examples known. One, measuring over fifteen inches in diameter, is inscribed "Day's / Pottery" (fig. 13) and was probably made to be displayed as a sign inside Day's workshop. Slip lines applied with a triple-spouted slip cup and combed at intervals complete the dish. The capital letters of the inscription were worked with a tool and have a feathery appearance, recalling that the term *feathering* was another word for *combing*. Combed slip is not typically associated with New Jersey or Connecticut earthenware, but it is seen frequently on slipware made in eastern Pennsylvania and in Philadelphia, so Day was perhaps familiar with the Philadelphia practice before he moved to Norwalk.[21] Day's dynamic script and the variety of decorative effects he incorporated into the dish's slip design offer unquestionable proof of his artistry and craftsmanship. The ornamental capital letters differ from the script on later Norwalk ware, which is closer in appearance to writing on paper. This, and Day's acquisition of the Hoyt pottery in 1796, indicate the sign was made before 1800.[22]

A second dish from the same time is a rectangular form ornamented with Day's conjoined slip-script initials AD (fig. 14).[23] Surrounding the initials is a wreath of wavy, looping lines; copper oxide glaze was applied liberally both to the letters and the infill decoration in order to add green to the

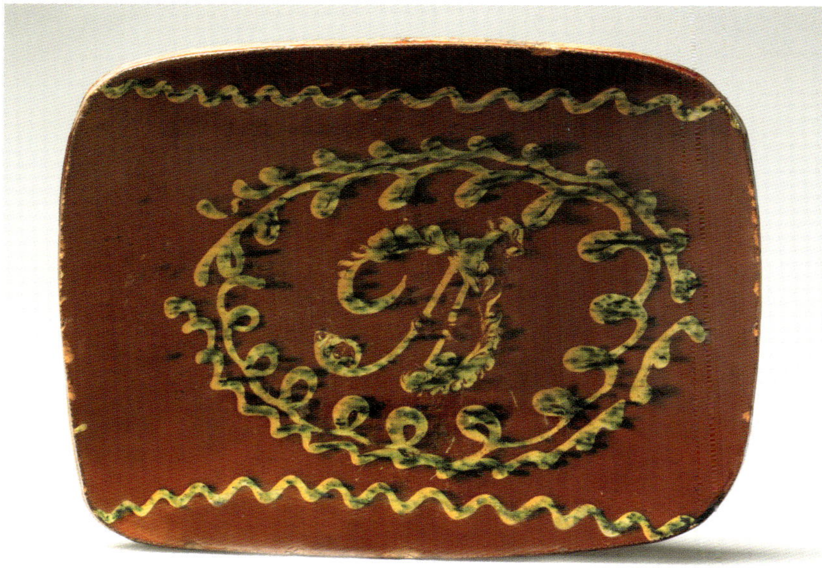

Figure 14 Dish, attributed to Absalom Day (1770–1843), Norwalk, Connecticut, 1794–1800. Slip-decorated earthenware. W. 16½". Inscribed on obverse: "AD" (conjoined initials). (Courtesy, The Metropolitan Museum of Art, Gift of Mrs. Robert W. de Forest, 1933, 34.100.42.)

design after firing. The use of copper oxide is also more commonly associated with Pennsylvania earthenware, but the Wintons found numerous sherds with green glaze during their excavations of the waster at the Hoyt pottery site.[24] Whether Day introduced Hoyt to the use of copper oxide glaze or was taught its use by Hoyt is unknown. These two dishes are the only slip-script wares that are firmly linked to Day, and both have inscriptions that refer to him. Slip-script pottery might not have been part of Day's regular production but instead was reserved for advertising and personal use, which could account for the lack of additional examples.

Figure 15 Dish, maker unidentified, Huntington, New York, probably 1851. Slip-decorated earthenware. D. 12¾". Inscribed on obverse with double-quill slip cup: "America" (Courtesy, Historic New England.) The inscription refers to the yacht *America*, which won the Royal Yacht Squadron's 100-Pound Cup sailing race around the Isle of Wight in 1851. Thereafter the race has been known as the America's Cup.

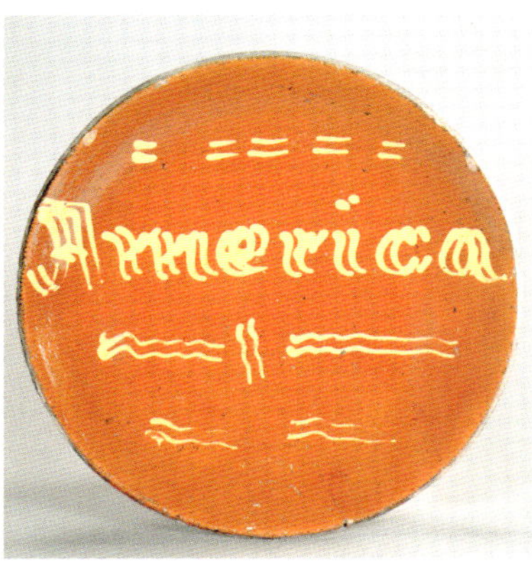

Day's influence on other potters in Norwalk was long-lasting. The Gregory and Betts families had settled in Norwalk in the seventeenth century, and the Norwalk landmarks Betts' Island, Betts Hill, and Gregory Point remain as evidence of their early association with the town. Born in Norwalk, John Betts Gregory (1782–1842) was apprenticed to Absalom Day, and it was in Day's shop that he probably learned the slip-script technique. Their association underscores the professional and familial connections that would spread the slip-script style. Sometime before 1807 Gregory "had a good business offer from Huntington, L. I. which he accepted."[25] This corresponds to the period when Samuel J. Wetmore operated a pottery in Huntington. Which pottery employed Gregory in unknown, but Huntington's proximity to Norwalk and the strong commercial exchange between the two towns explains why a talented young Norwalk potter was recruited by a Huntington pottery and how slip-script earthenware dishes came to be made in Huntington in the 1850s (fig. 15).

Gregory returned to Norwalk about 1809 and then moved to Clinton, New York, a farming and mercantile center in Oneida County and the home of Hamilton College. In 1810 he purchased the pottery on College Street established by Connecticut native Erastus Barnes (1773–1807); four years later he acquired Barnes's former house.[26] In May 1817 he announced in a

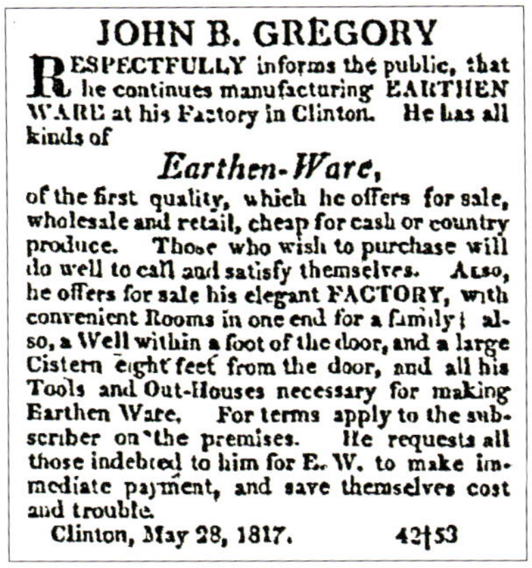
JOHN B. GREGORY

RESPECTFULLY informs the public, that he continues manufacturing EARTHEN WARE at his Factory in Clinton. He has all kinds of

Earthen-Ware,

of the first quality, which he offers for sale, wholesale and retail, cheap for cash or country produce. Those who wish to purchase will do well to call and satisfy themselves. Also, he offers for sale his elegant FACTORY, with convenient Rooms in one end for a family; also, a Well within a foot of the door, and a large Cistern eight feet from the door, and all his Tools and Out-Houses necessary for making Earthen Ware. For terms apply to the subscriber on the premises. He requests all those indebted to him for E. W. to make immediate payment, and save themselves cost and trouble.

Clinton, May 28, 1817. 42†53

Figure 16 Announcement by John B. Gregory dated May 28, 1817, published in the *Columbian Gazette* (Utica, N.Y.), June 3, 1817.

Utica, New York, newspaper (fig. 16) that he "continues manufacturing EARTHEN WARE at his Factory in Clinton." The advertisement also offered for sale his "elegant FACTORY, with convenient Rooms at one end for a family" and "all his Tools and Out-Houses necessary for making Earthen Ware."[27] Gregory did not put his house up for sale, so presumably he intended to remain in Clinton. If he had intended to give up potting, however, his plans changed—perhaps because the pottery did not sell—because he continued to manufacture in Clinton for another fourteen years.

Gregory must have made a significant amount of utilitarian earthenware in the twenty-three years he worked in Clinton, but only two signed dishes from this time are known. The first is the earliest dated example of Norwalk-style slip-script earthenware (fig. 17). Inscribed in manganese slip, the dish might have been displayed in Gregory's shop and perhaps was inspired by the memory of his former master's large advertising dish. Gregory's distinctive script can be identified by the capital letters B, D, and G, which are seen on several signed and attributed works.

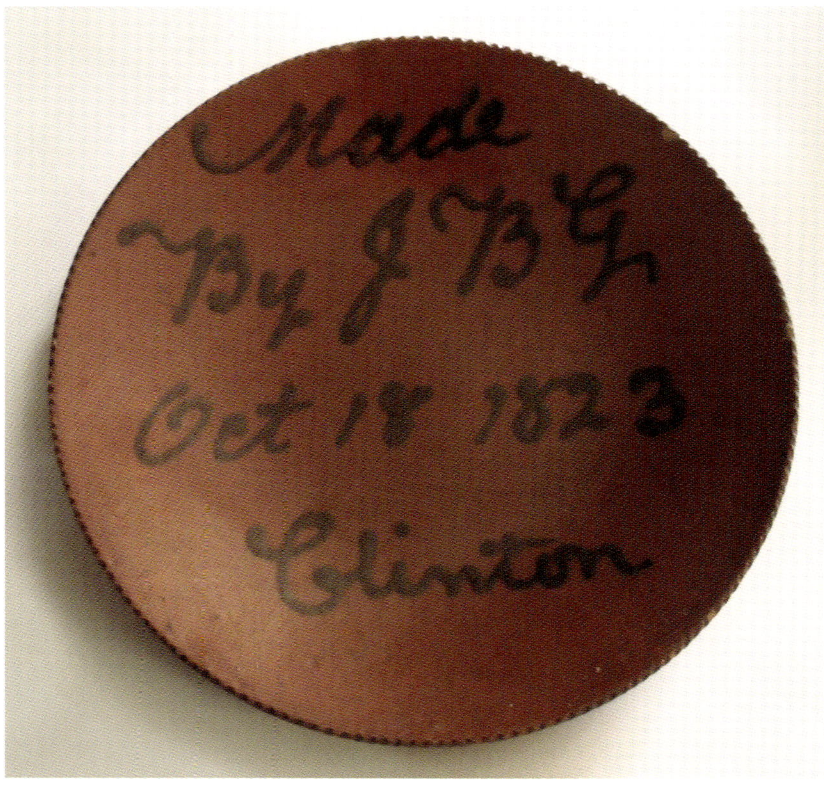

Figure 17 Dish, John Betts Gregory (1782–1842), Clinton, New York, 1823. Slip-decorated earthenware. D. 15⅝". Inscribed on obverse: "Made / By JBG / Oct 18 1823 / Clinton" (Courtesy, Clinton Historical Society.)

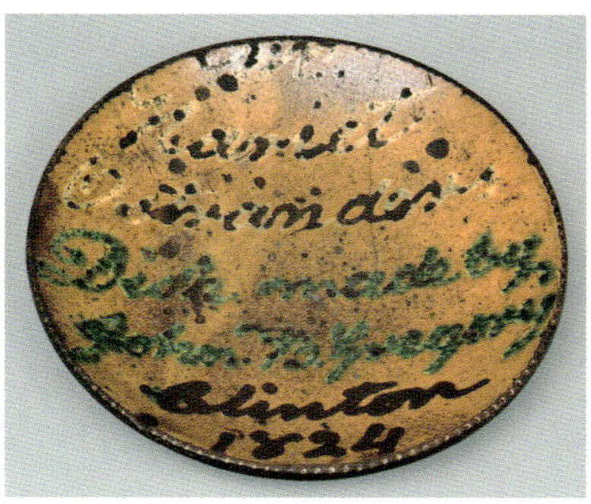

Figure 18 Dish, John Betts Gregory (1782–1842), Clinton, New York, 1824. Slip-decorated earthenware. D. 10½". Inscribed on obverse: "Harriet / Ostrander / Dish made by / John B Gregory / Clinton / 1824" (Courtesy, Crocker Farm.)

The following year, Gregory made a dish for Harriet Gregory Ostrander (1790–1832; fig. 18). Also born in Norwalk, Harriet was related to John; they were probably cousins. She and her husband, Captain William Ostrander, lived in Albany, New York, where William ran commercial sloops on the Hudson River. The inscription, written in white slip and manganese oxide draws at least as much attention to the dish's maker as to its recipient. Gregory's characteristic capital letters, the copper oxide, and inscriptions written in different colored slips differentiate his work from slip-script wares made later in Norwalk. His reference to this as a "Dish," a term that would appear frequently on Norwalk slip-script ware made later, is the earliest example of an American pottery form identified in this way.[28]

"Why / Will you / Die" is inscribed on the third dish attributed to Gregory's Clinton years (fig. 19). This chilling text has eluded interpretation, leaving questions about the nature of Gregory's personality and why he would create the dish. Gregory was remembered as a "devout Methodist" whose Clinton home was a gathering place for itinerant preachers and believers from 1819 to 1827, as Day's house had been in Norwalk.[29] Gregory expressed his religiosity and his knowledge of the Bible through his work. The inscription is taken from Ezekiel 18:31: "Cast away from you all your transgressions, whereby ye have transgressed; and make you a new heart and a new spirit: for why will ye die, O house of Israel?" The verse is a caution against spiritual death through sin. However, Gregory chose the most dismaying part of the verse to inscribe on the dish. By taking it out of its larger context, he seems to pose a question that most would be hard-pressed to answer if they considered it at all. But this fragment of a biblical passage would have been recognized by others who knew the complete verse. Gregory's combination of white slip, white slip accented with copper oxide, and iron oxide adds power to the inscription's meaning. Writing the word "Die" in iron oxide emphasizes the literal connotation of the word, its explicit meaning reinforced by the black letters.

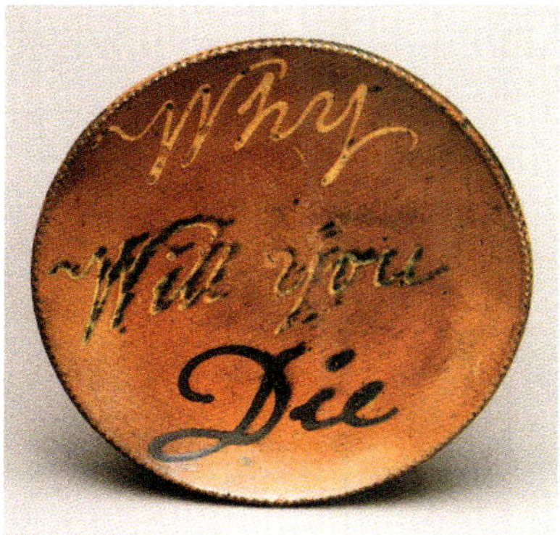

Figure 19 Dish, attributed to John Betts Gregory (1782–1842), Clinton, New York, 1823–1831. Slip-decorated earthenware. D. 11¾". Inscribed on obverse: "Why / Will you / Die" (Courtesy, George R. Hamell.)

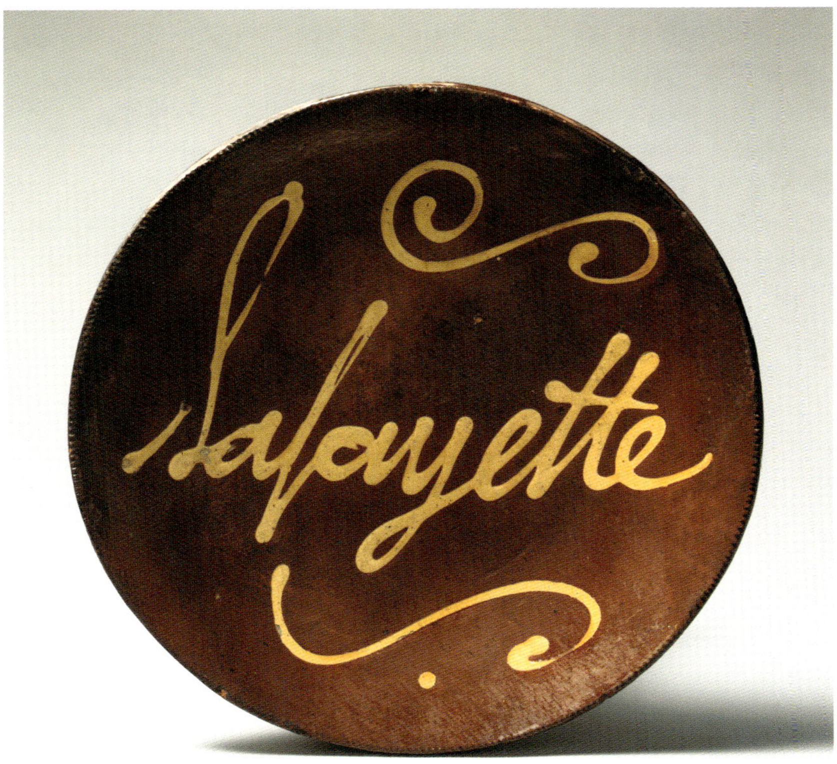

Figure 20 Dish, attributed to Day's Pottery or Smith Pottery, Norwalk, Connecticut, ca. 1824. Slip-decorated earthenware. D. 12¼". Inscribed on obverse: "Lafayette" (Courtesy, The Metropolitan Museum of Art, Gift of Mrs. Robert W. de Forest, 1933, 34.100.223.)

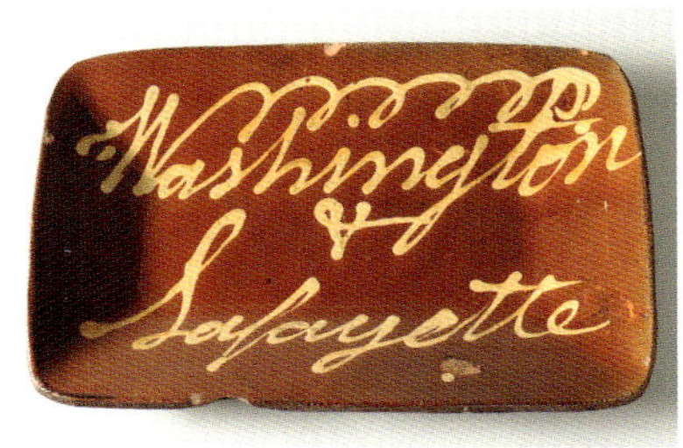

Figure 21 Dish, attributed to Day's Pottery or Smith Pottery, Norwalk, Connecticut, ca. 1824. Slip-decorated earthenware. D. 15½". Inscribed on obverse: "Washington / & / Lafayette" (Courtesy, Lewis Scranton.)

The earliest identified Norwalk commercial slip-script earthenware was probably made in 1824, when the marquis de Lafayette visited the town during his American tour. At the Norwalk Hotel, Lafayette "received a throng of Norwalk's best people in its parlors. The town went wild with enthusiasm and erected a triumphal arch on the bridge bearing in letters spelled in lights, 'Welcome Lafayette.'"[30] It was probably Day's Pottery that capitalized on the affection that Norwalk's citizens displayed for the last living Revolutionary War general by making dishes inscribed "Lafayette" and "Washington / & / Lafayette" (figs. 20, 21).[31] These dishes have been found in sufficient numbers to indicate they were popular souvenirs of the general's visit.[32]

In 1825 Absalom Day transferred control of his Norwalk pottery to his nephew Asa E. Smith, who had been Day's apprentice beginning in 1812. Smith presumably continued working with his uncle until he took over the shop. The reason Day stopped potting and directed his energies to his farm was likely ill health. Many potters suffered physical ailments associated with breathing clay dust and working with lead glaze, and at age fifty-five Day might have been unable to continue working. From November 1827 through February 1829, an advertisement for "Trotter's Genuine Pectoral of Horehound" carrying Day's written endorsement of the product appeared more than two hundred times in New York City's *Commercial Advertiser*. "For about four years," Day wrote, "I was afflicted with a violent cough, attended with all the symptoms of a confirmed pulmonary complaint. So great was my debility, that I was unable to attend to my business or public duties."[33] The appearance of Day's symptoms corresponds to the period shortly before he transferred the pottery to his nephew.

Probably encouraged by the success of the Lafayette dishes, Asa E. Smith expanded production, adding other inscriptions to his wares. To do so required the skills of a literate potter who was expert in the drape-molding process and the use of a coggling wheel. In 1686 English naturalist Robert Plot described the slip used by potters in Staffordshire, England, to decorate earthenware as being composed of clays that are "looser and more friable" than "throwing clays." Slip clay was "mixed with water they [the potters] make into a consistence thinner than a syrup, so that being put into a bucket it will run out through a quill; this they call slip, and is the substance wherewith they paint their wares."[34]

Writing with slip required that it be viscous enough to flow easily and adhere to the surface, but not so thin that it ran after it was applied. Great skill was required both to write with a single-quill slip cup in an attractive Spencerian script that fit comfortably on the dish and to avoid errant drips. The decorator needed to visualize the completed inscription before starting as there was only one opportunity to do it correctly. Misspellings abound, indicating that such errors did not automatically require a piece to be discarded. From the start, the inscriptions show little hesitation and appear to have been written quickly and confidently by a practiced hand. This decorator's self-assurance is seen in the S-scrolls above and below "Lafayette" and the repeating loops of slip trailing from the capital W in "Washington." The

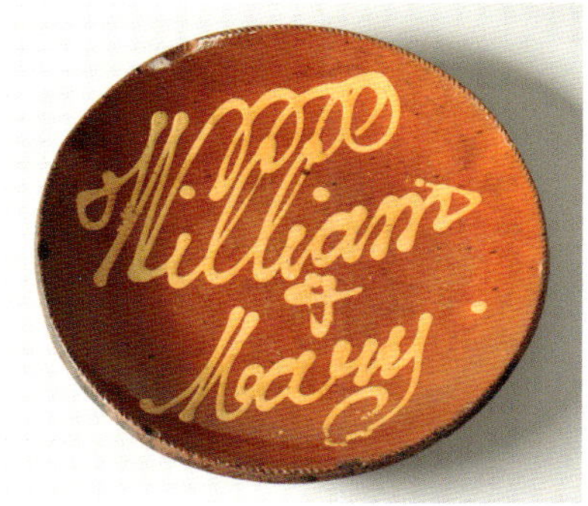

Figure 22 Dish, attributed to the Smith Pottery, Norwalk, Connecticut, 1825–1830. Slip-decorated earthenware. D. 11⅛". Inscribed on obverse: "William / & / Mary" (Photo, Skinner, Inc.)

capital W also appears on dishes inscribed "William / & / Mary" (fig. 22) commemorating the dual sovereigns who signed the English Bill of Rights that inspired resistance to British rule in North America.[35]

That same script is seen on Norwalk dishes made over the course of many years. The Wintons wrote, "It is stated that this workman was a man named Chichester." They did not identify their source, but the claim persisted and has evolved into a common belief that Henry Chichester was the decorator. Two Henry Chichesters lived in Norwalk in this period: Henry Chichester Sr. (1762–1849), who was born in Huntington, Long Island, and married Deborah Hoyt; their son Henry Jr. (1799–1847) was a potter who before 1825 made buttons, doorknobs, and earthenware in Norwalk in partnership with his brother-in-law James Quintard under the name "Chichester & Quintard." Henry Jr., therefore, might be a candidate. However, no evidence that either Henry Sr. or Henry Jr. had worked for the Smith Pottery has been found. Moreover, a Norwalk dish dated 1854 was made several years after both men had died.[36] As satisfying as it would be to identify the decorator with confidence, the inability to do so does not diminish this potter's achievement or the significance of his contribution to the history of Norwalk pottery.

Gifting pottery, whereby parents gave pottery to their children when they were young or to assist them in setting up households as they approached adulthood, was probably more common in early America than is generally realized. Norwalk slip-script dishes bearing names and initials outnumber other inscriptions by a wide margin. These dishes probably represent a small fraction of the total number made, indicating that the practice of making personalized pottery to commemorate births, marriages, and other milestones in life was fairly common in Norwalk. Slip inscriptions fall into several categories: names and initials, foods, place-names, biblical texts, patriotic sentiments, aphorisms, and names of saints. Numerous dishes with inscriptions such as "Nancys / Dish," "Lucys Dish," and "Sarahs Dish"

Figure 23 Dish, attributed to the Smith Pottery, Norwalk, Connecticut, 1825–1850. Slip-decorated earthenware. W. 11¾". Inscribed on obverse: "Nancys / Dish" (Courtesy, Yale University Art Gallery, 1931.1788.)

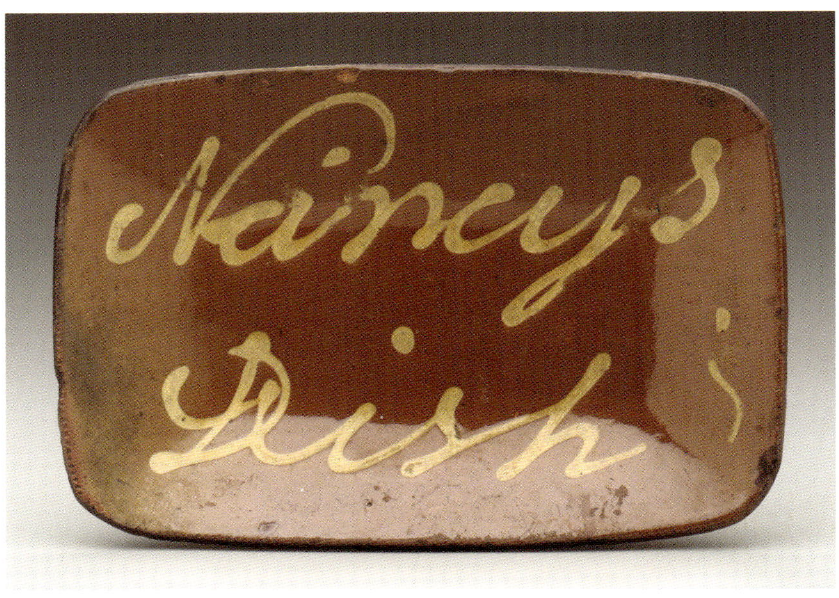

(fig. 23) were likely made for children, similar to the personalized dishes and bowls that parents purchase for their children today.

Others bear only the initials of individuals, including six that are inscribed with both "CB" and "WB"; one each lettered "HB," "NB," and "SB"; and three initialed "DE." The reason for the duplicates and the surnames involved are not known, but it is possible they were made for members of the same two families. Full names, such as "Angeline Cleveland" and "Finetta Wheeler" mentioned earlier, appear only occasionally. A dish inscribed "Francis / Smith / Norwalk" (fig. 24) might identify Francis H. Smith, the infant son of Asa Smith's brother Ward. Born in New York City in 1850, Francis died at age two months and was buried in Norwalk. If the dish was made to celebrate the child's birth, its significance soon changed to become a poignant reminder of the family's loss.[37]

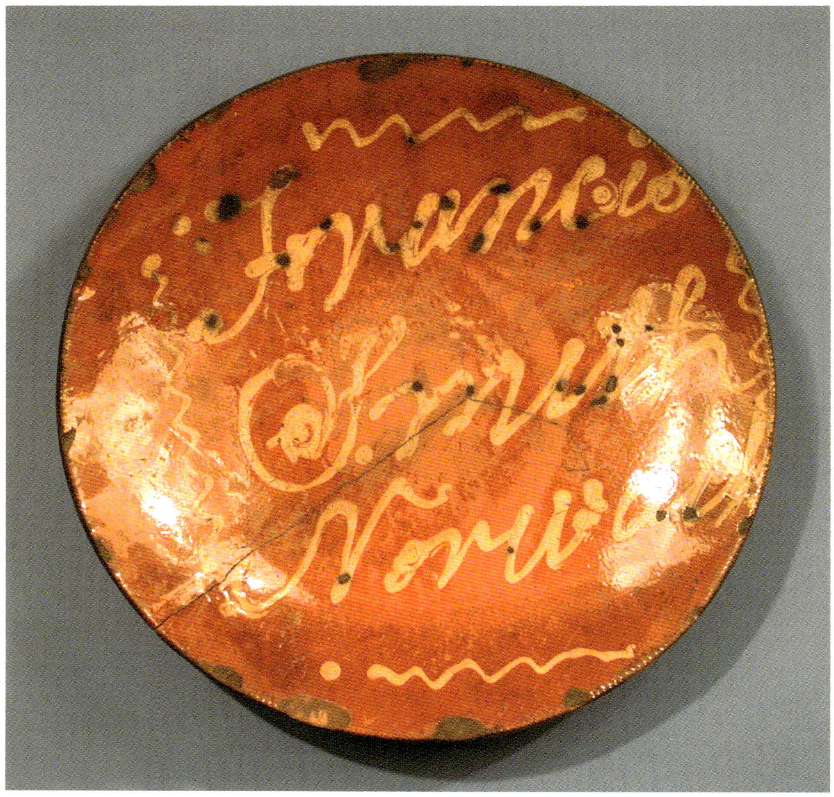

Figure 24 Dish, attributed to the Smith Pottery, Norwalk, Connecticut, 1825–1850. Slip-decorated earthenware. D. 14¼". Inscribed on obverse: "Francis / Smith / Norwalk" (Courtesy, The Connecticut Historical Society, 1961.26.1.)

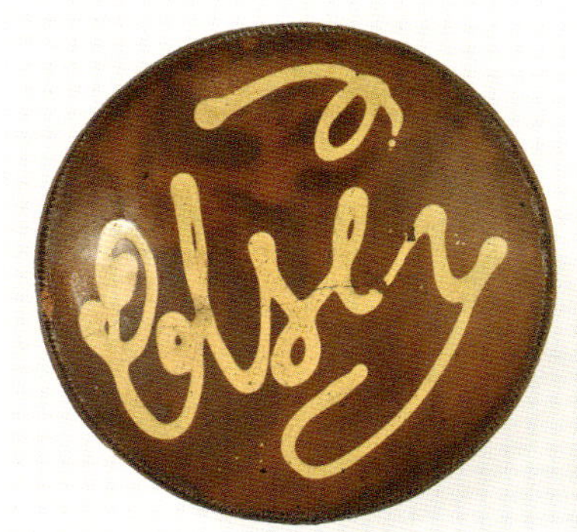

Figure 25 Dish, attributed to the Smith Pottery, Norwalk, Connecticut, 1825–1850. Slip-decorated earthenware. D. 8". Inscribed on obverse: "Elsey" (Courtesy, Pook & Pook, Inc.)

Seven dishes made for individuals whose first names start with the letter E have script that differs somewhat from the examples discussed thus far. One of two inscribed "Elsey" (fig. 25) typifies the others with its more vertical script and a capital E that begins with a large loop.[38] None of the seven dishes includes a surname, which suggests they were made for children. The infill motifs of looping and coiled lines are bolder than the more restrained straight and wavy lines and S-scrolls seen on the majority of Norwalk pieces. It is noteworthy that on both of the "Elsey" dishes the decorator did not leave sufficient room for the descending letter y, which therefore had to be written horizontally so it would fit on the dish. This anomaly might indicate the hand of a second decorator. However, the same capital E appears on a dish inscribed "Remember / thy Last / End."[39] If two potters were employed by Smith to inscribe dishes, this would be evidence that inscribed ware was not just a sideline but an important part of the pottery's production.

Earthenware dishes were used for preparing and serving foods, and several inscriptions document the foodways of the town in the second quarter of the nineteenth century. Native Americans had harvested oysters from Long Island Sound's rich oyster beds and dug clams in coastal areas for millennia. In 1721 the town prohibited anyone other than its residents from "rakeing and gathering of Oysters within ye harbours, coves, or any other place appertaining and being within the limits of our township."[40] Norwalk's numerous oystermen harvested the shellfish, which, along with clams, provided food and income for many families. The importance of shellfish to

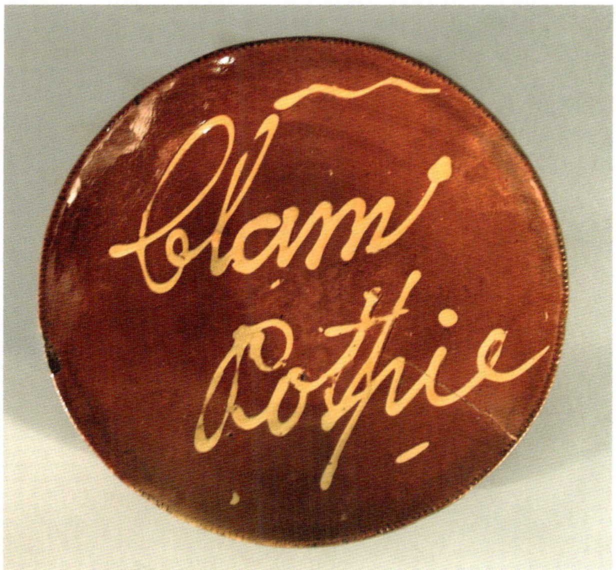

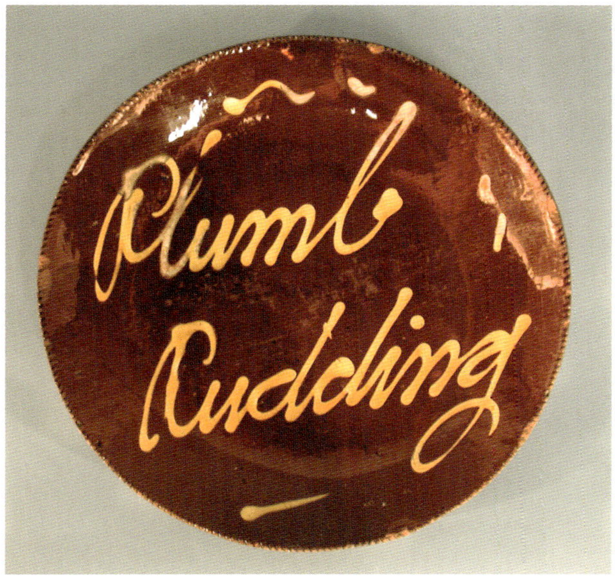

Figure 26 Dish, attributed to the Smith Pottery, Norwalk, Connecticut, 1825–1850. Slip-decorated earthenware. D. 10". Inscribed on obverse: "Clam / Potpie" (Courtesy, The Connecticut Historical Society.)

Figure 27 Dish, attributed to the Smith Pottery, Norwalk, Connecticut, 1825–1850. Slip-decorated earthenware. D. 13". Inscribed on obverse: "Plumb / Pudding" (Courtesy, The Connecticut Historical Society.)

the local diet and economy is preserved on dishes inscribed "oysters & Clams" (or "Clams & oysters") and "Clam Potpie" (fig. 26). "Plumb Pudding" (fig. 27) and "Apple Pie" offer glimpses of English dishes that colonists brought to America and the fruits grown in local orchards. The unglazed bottoms of some dishes have been blackened by fire, but others are not. Because the dishes were relatively inexpensive and decorative, some owners did not feel compelled to use them for preparing food but instead chose to use them as serving dishes or for display.

Absalom Day directed the spiritual lives of many South Norwalk residents, which accounts for numerous dishes inscribed with religious sentiments and the names of saints. The Bible had been a powerful force in Connecticut since the earliest period of settlement, and dishes bearing biblical passages and related epigrams found a strong market in Fairfield County. The inscriptions "forgive / all injuries" and "Pray for / the Living / & the / Dead" (figs. 28, 29) are two of the seven acts of corporal mercy that Christians are expected to perform.

Figure 28 Dish, attributed to the Smith Pottery, Norwalk, Connecticut, 1825–1850. Slip-decorated earthenware. D. 14½". Inscribed on obverse: "forgive / all injuries" (Courtesy, Historic New England.)

Figure 29 Dish, attributed to the Smith Pottery, Norwalk, Connecticut, 1825–1850. Slip-decorated earthenware. D. 14½". Inscribed on obverse: "Pray for / the Living / & the / Dead" (Courtesy, Sotheby's.)

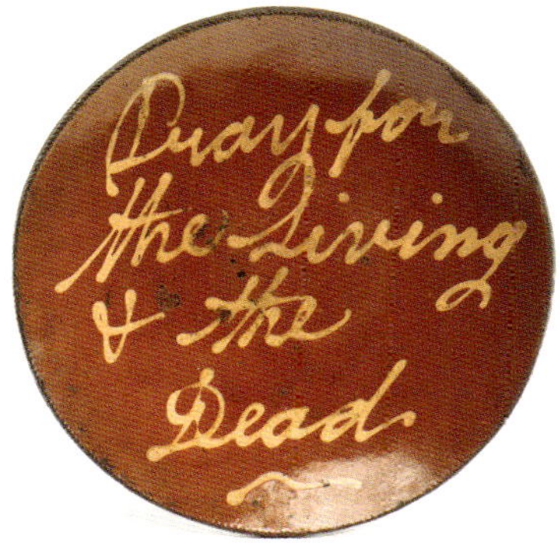

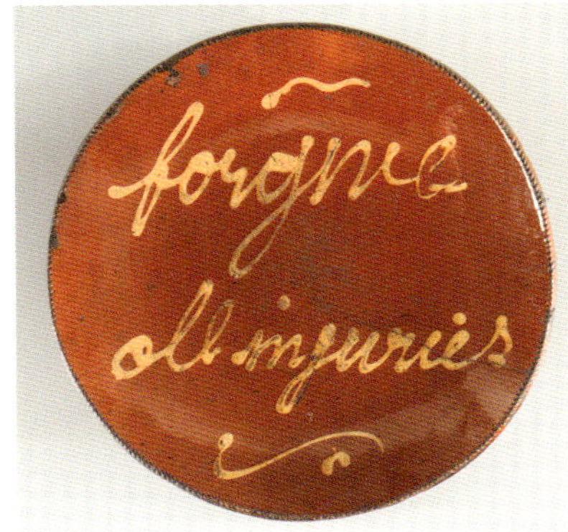

Figure 30 Dish, attributed to the Smith Pottery, Norwalk, Connecticut, 1825–1850. Slip-decorated earthenware. D. 12". Inscribed on obverse: "Honour / the / humane" (Courtesy, The Art Institute of Chicago, Gift of Mrs. John Worthy, 1918.284.)

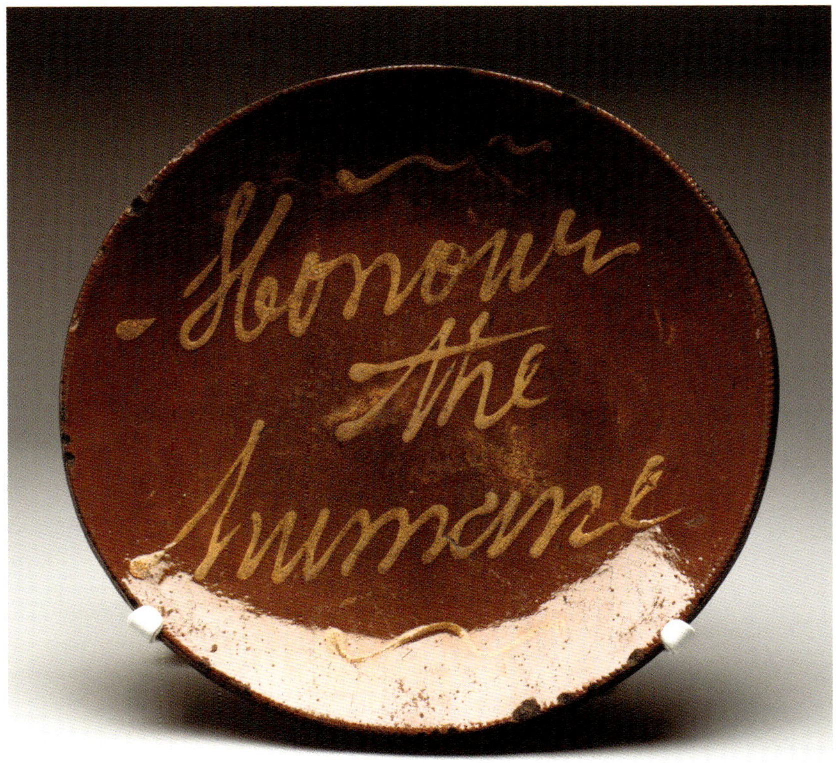

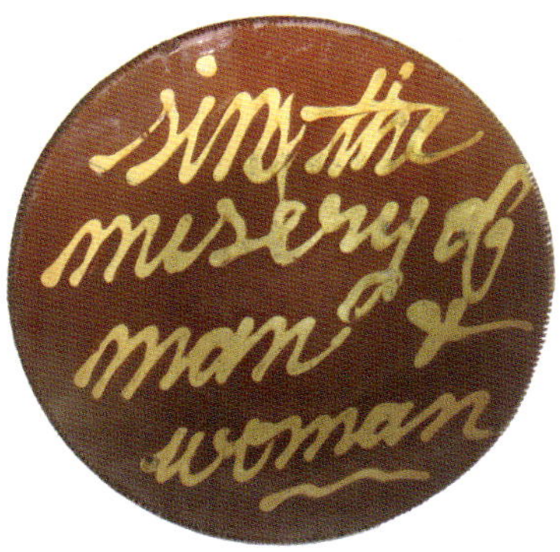

Figure 31 Dish, attributed to the Smith Pottery, Norwalk, Connecticut, 1825–1850. Slip-decorated earthenware. D. 10". Inscribed on obverse: "sin the / misery of / man & / woman" (Courtesy, Sam Herrup.)

Two dishes inscribed "Honour / the / humane" (fig. 30) served as reminders to acknowledge compassionate behavior in others and incorporate the attribute into one's life. While "sin the / misery of / man & / woman" (fig. 31) is not from scripture, the link between sin and suffering is found throughout the Bible.

Customers of the Smith Pottery could order dishes inscribed with the names of saints who served as models for a Christian life. The number of extant examples is large, with those inscribed "st. / francis / xavier" and "st / Remigius" (figs. 32, 33) typifying most. Among the favorites were martyred saints, two of whom had died or whose remains were rediscovered about the time that Connecticut was settled: "st. / fidelas / Martyr" commemorates the man who died at the hands of Calvinists in 1622 (fig. 34),

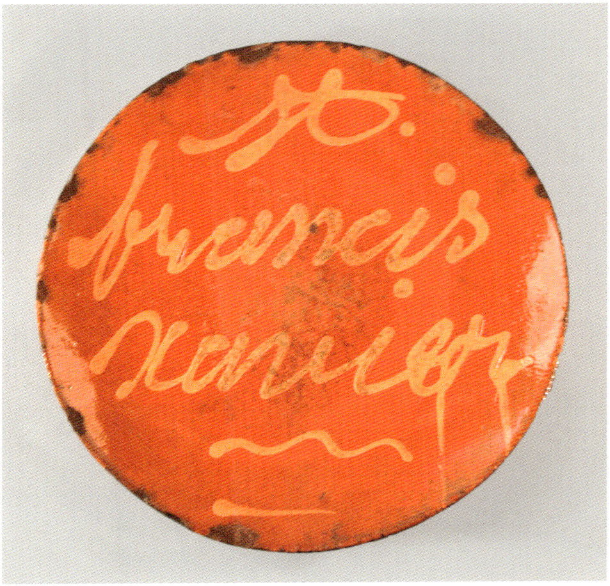

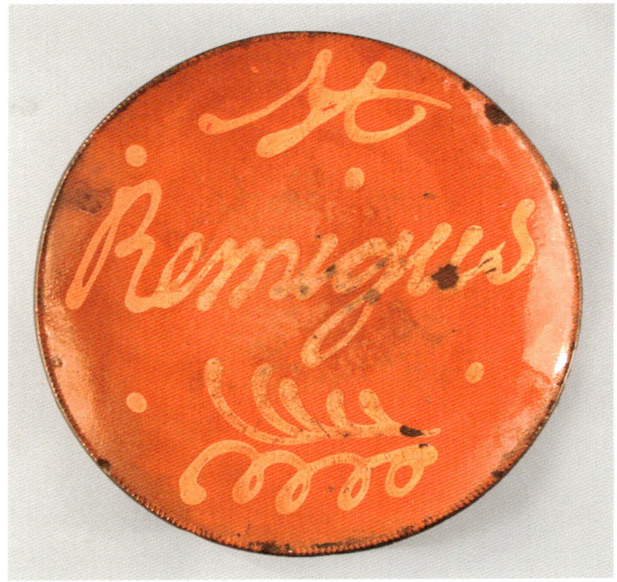

Figure 32 Dish, attributed to the Smith Pottery, Norwalk, Connecticut, 1825–1850. Slip-decorated earthenware. D. 11". Inscribed on obverse: "st. / francis / xavier" (Courtesy, The Connecticut Historical Society, 1945.1.1368.)

Figure 33 Dish, attributed to the Smith Pottery, Norwalk, Connecticut, 1825–1850. Slip-decorated earthenware. D. 11". Inscribed on obverse: "st / Remigius" (Courtesy, The Connecticut Historical Society, 1945.1.1369.)

and "St Martina / Virgin & / Mayrter" died in 226; her burial site was discovered in 1634. By preserving the names of individuals who gave their lives for their faith, the dishes can be seen as an extension of the Methodist Church's ministry in Norwalk.[41]

John Betts Gregory left Clinton and returned to Norwalk in 1831. He purchased Half Mile Island (now Canfield Island) and built kilns for burning earthenware and stoneware in the "cholera year" of 1832. He apparently resumed making slip-script pottery on his return. Three dishes attributed to

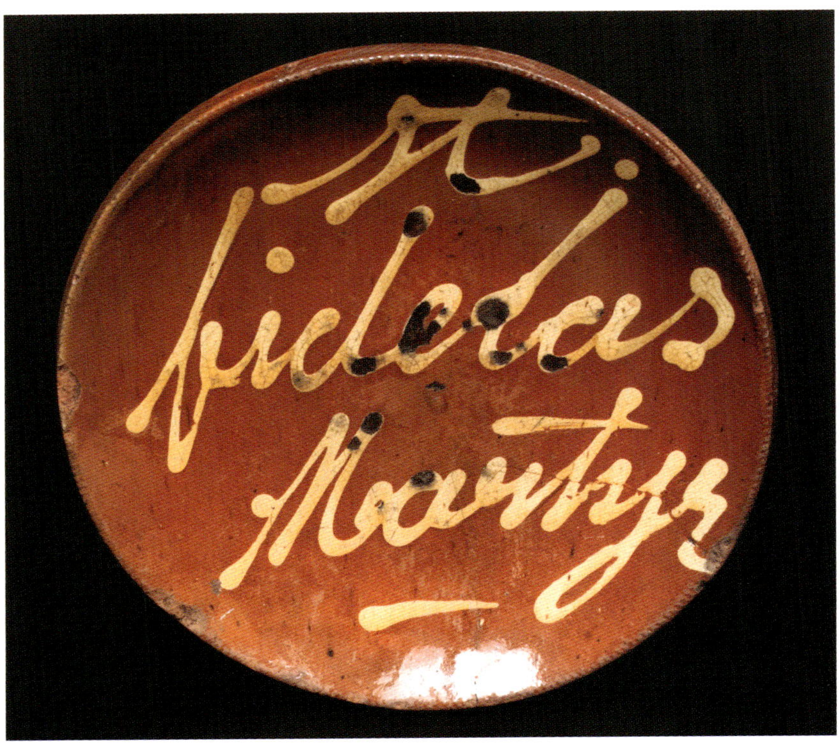

Figure 34 Dish, attributed to the Smith Pottery, Norwalk, Connecticut, 1825–1850. Slip-decorated earthenware. D. 10¼". Inscribed on obverse: "st. / fidelas / Martyr" (Sold at Crocker Farm, March 14, 2015, lot 113.)

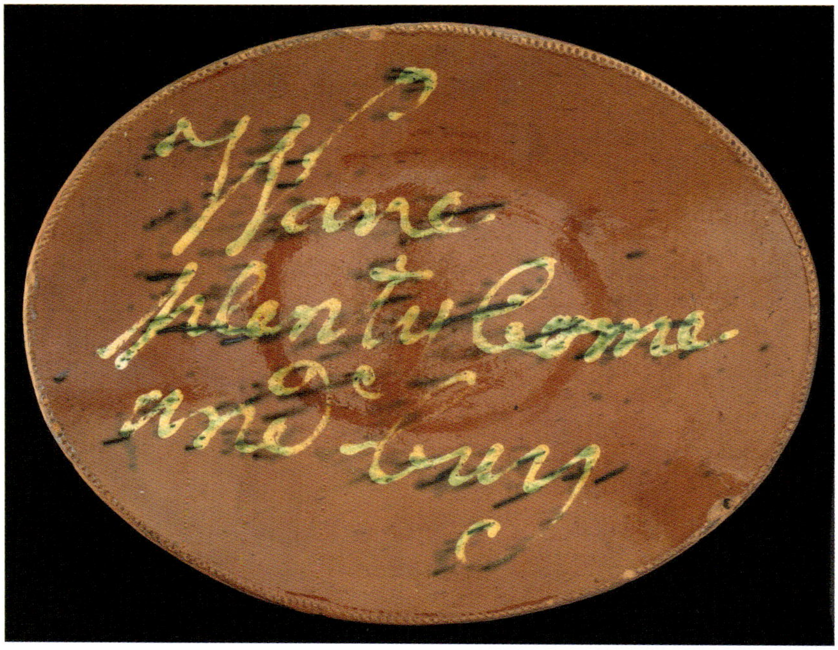

Figure 35 Dish, attributed to John Betts Gregory (1782–1842), probably Norwalk, Connecticut, 1832–1840. Slip-decorated earthenware. L. 16⁵⁄₁₆". Inscribed on obverse: "Ware / plenty Come / and buy" (Courtesy, The Colonial Williamsburg Foundation.)

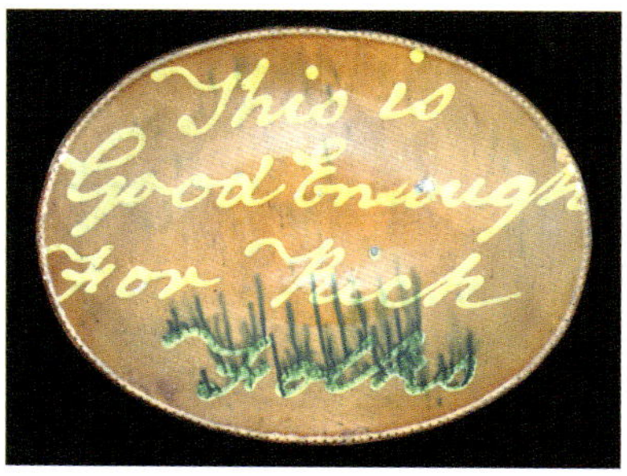

Figure 36 Dish, attributed to John Betts Gregory (1782–1842), probably Norwalk, Connecticut, 1832–1840. Slip-decorated earthenware. L. 14¼". Inscribed on obverse: "This is / Good Enough / For Rich / Folks" (Courtesy, Nancy and Gary Stass.)

Gregory that may have been made in Norwalk include an oval dish inscribed "Ware / plenty Come / and buy" (fig. 35). Featuring a flag on the capital W that identifies his hand, Gregory's characteristic streaks of copper oxide give this inscription vitality that is lacking in his Clinton works. This advertisement in clay may have hung in Gregory's Norwalk pottery to demonstrate his skill and the quality of his ware.

Gregory inscribed another oval dish "This is / Good Enough / For Rich / Folks" (fig. 36). Vertical streaks of copper oxide give the word "Folks" surprising dimensionality and depth that are absent on ware from the Smith Pottery. This inscription recalls "The best is not too good for you" inscribed on slipware made in the eighteenth century by Staffordshire potters. Gregory's sentiment, however, has an edgy ring to it, as if he were scolding the wealthy for not patronizing his pottery or for believing that his products were fit only for the middle class. Evidence of Gregory's distinctive personality may support this. In Clinton he was remembered as "quite a recluse, being seldom seen outside of his own premises," but a "genial soul" who "loved to scatter jokes and bits of humor among old and young who came to inspect his work or buy his wares." It is said that Gregory often sang while he worked, one of his favorite spirituals being "The Potter and His Clay." The lyrics were inspired by the same biblical passage in Romans that the Reverend Coles was reminded of as he watched Absalom Day at work:

> Behold the potter and the clay!
> He forms his vessels as he please.[42]

Celebrating the potter's independence, the verse seems to have been important to this man who spoke his mind directly and freely through his work.[43]

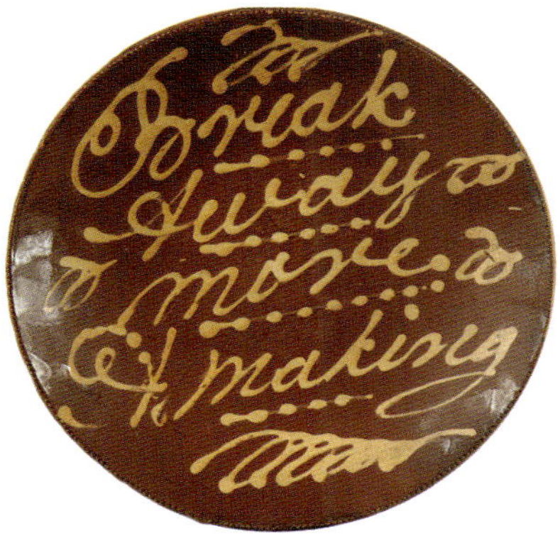

Figure 37 Dish, attributed to the Smith Pottery, Norwalk, Connecticut, 1825–1850. Slip-decorated earthenware. D. 13¼". Inscribed on obverse: "Break / Away / more / making" (Courtesy, Pook & Pook, Inc.)

Norwalk ware was sold at the potteries, it was shipped on sloops to sell in coastal Connecticut communities, in Massachusetts, and in New York City, and it was carried into the surrounding countryside by peddlers. The inscriptions "Break / Away / more / making" (fig. 37) and "Cheap Ware" refer to the affordability of Smith's ware, with "Cheap" meaning inexpensive, not of low quality. Making use of the word "cheap" must have been an effective marketing tool. Two dishes bear the inscription "pony / up the / Cash" (fig. 38), which might be a playful encouragement of sales. Inscribed Norwalk dishes average about 11 inches in diameter, but they range from 8 inches to more than 14 inches. A price list for "brown earthenware" published in 1864 by A. E. Smith's Sons lists 11-inch "baking dishes" priced at $1.00 per dozen, or about 8 cents each.

The Smith Pottery appears on an 1836 drawing of Old Well on Water Street adjacent to docks at the south end of Norwalk harbor (fig. 39). A detail at the bottom of the drawing shows the pottery house on the left, a gable-fronted building on the right that may have housed a retail shop, and smoke labeled "yellow ware" billowing from a single bottle kiln. Asa E. Smith and Noah Selleck were in partnership under the name Selleck & Smith from 1837 to 1843. This was the first of several organizational changes the Smith firm underwent in succeeding years. Absalom Day died in 1843, the year that his son Noah and Asa Smith formed the firm Smith & Day. This partnership continued until 1848, when Smith and his son Theodore, who managed the pottery's wareroom at 38 Peck Slip in New York City, formed A. E. Smith & Sons. From about 1850 until 1860, Norwalk stoneware stamped "A.E. Smith & Sons, Manufacturers, 38 Peck Slip, N.Y." was sold in New York City. The importance of that market for Smith's ware is documented by a dish inscribed "Newyork / City" (fig. 40).

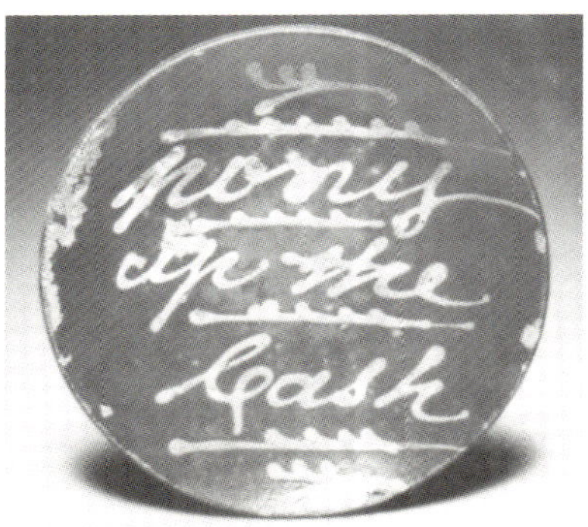

Figure 38 Dish, attributed to the Smith Pottery, Norwalk, Connecticut, 1825–1850. Slip-decorated earthenware. D. 10½". Inscribed on obverse: "pony / up the / Cash" (Courtesy, Lewis Scranton.)

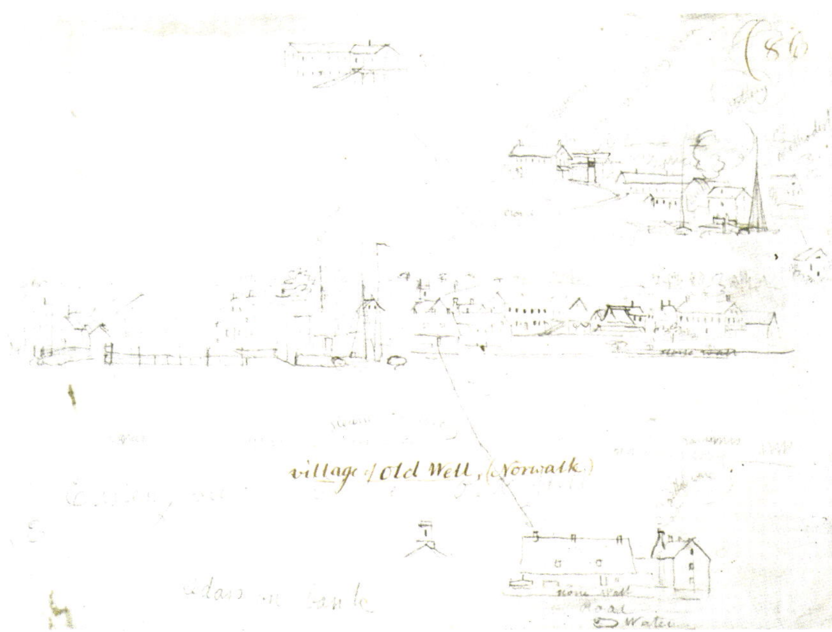

Figure 39 *Village of Old Well*, John Warner Barber (1798–1885), Norwalk, Connecticut, 1836. Graphite on paper. 7½" x 9⅞". (Courtesy, Connecticut Historical Society, 1953.5.212.)

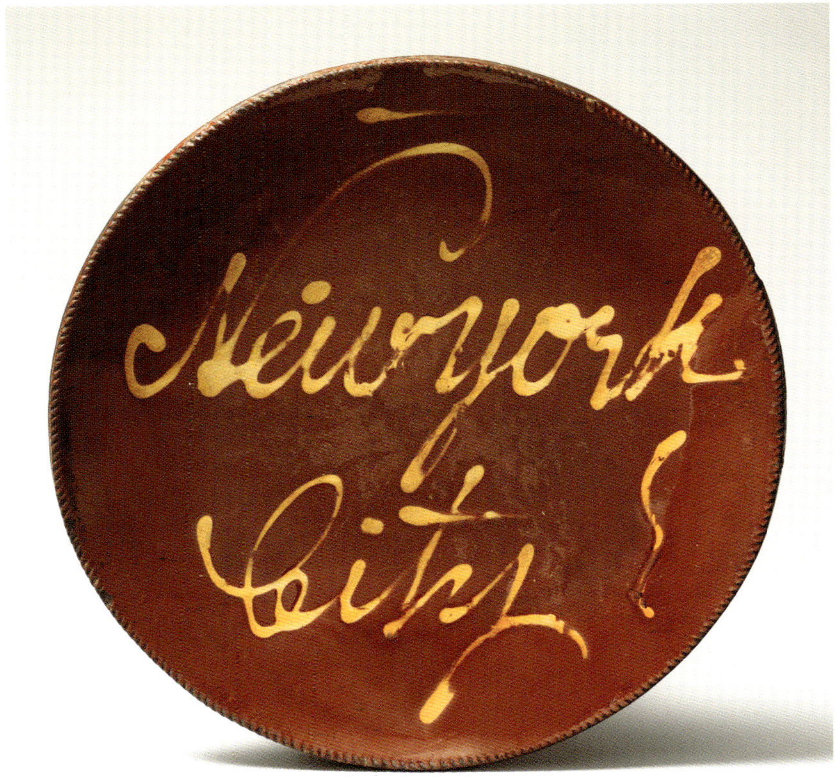

Figure 40 Dish, attributed to the Smith Pottery, Norwalk, Connecticut, 1825–1850. Slip-decorated earthenware. D. 13". Inscribed on obverse: "Newyork / City" (Courtesy, The Metropolitan Museum of Art, Gift of Katharine Prentis Murphy, 1948, 48.124.)

Nationalistic inscriptions such as "United / States" (fig. 41) bring to mind the early "Lafayette" and "Washington & Lafayette" dishes. In 1851 Lajos Kossuth (1802–1894), a leader of Hungary's struggle for independence from Austria, arrived in the United States to tour the nation. Kossuth was the first

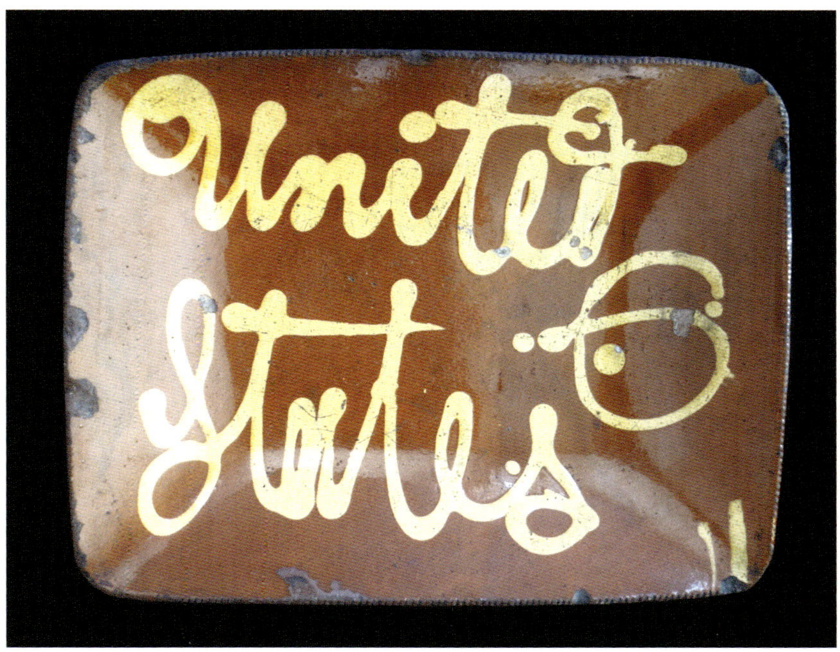

Figure 41 Dish, attributed to the Smith Pottery, Norwalk, Connecticut, 1825–1850. Slip-decorated earthenware. L. 12¼". Inscribed on obverse: "United / States" (Courtesy, Nancy and Gary Stass.)

foreign dignitary to address a joint session of Congress since Lafayette, and he was feted at the White House in 1852. Unlike Lafayette, Kossuth did not visit Norwalk, but popular support for him and Hungary's bid for self-rule was strong, prompting Smith to produce a dish inscribed "Kossuth / 1852" (fig. 42). This being the only known example of this dish suggests it did not gain wide popularity.

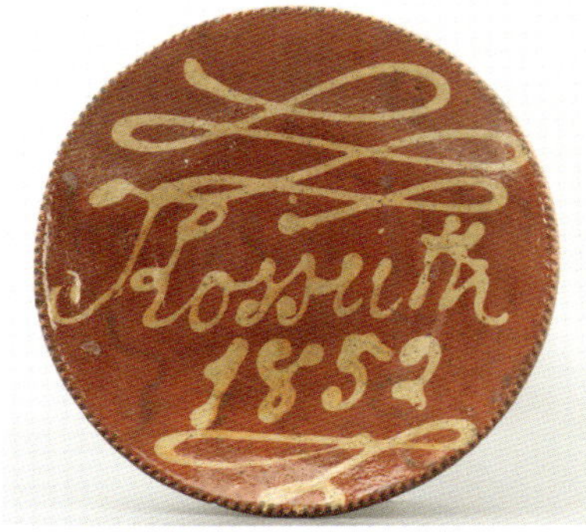

Figure 42 Dish, attributed to the Smith Pottery, Norwalk, Connecticut, probably 1852. Slip-decorated earthenware. D. 13". Inscribed on obverse: "Kossuth / 1852" (Courtesy, Historic New England; Beauport, Sleeper-McCann House.)

Potters occasionally made personal items for their own amusement or as gifts for family or friends. Known as "end of the day" pieces, they can be sculptural works that were fashioned from clay into animal or human forms. They are often a significant departure from the kind of ware the potter made day in and day out. A Norwalk dish bears the only inscription referencing the work of potters: "This is / the last / for to / day" (fig. 43). The dish bridges production work and novelty, making it possibly the earliest known example of this ware. A standard form produced by the pottery was given a personal message that brings the long days and repetitive work of potters down to a very personal level.

The 1850s were especially rewarding for Asa E. Smith. In 1850, when his occupation was listed as "potter," the value of his real estate was $9,000. Ten years later, now owner of a "Stone Ware Factory," his real estate was valued at $45,000 and his personal estate $15,000. His income grew substantially as his firm's stoneware found a growing market not just locally but also in New York City and elsewhere in the United States, where Smith's stoneware was shipped by rail. His success over decades might be attributed to a belief that "Honesty / is the Best / of polacy," which is inscribed on one of his firm's dishes (fig. 44).

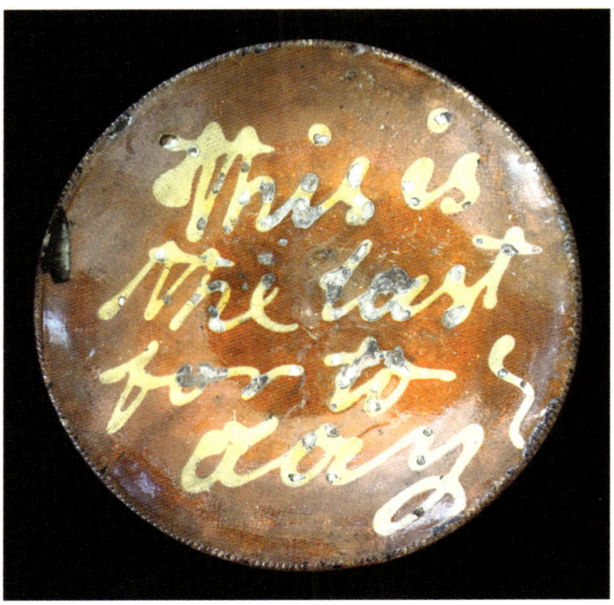

Figure 43 Dish, attributed to the Smith Pottery, Norwalk, Connecticut, 1825–1850. Slip-decorated earthenware. D. 12". Inscribed on obverse: "This is / the last / for to / day" (Courtesy, Nancy and Gary Stass.)

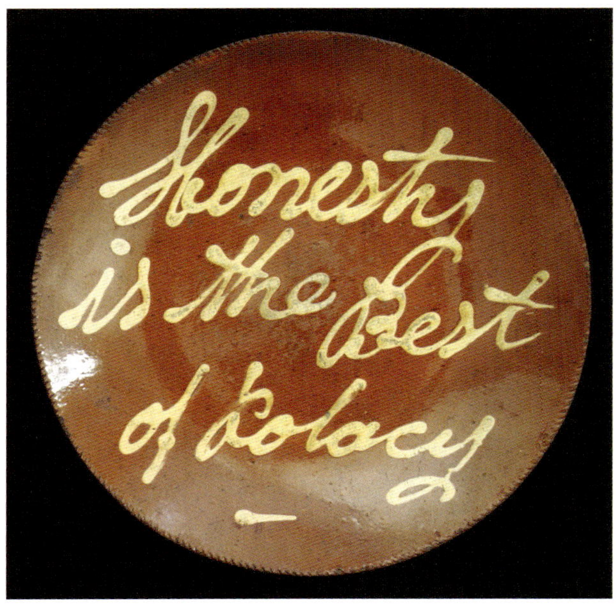

Figure 44 Dish, attributed to the Smith Pottery, Norwalk, Connecticut, 1825–1850. Slip-decorated earthenware. D. 12¼". Inscribed on obverse: "Honesty / is the Best / of polacy" (Courtesy, Nancy and Gary Stass.)

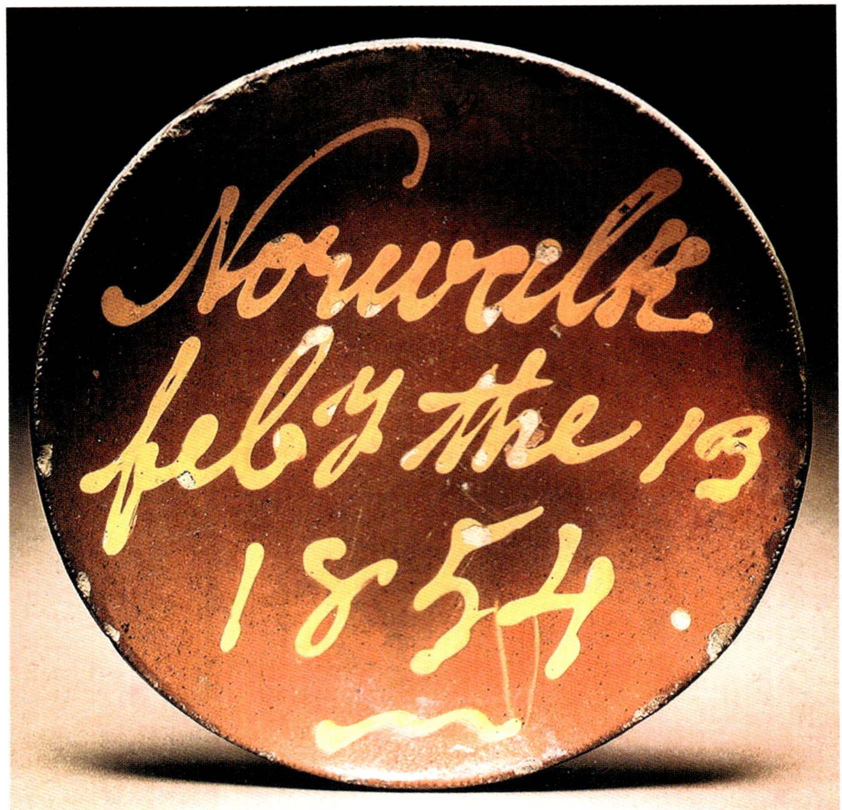

Figure 45 Dish, attributed to the Smith Pottery, Norwalk, Connecticut, 1825–1850. Slip-decorated earthenware. D. 18". Inscribed on obverse: "Norwalk / feb^y the 13 / 1854" (Courtesy, Norwalk Historical Society.)

On February 14, 1854, Smith advertised in the *New-York Daily Tribune* his "desirable country residence" on Norwalk's East Avenue for sale, one day after the dish that serves as the keystone for identifying Norwalk slip-script earthenware was made (fig. 45).[44] "Norwalk / feb^y the 13 / 1854" was inscribed by the artisan who had lettered dishes for thirty years. That the manufacture of this dish and Smith offering his house for sale occurred a day apart seems beyond coincidence. Perhaps the dish held special significance for Smith by signaling the end of his pottery's production of slip-script earthenware and the firm's emphasis on marketing stoneware regionally. Norwalk is given prominence in the inscription, acknowledging the place whose residents had supported Smith, Absalom Day, and John Betts Gregory for more than fifty years. If the dish was among the last of its kind made by Smith's pottery, it was a fitting tribute to the town and to the potters whose earthenware carried messages that would continue to resonate nearly two centuries later.[45]

ACKNOWLEDGMENTS For their assistance with the preparation of this article, I sincerely thank Anthony Butera, Brian Cullity, Robert Hunter, Suzanne Katz, Angelika Kuettner, Melissa Maichele, Lewis Scranton, Nancy and Gary Stass, and the High Museum of Art.

1. "Antiques Auction," *Norwalk Bulletin*, September 2, 1953.
2. Andrew L. Winton and Kate Barber Winton, "Norwalk Potteries," *Old-Time New England* 24 (1934): 75–92.

3. Ibid., p. 80.

4. Ibid., pp. 90–91.

5. Albert Hastings Pitkin, *Early American Folk Pottery, Including the History of Bennington Pottery* (Hartford, Conn.: Case, Lockwood & Brainard Co., 1918), p. 92.

6. Edwin Hall, *The Ancient Historical Records of Norwalk, Conn.* (Norwalk, Conn.: James Mallory & Co., 1847), p. 173.

7. A reference to "Norwalk Ware" appeared in an advertisement placed by New Haven, Connecticut, merchant Ebenezer Huggins in the *Connecticut Herald*, June 9, 1812. In 1822 the schooner *Galen* docked in Boston to sell "a small quantity of Norwalk WARE; viz. Baking Dishes, Butter Pots, Wash Bowls, Jars, &c." *The Repertory*, January 10, 1822. Dishes inscribed "City of Troy" and "New York City" indicate that the ware was also made on order or on speculation for New York customers, though Fairfield County remained the primary market. The "City of Troy" dish was brought to the author's attention by Kathryn T. Sheehan in an email, February 14, 2015.

8. Inscribed earthenware was made beginning in the late eighteenth century by several Pennsylvania-German potters who mastered the sgraffito technique of scratching images and inscriptions through a contrasting coating of slip to reveal the darker clay body beneath. This ware employs a different technique: the inscriptions are typically located on the perimeter of the dish and they often address relationships between men and women. These differences, along with the geographical and cultural distance between New Englanders and German Americans, distinguish this Pennsylvania pottery tradition from that of Norwalk.

9. Sotheby's, *Important Americana*, sale NO7025, October 9, 1997, lot 159.

10. Kevin Wright, "Potters along the Hackensack River (Bergen County) 1805–1867," www.bergencountyhistory.org/Pages/BCHSPottery.html (accessed January 10, 2015).

11. Stanley L. Engerman and Robert E. Gallman, "The Emergence of a Market Economy before 1860," in *A Companion to 19th-Century America*, edited by William L. Barney (Malden, Mass.: Wiley-Blackwell, 2006), p. 133.

12. Angeline Scott Donley, "The Homely Pottery of Old New England," *Country Life in America* 19 (1910): 46.

13. 1850 Federal Census for Norwalk, Fairfield, Connecticut, roll M432_38, p. 143B, image 292.

14. Collection of the Stamford Historical Society, Stamford, Connecticut.

15. Edwin Atlee Barber, *The Pottery and Porcelain of the United States* (New York: G. P. Putnam's Sons, 1893), p. 181. Several men in the 1850 Norwalk census listed their occupations as "button maker," suggesting that Norwalk was a center of button production.

16. Charles A. Philhower, *Brief History of Chatham* (New York: Lewis Historical Publishing Co., 1914), p. 10.

17. Winton and Winton, "Norwalk Potteries,"p. 85.

18. *Republican Journal*, February 4, 1797.

19. Francis Asbury, *The Journal of the Reverend Francis Asbury*, 3 vols. (New York: N. Bangs and T. Mason, 1821), 3:65. Day built "a substantial house in the Georgian style of architecture" that was razed in 1876 to make way for a business block. [Samuel Richards Weed], *Norwalk after Two Hundred & Fifty Years: An Account of the Celebration of the 250th Anniversary of the Charter of the Town, 1651—September 11th—1901* (South Norwalk, Conn.: C. A. Freeman, 1901), p. 302.

20. Coles continued: "One thing more I observed: the potter made no vessel for the purpose of destroying it. This fact, I thought, might convince the most skeptical, if he wished to be convinced, that if capricious man never made a vessel for the purpose of breaking it to pieces, the just and holy God never created a single human soul for no other object than to show forth his power in destroying it." Rev. George Coles, *My First Seven Years in America* (New York: Carlton & Phillips, 1852), p. 48.

21. Barbara H. Magid and Bernard K. Means, "In the Philadelphia Style: The Pottery of Henry Piercy," in *Ceramics in America*, edited by Robert Hunter (Hanover, N.H.: University Press of New England for the Chipstone Foundation, 2003), pp. 47–86.

22. A small raised heart is molded into the back of the dish, which has short parallel lines incised around the perimeter and crudely fashioned letters "AW" in the center. It would be reasonable to assume that these are the initials of the potter who made the dish. Transporting clay from New Jersey, marketing his ware, and preaching took Day away from Norwalk for weeks at a time, and he would have relied on other potters to run the pottery in his absence. Still, it seems unlikely that Day would have entrusted an employee to make this sign for his own pottery. The names of the individuals who are known to have apprenticed to Day do not match

these initials, and the only identified Connecticut potter having those initials is Ashbel Wells Jr. of Hartford, who opened a store in Hartford in 1787—predating Day's Norwalk work. Melissa Maichele, email to author, January 5, 2015. For information about Ashbel Wells Jr., see William DeLoss Love, *The Colonial History of Hartford Gathered from the Original Records* (Hartford, Conn.: William DeLoss Love, 1914), p. 177.

23. The practice of using conjoined initials was common in the American colonies in the eighteenth century. Day's initials resemble conjoined initials carved on the gravestones of six members of the Day family who died between 1802 and 1815 and are buried in Hillside Cemetery in Chatham, New Jersey. Day's travels may have taken him to Chatham to visit family or to attend the funerals of some of these relatives where he saw the gravestones and appropriated the motif.

24. Winton and Winton, Norwalk Potteries, p. 83.

25. Rev. Charles M. Selleck, *Norwalk* (Norwalk, Conn.: Charles M. Selleck, 1896), p. 31.

26. William C. Ketchum Jr., *Potters and Potteries of New York State, 1650–1900* (Syracuse, N.Y.: Syracuse University Press, 1987), p. 271.

27. *Columbian Gazette* (Utica, N.Y.), June 3, 1817.

28. An advertisement in the *Philadelphia Inquirer* in 1839 offered for sale pie plates "made expressly to stand the fire in baking"—a spurious claim given that the heat of the kiln in which earthenware was fired was far greater than a household oven, so baking food in earthenware dishes would not harm them. Savory and sweet pies could feed a family at little expense, so these baking wares were common in American households.

A dish inscribed "E Selby / Conover" may document a previously unrecorded partnership between Edward Selby and Abraham Conover, who were potters in Poughkeepsie, New York. Copper oxide was applied liberally to both the inscription and the infill decoration on the dish, recalling Gregory's practice. Harriet Ostrander lived in Albany County, so it is conceivable that Gregory secured other commissions in the area. However, why Poughkeepsie potters would commission an outsider to make this promotional dish is unclear, and an attribution to Gregory remains tentative. See Ketchum, *Potters and Potteries of New York State*, pp. 116–17, and Sotheby's, *Americana*, sale NO8158, January 22, 2006, lot 265.

29. Rev. A[mos] D[elos] Gridley, *History of the Town of Kirkland, New York* (New York: Hurd and Houghton, 1874), pp. 107, 165.

30. Connecticut Daughters of the American Revolution, Norwalk Chapter, *The Colonial and Revolutionary Homes of Wilton, Norwalk, Westport, Darien and Vicinity* (Norwalk, Conn.: DAR, Norwalk Chapter, 1900), pp. 44–45.

31. A dish inscribed "Lafayette & Jackson" (private collection) links Lafayette, a hero of the Revolutionary War, and Andrew Jackson, who was celebrated as a hero in the War of 1812.

32. Earthenware dishes having molded profiles of Lafayette and Washington are attributed to Bergen County, New Jersey, potter Henry Van Saun (d. 1829). Wright, "Potters along the Hackensack River (Bergen County) 1805–1867."

33. *Commercial Advertiser* (New York, N.Y.), November 29, 1827. Day's endorsement, dated October 1827, continues: "my friends all thought me in decline; and after taking every medication recommended, without effect, I was resigned to my fate, and looked forward to the grave as the only place of relief. . . . [T]he first bottle afforded relief; and by continuing to use it, I have been restored to my usual health and strength." No evidence that Day returned to making pottery after 1825 has been found, however.

34. Robert Plot, *The Natural History of Stafford-shire* (Oxford, Eng., 1686), p. 122. Plot was keeper of the Ashmolean Museum and professor of chemistry at Oxford University. The process of turning, decorating, and burning earthenware has little changed from how he describes it.

35. Tin-glazed earthenware dishes bearing images of King William and Queen Mary were made in England from the time Mary ascended to the throne in 1689 to her death in 1694. The couple may not have been commemorated on ceramics again until Norwalk potters did so.

36. For information about Henry Chichester Jr.'s activities, see Herbert Armstrong Poole, *The Genealogy of John Lindsley (1845–1909) and His Wife Virginia Thayer Payne (1856–1941) of Boston, Massachusetts* (Milton, Mass.: Herbert Armstrong Poole, 1950), p. 263.

37. "Connecticut Deaths and Burials, 1772–1934," database, FamilySearch (https://familysearch.org/ark:/61903/1:1:F7KF-C4R [accessed March 27, 2015]), Francis H. Smith, May 20, 1850; citing reference 368; FHL microfilm 3,355.

38. The names on the other dishes are "Enos," "Enoch," "Elennor," and two dishes are inscribed "Emma."

39. This inscription is from Ecclesiasticus 7:40: "In all thy works remember thy last end, and thou shalt never sin." The verse refers to the common belief that the sins one committed in life will be tallied on Judgment Day. A variation of this inscription, "REMEMBER THY END TRULY," is written on a seventeenth-century Metropolitan ware pot. John Eliot Hodgkin and Edith Hodgkin, *Examples of Early English Pottery Named, Dated, and Inscribed* (London, Eng., 1891), p. 53.

40. Edwin Hall, *The Ancient Historical Records of Norwalk, Conn., with a Plan of the Ancient Settlement, and of the Town in 1847* (Norwalk, Conn.: James Mallory & Co., 1847), p. 110.

41. "St. Eulogias," a misspelling of St. Eulogius of Córdoba, is inscribed on another dish.

42. Gridley, *History of the Town of Kirkland*, p. 165; Samuel W. Durant, *History of Oneida County, New York* (Philadelphia, Pa.: Everts & Fariss, 1878), p. 464; Isaac Watts, *Hymns and Spiritual Songs* (London, Eng.: Printed for W. Strahan et al., 1773), p. 92.

43. This trait is seen on a dish inscribed "For / Hansom / Molly" (Connecticut Historical Society). Considering the mores of the time, Gregory's observation seems surprisingly bold.

44. *New-York Daily Tribune*, February 14, 1854. No evidence that Smith sold the residence in which he had lived since at least 1847 has been found. Census lists for 1860 and 1870 show Smith living among the same East Avenue neighbors as earlier.

45. The recent discovery of Norwalk slip-script earthenware in archaeological contexts in Brooklyn, Manhattan, and Staten Island provides additional evidence of the wide appeal of Asa E. Smith's ware. More such finds are inevitable and will expand our understanding of the popularity of this distinctive pottery and the meaning it held for customers living far from where it was made.

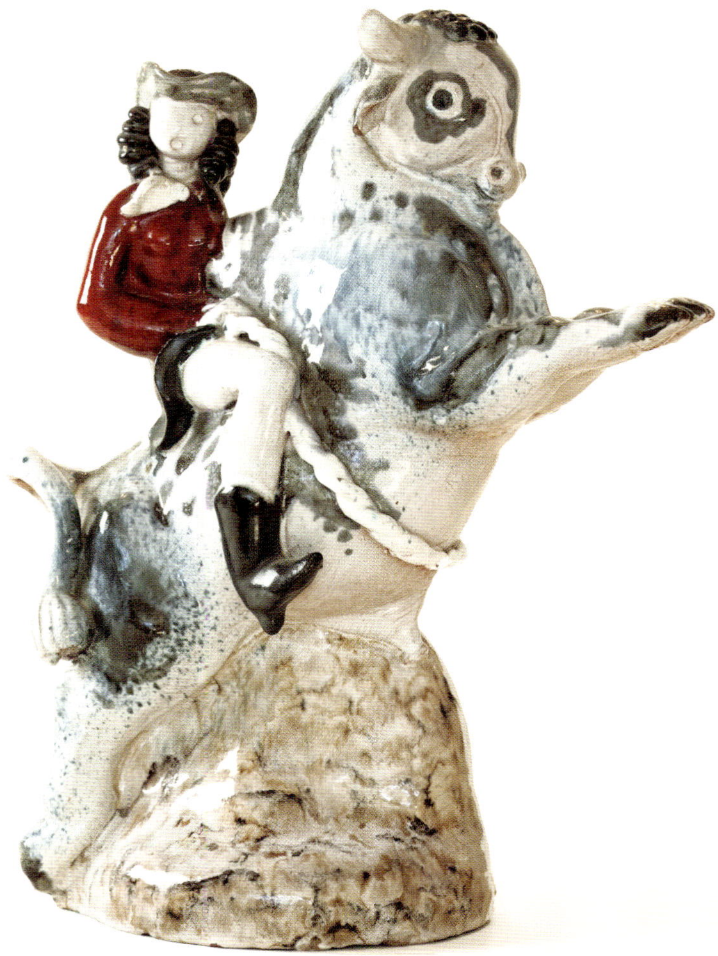

Figure 1 Russell Aitken, *Wyoming Europa*, 1939. Glazed earthenware. H. 12". (Private collection; photo, Randl Bye.)

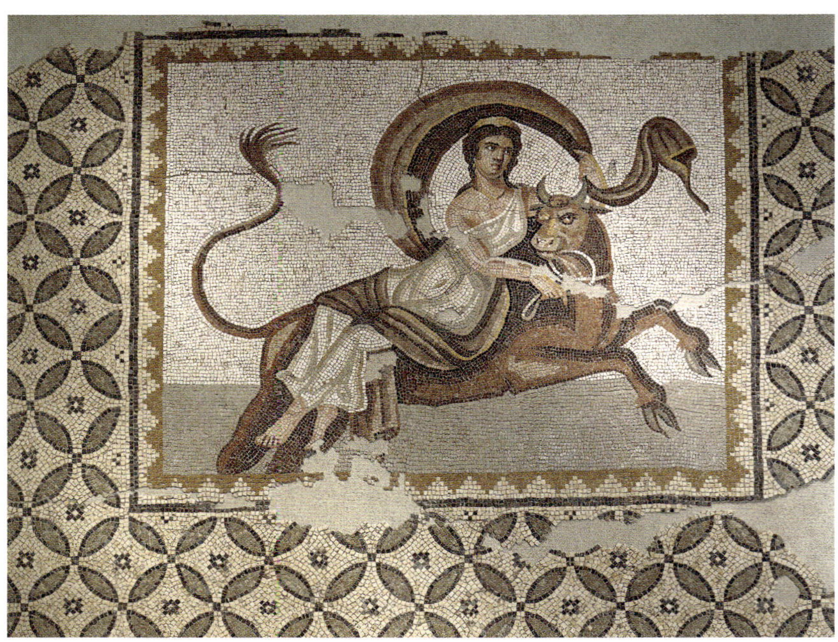

Figure 2 Tile mosaic, *Europa Abducted by Zeus Transformed into Bull*, Byblos, Lebanon. Roman era, 3rd century A.D. (De Agostini Picture Library, G. Dagli Orti, Bridgeman Images.)

Tom Folk

The Allegory of Europa in Twentieth-Century American Sculpture

▼ IRENE ROOSEVELT AITKIN, the second wife of ceramic artist Russell Barnett Aitken (1910–2002), donated a collection of her husband's ceramic sculptures to the Cleveland Museum of Art in 2013. Among the items is a study collection of ceramic maquettes made by the artist and a scrapbook relating to his career in ceramics. Russell Aitken, a Cleveland native who graduated as valedictorian from the Cleveland Institute of Art, was considered a leading American ceramist. One of the ceramic maquettes was for his *Wyoming Europa* (fig. 1). The maquette, 12 inches high, is marked only with his initials, RBA. The finished version, which is illustrated in the scrapbook on a page taken from an unidentified magazine, was recently discovered in the collection of the noted furniture designer Harold M. Schwartz in San Miguel de Allende, Mexico.

This article focuses on seven ceramic and two bronze sculptures dating from the 1920s to 1940 that all illustrate the theme of Europa and the Bull except for one, which depicts only Europa. All of the artists were American-born except for Carl Milles (1875–1955), who was from Sweden, and Valerie ("Vally") Wieselthier (1895–1945), from Austria, major European artists who had established careers before they came to the United States. All told, the nine sculptures show a progression in style and meaning. Aitken's *Wyoming Europa* is the penultimate version, where, in effect, we see the Americanization of Europa. The Cowan Pottery in Rocky River, a suburb of Cleveland, and the Art Academy of Cranbrook, in Bloomfield Hills, Michigan, are also significant in this development.[1]

The Myth of Europa as Allegory

The image of the mythological Europa has always been an allegory for the origins of Western, or European, civilization. In Greek mythology Europa, a woman of Phoenician origin, was the mother of King Minos of Crete. Zeus fell in love with her and adopted the form of a white bull in order to abduct her (fig. 2). When Europa mounted his back, Zeus flew over the Mediterranean and landed on the island of Crete. He then took human form and laid with her. Their offspring were Minos, Rhadamanthus, and Sarpedon. Minos became the king of Crete and the Minoan civilization was named after him.

The ritual of the taming of the bulls dates as far back as about 1500 B.C. in Crete. The Minoans fostered a bull culture, and bull vaulting and bull sacrifices were an integral aspect of their strange and still little-understood religious practices, involving both young men and women vaulting over charging bulls, as is depicted in Minoan frescoes. The American sport of bull riding, developed during the expansions of the western territories, has its parallels to the ancient Greek rites. In the American practice, a rider mounts the bull and with one hand tightly grips a flat braided rope tied around the bull. When the bull is released, it bucks, rears, and kicks, trying to throw off the rider. For the ride to count, the rider must stay atop the bull for at least eight seconds—what has been called the most dangerous eight seconds in sports.

The allegory of Europa and the Bull had considerable political symbolism throughout its history well into modern times. And its manifestation in works of art—in painting and sculpture—reflected the prevailing political and social milieu. Europa and the Bull was the subject of many old master paintings in the sixteenth, seventeenth, and eighteenth centuries (fig. 3). Various incarnations of Europa, sometimes with the bull and sometimes as a stand-alone figure, were depicted with allegorical images of Asia, Africa, and America beginning in the sixteenth century (fig. 4). In the eighteenth century the Meissen factory, the first factory to produce hard-paste porcelain in Europe, introduced its version of this subject. The figure was subsequently copied by other European and British ceramic factories and produced into modern times (fig. 5).

Figure 3 Maerten de Vos (1532–1603), *The Rape of Europa*, ca. 1590. Oil on oak panel. 52⅝" x 68¹¹⁄16". (Museo de Bellas Artes, Bilbao, Spain, Bridgeman Images.)

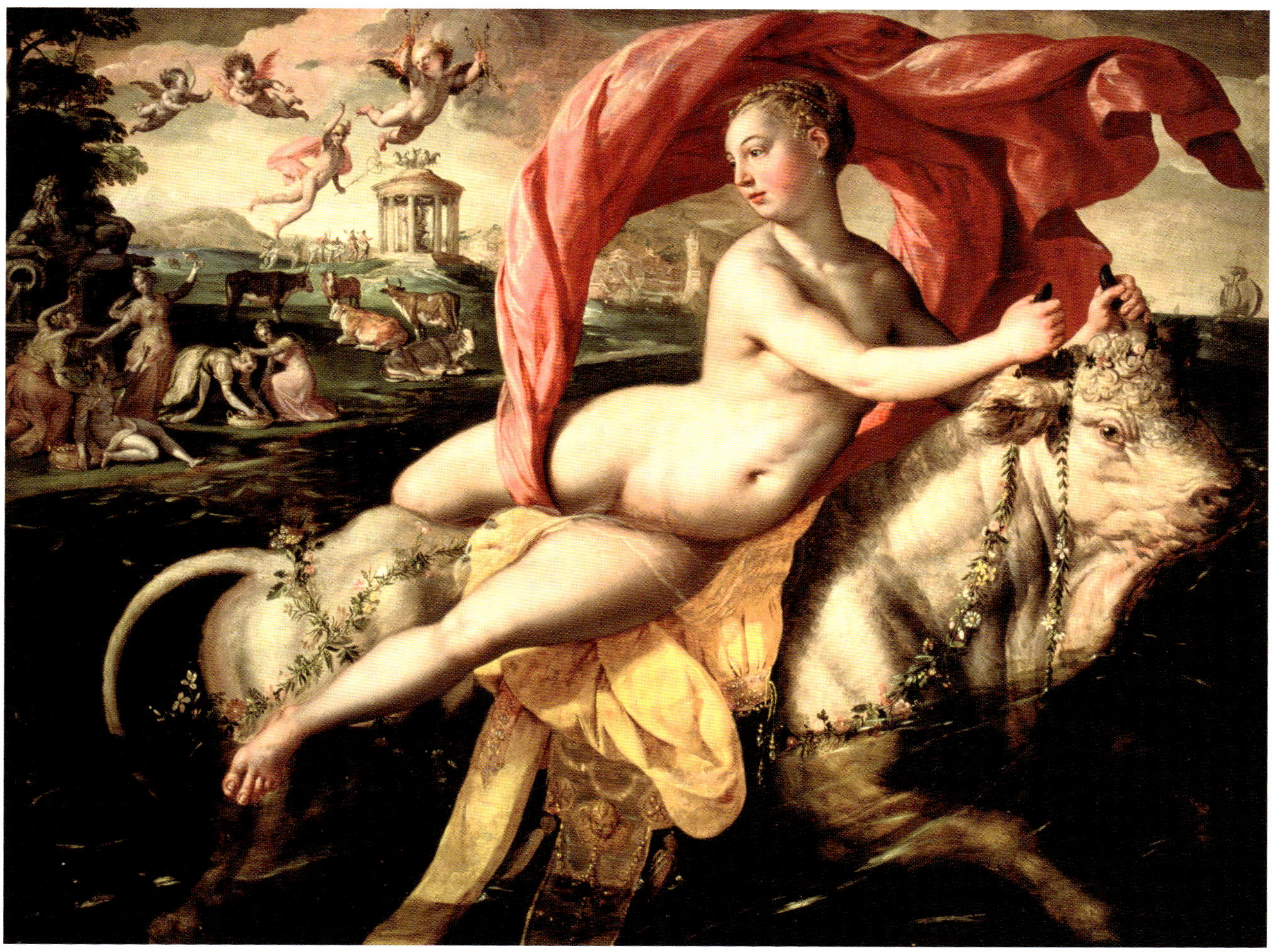

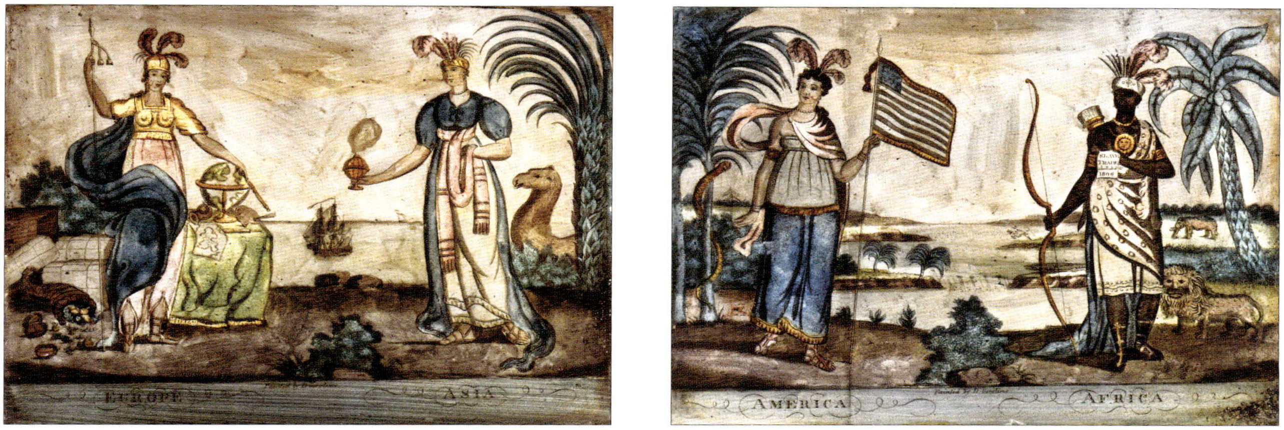

Figure 4 D. Gerolina, (*left*) *Europe and Asia* and (*right*) *America and Africa*, England, ca. 1807. Reverse painting on glass. 14" x 10". (Private collection.)

Figure 5 Candlestick depicting Europa and the Bull, Charles Gouyn's Factory, London, England, 1750–1760. Soft-paste porcelain. H. $7\frac{11}{16}$". (© Victoria and Albert Museum, London.)

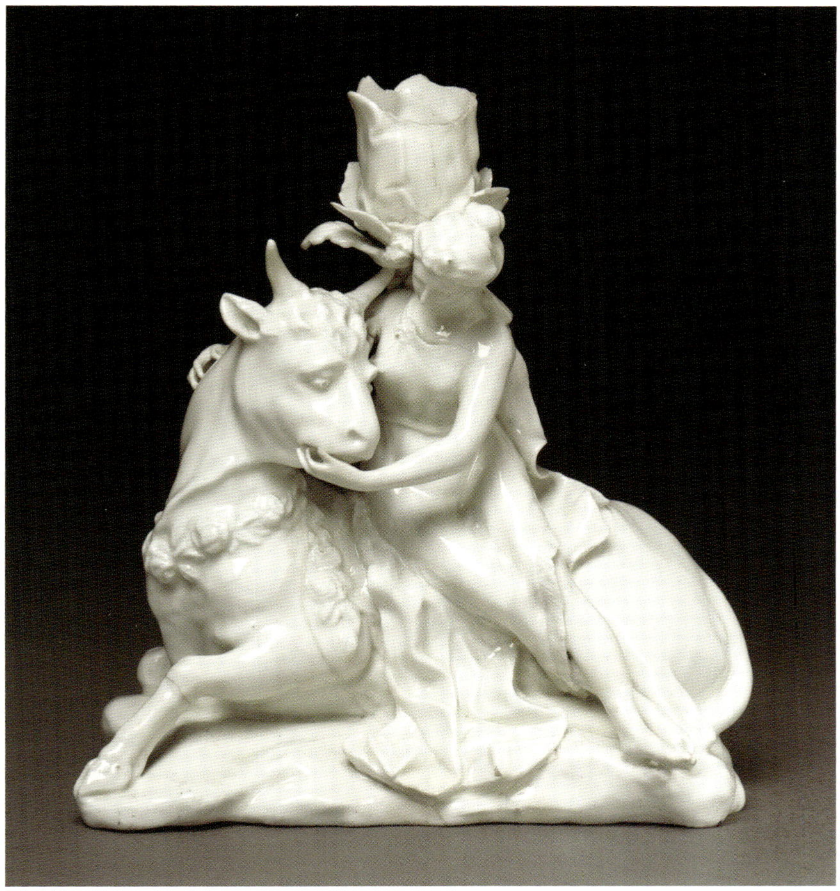

In 1939 Americans increasingly feared that the United States might be called to defend Western civilization as the conflict in Europe grew. It was at this time that Aitken transformed the essence of an ancient Cretan myth into something that referenced the mythology of the American West, *Wyoming Europa* (see fig. 1). In Aitken's portrayal, the rider is a cowgirl who wears a cowboy hat and western boots. This Europa does not even ride sidesaddle, as women typically did. She is full of American confidence and has the bull under control. Although *Wyoming Europa* can be seen as a barometer relating to America's concern for the events in Europe, its mood is jocular, not fearful.

The same year Aitken created *Wyoming Europa*, the World's Fair opened in New York. Situated historically between the Great Depression and World War II, the fair proved to be a microcosm of the rapid changes occurring in the larger world. With the theme "Building the World of Tomorrow," it was opened by President Franklin D. Roosevelt on April 30, 1939, and was hailed as the greatest international exposition in history. Sixty foreign nations participated—Germany was the only major power that did not.

Carl Milles

Sweden, which did not participate in either world war, was nevertheless affected by the New York stock market crash of 1929 and experienced a financial depression in 1930 and 1931.[2] The country's economic problems seemed not to have much effect on its leading sculptor, Carl Milles, however.[3]

Milles is probably best remembered in the United States for two bronze sculptural fountains (ca. 1934–1937) that he created to enhance the grounds

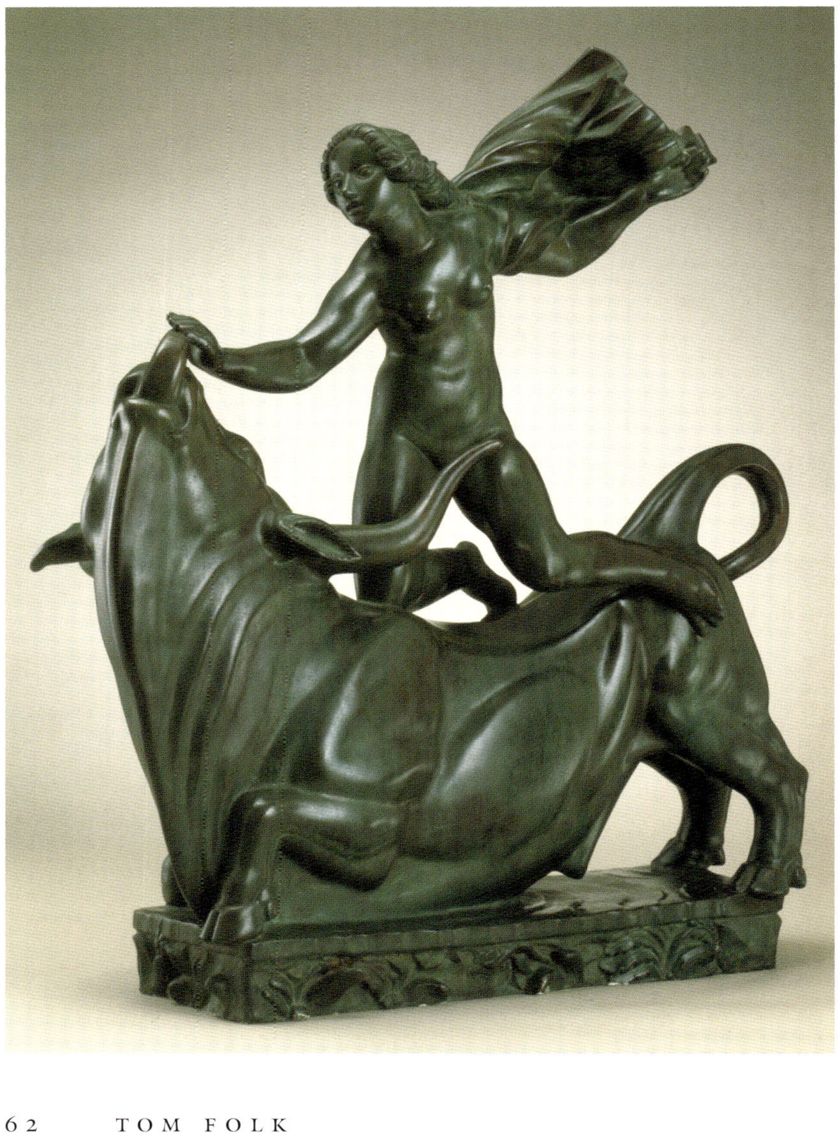

Figure 6 Carl Milles, *Europa and the Bull, Study for Large Fountain*, 1926. Bronze. H. 31". (Cranbrook Foundation, gift of George Gough Booth and Ellen Scripps Booth; archive photo, courtesy Cranbrook Art Museum.)

of the Art Academy buildings of Cranbrook. The buildings, designed by Finnish architect Eliel Saarinen, and Milles's sculptures present an exemplary—and unusual—unity of design. Milles had studied from 1897 to 1904 in Paris, where he came under the influence of Auguste Rodin, as did many young sculptors at that time. But later, while Milles was living in Munich, the late-nineteenth-century German sculptor Adolf von Hildebrand had a more substantial and long-lasting influence.

The Cranbrook Art Academy wanted Milles to join the faculty as a resident artist in the early twenties, when Milles was first establishing his reputation; in time, the school became more aggressive in the effort to bring him to America and offered him a professorship.

Milles was well known in the United States before his appointment at Cranbrook. In 1929 he was in a group exhibition of European sculpture at the Metropolitan Museum of Art, and in 1931 the City Museum of Saint Louis mounted his first comprehensive exhibition, which then traveled to the Detroit Institute of Arts, the Cleveland Museum of Art, and the Brooklyn Museum. Milles had first visited Cranbrook in 1929 and received the position of professor of sculpture in 1931. Cranbrook treated Milles well, well beyond how other resident artists were treated. In fact, the school did not even require him to offer formal instruction; modeling, casting, and other sculptural practices were offered by other instructors, and Milles would lecture advanced students while criticizing their work.[4] In 1951 he retired from Cranbrook and returned to Sweden.

Milles had created a bronze Europa sculpture, commissioned in 1919, for the marketplace in Halmstad, Sweden. Conceived in Sweden, his Europa is probably the most European. His reductive figures display modern tendencies but they also indicate a reluctant interest in Art Deco. Cranbrook has a replica of this Europa sculpture that was cast in 1926 (fig. 6) and is 31 inches tall. Cranbrook also has a larger, outdoor version, also cast in 1926, which is 112 inches tall and positioned at the top of a series of descending pools, where bronze nude male tritons support fish that squirt jets of water. Milles's figure of Europa is kneeling precariously on the bull's back, the moment of conception suggested by her touching the bull's elongated tongue. The bull is depicted with its front legs low to the ground and its hind legs elevated. Art historian Joan Marter explains, "The animal is not shown leaping forward, as in versions of this subject by other artists . . . rather, the short front legs have collapsed beneath the animal."[5]

Paul Manship

The American sculptor Paul Manship is considered by many to be the greatest Art Deco sculptor of all, best remembered for his enormous 1934 *Prometheus* in Rockefeller Center, the most well-known Art Deco sculpture in the world.[6] As a style, Art Deco relates to the 1925 Exposition Internationale des Art Décoratifs et Industriels Modernes in Paris, though the term was coined in the 1960s by curators and art dealers. It is characterized by clean, streamlined, geometric forms with much stylization or conventionalization. Manship's sculptures from the 1910s, such as his *Indian Hunter* and accompanying sculpture *Pronghorn Antelope* of 1914, are already very much in the Art Deco style.

After becoming the assistant of noted American sculptor Solon Borglum in New York City, Manship won the Prix de Rome in 1909, which enabled him to spend three years at the American Academy in Rome. Returning to New York in 1912, he gradually developed his quintessential Art Deco style. Manship moved to Paris in 1921 and set up a studio where the French

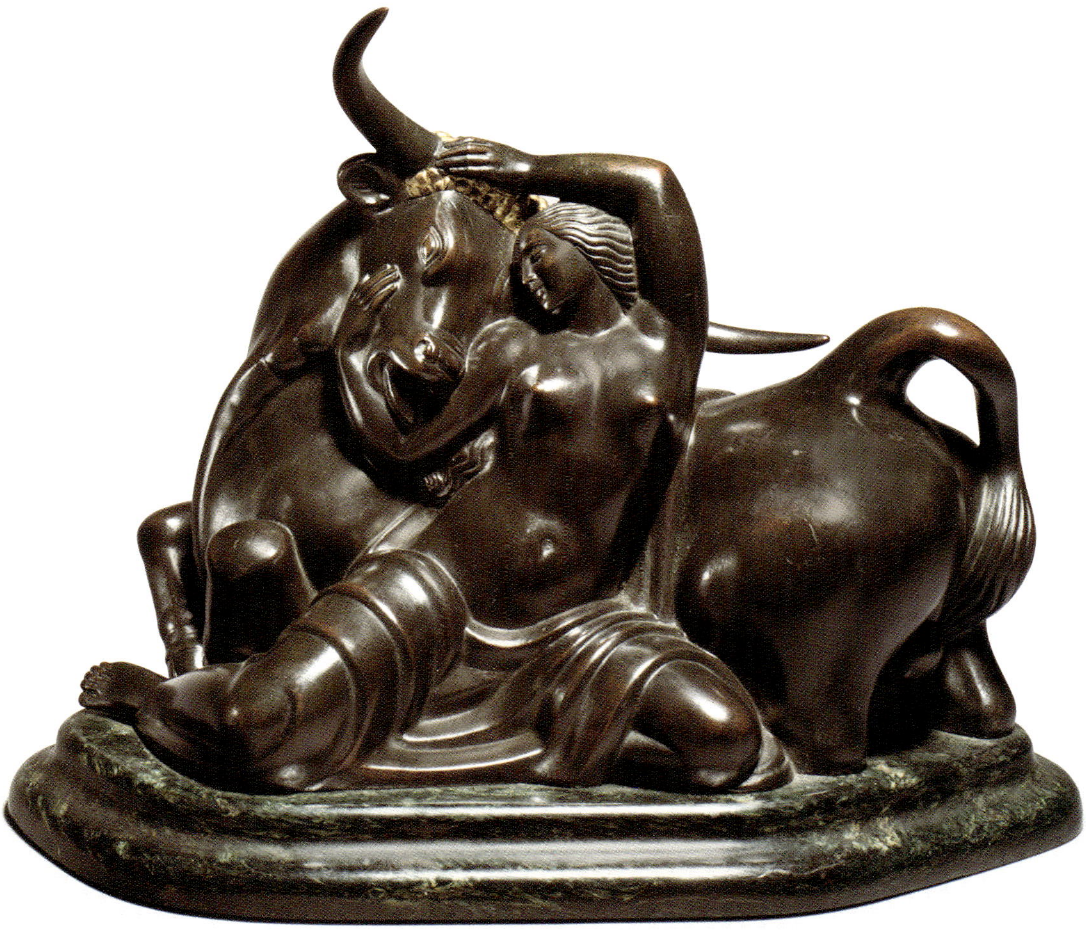

Figure 7 Paul Manship, *Europa and the Bull*, 1924. Bronze. H. excluding marble base 9½". (© Christie's Images Limited 2016.)

painter of decorative themes Jean Dupas had worked. One of Manship's most iconic Art Deco sculptures is his 1924 *Europa and the Bull* (fig. 7). This bronze sculpture displays the sensuality that was lacking in Milles's version. The partially clad Europa embraces the reclining bull. But here she is kneeling in front of him, not on top of him, as in the Milles work. The exposed tongue of the bull may represent the subtle influence of the Milles version.

Although Manship usually worked in bronze, he occasionally sculpted in terra cotta, and in 1928 he created a ceramic version of Europa for the Cowan Pottery in Rocky River, Ohio (fig. 8). This figure of Europa apparently came from one of the same molds used for his bronze version. Manship met ceramist Guy Cowan at the Exposition of Art in Trade at Macy's in New York, where there was a substantial amount of Cowan pottery on display.[7] *Europa* is the only known Manship design created for the Cowan Pottery, and the example in its museum might be the sole extant version.

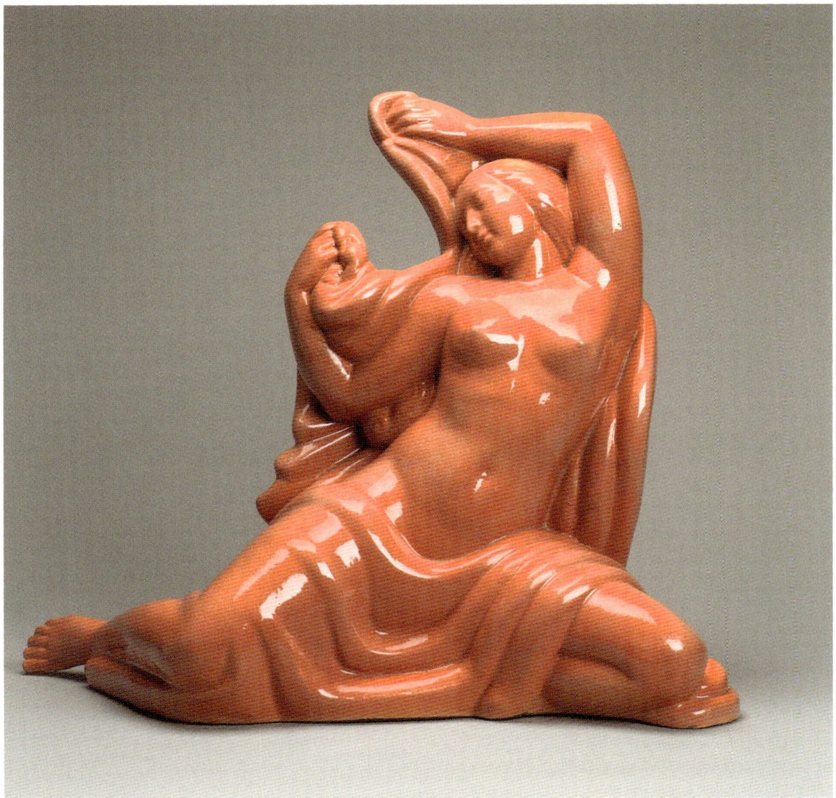

Figure 8 Paul Manship, *Europa*, for the Cowan Pottery Studio, Rocky River, Ohio, ca. 1928. Earthenware with "terra cotta" glaze. H. 15". (Cowan Pottery Museum, Rocky River Public Library.)

Waylande Gregory

Guy Cowan was one of the leading figures in American ceramics.[8] His family had been in the pottery business for generations and he was one of the first graduates of Alfred, the New York School of Clay Working and Ceramics. In 1913 he opened the Cleveland Tile and Pottery Company, which he closed in 1917 to enter the service in World War I. In 1919 he established a pottery at Rocky River near Cleveland, naming it the Cowan Pottery Studio. Starting in 1925 he produced limited-edition ceramic sculptures that were exhibited in fine art venues, such as the annual exhibitions presented by the Pennsylvania Academy of the Fine Arts in Philadelphia.

Waylande Gregory was Cowan's leading sculptor and designer, arriving in 1928 to become the company's only full-time employee.[9] Born in Kansas in 1905, Gregory had been the assistant of Chicago sculptor Lorado Taft, who was noted for his contributions to the World's Columbian Exposition in 1893. While at Cowan, Gregory designed some of its finest ceramic sculptures, including *Salome* (1928), *Nautch Dancer* (1930), and *Burlesque Dancer* (1930). The Cowan Pottery closed at the end of 1931 due to financial problems stemming from the Great Depression. Gregory immediately started teaching at Cranbrook in early 1932, as a resident artist, and he developed the ceramics program there. Although Gregory's arrival was greeted at Cranbrook with great expectation, he was not treated as grandly as Milles had been. Gregory had a full-time job at Cowan, but the situation at Cranbrook was very different. After negotiating his residency with Cranbrook's resident architect Eliel Saarinen, Gregory seemed to be communicating more with George Booth, Cranbrook's benefactor, once he arrived.

Setting up a ceramics program proved to be very expensive, costing $2,500. Gregory hired a contractor to build a kiln. This kiln gave Gregory an opportunity to create much larger pieces than he had at Cowan (up to 30 inches high). He also acquired a low-temperature kiln for smaller sculptures. However, Gregory did not want to understand that Booth expected to be compensated for the investment, which he viewed as a loan. At Cranbrook, Gregory and his wife were provided with complimentary lodgings, but all other income had to come from the sale of artworks and from teaching students on his own. The fee for the spring 1932 term, which included twelve two-hour classes, was $35, and he advertised his classes with elegant brochures printed at the Cranbrook shop (fig. 9). Because he was charismatic, he had a following of students, but since he was at Cranbrook for only about eighteen months, he did not have enough time to develop many followers in the field of ceramics. (The exception was Howard Whalen, who was his assistant at Cranbrook and who later opened his own pottery in California.)

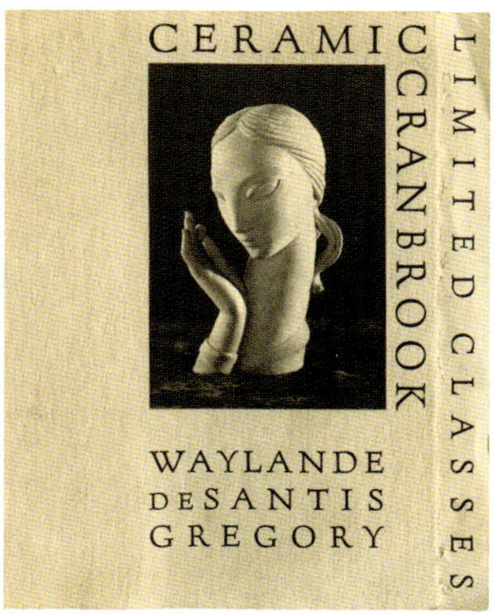

Figure 9 Cranbrook brochure, 1932, showing Waylande Gregory's *Girl with Olive*, 1932.

In January 1933 Gregory renegotiated his contract at Cranbrook but not without listing several complaints. He found Cranbrook far removed from cultural centers, making it difficult to meet clients there and harder to find new students. And he was furious that when Raymond Hood, the chief designer of New York City's Art Deco skyscrapers and a possible business connection, visited Saarinen, Saarinen failed to make an introduction.[10] It seems that mounting tensions between Gregory and the administration at Cranbrook may have been fueled by jealousy on the part of Milles, since Gregory's career was certainly on the rise. In fact, in a 1968 interview, Finnish-born Maija Grotell, who would later replace Gregory at Cranbrook as resident ceramic artist, recalled, "And everyone said to me, 'Do not speak to Carl [Milles] about ceramic sculpture. Don't dare speak to him about it.'"[11] Although Milles sculpted in bronze and Gregory in clay, Grotell's odd comment suggests a professional rivalry between the two. It is interesting that Gregory's replacement as Cranbrook's resident ceramist was a vessel maker, not a ceramic sculptor who might compete with Milles.

By March 1933 there was obvious friction mounting between Gregory and the Cranbrook management team of Booth, Saarinen, and Richard Raesman, the executive secretary. Hostilities exploded on March 4, a national bank holiday. Because the students were away and the Depression was still in full swing, Booth decided to close the Craft Shops and turn off the electricity and heat to save money. Gregory, who had a kiln loaded with major work, turned the electricity back on. While the sculptures were being fired, Booth had the electricity turned off again, damaging most of the pieces left in the kiln, including Gregory's *Europa and the Bull*. Gregory left Cranbrook in June without making peace. Gregory sued Booth, but the settlement, which came two years later, turned out to be a wash, as the value of Gregory's destroyed sculptures were balanced against the money he owed Booth for the two kilns, which he had never attempted to pay back.

Gregory re-created the destroyed *Europa and the Bull* (fig. 10), which was exhibited at his one-man exhibition at the Montclair Museum in 1934.[12] It was purchased in December 1941 by the Syracuse Museum of Art. As Gregory described it, "the sculpture is planned to be partially glazed in the natural texture of the clay. This brings about a splendid play of contrasts in the

Figure 10 Waylande Gregory, *Europa and the Bull*, ca. 1934. Earthenware. H. 23". (Everson Museum of Art, museum purchase.)

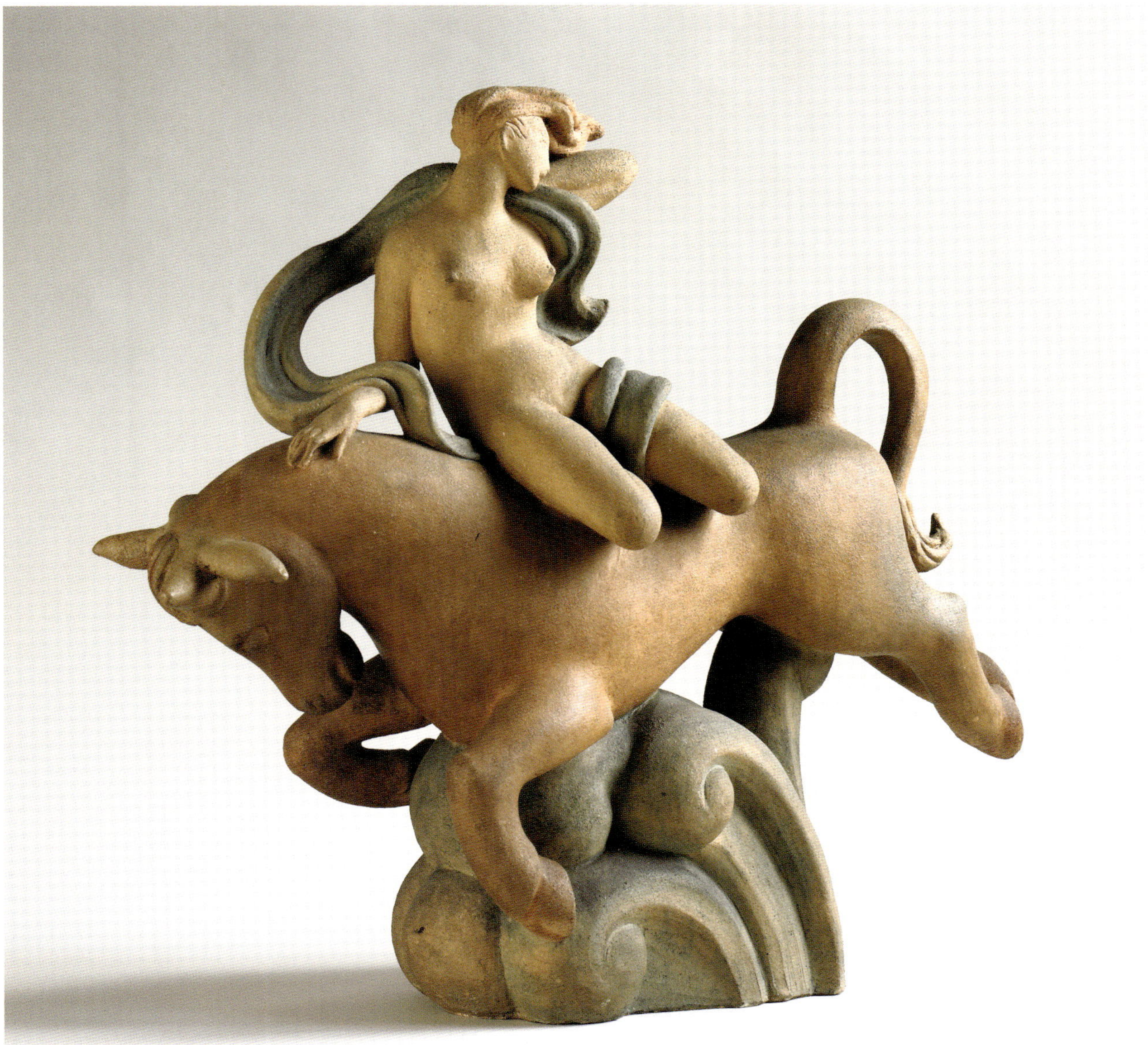

form of synthesis, texture contrasts and color relation."[13] The piece seems to be the most classically Art Deco version in ceramics. Although the kneeling Europa is on the back of the bull, as in the Milles version (see fig. 6), the sculpture overall shows the influence of Manship's second version of Europa and the Bull, *Flight of Europa*, of 1925 (Smithsonian American Art Museum), in which the belly of the bull is supported by dolphins. In Gregory's version, the bull is supported by waves. Gregory was much enamored of Manship's sculpture, and the two became friends within a few years. After Manship's death, Gregory wrote about the influence of ancient art on Manship's sculpture: "It was the exposure at first hand of the glories of ancient Hellenistic and Roman sculptures which started the crystallization of this style [Art Deco]."[14]

Valerie Wieselthier

A major shift in depictions of Europa in the United States occurred following the introduction of the more playful style of Vally Wieselthier, especially after she visited Cleveland in 1929 and influenced young American ceramists such as Viktor Schreckengost and Russell Aitken. Changes regarding the Jewish population in Austria no doubt encouraged her to move to the United States. Gregory's classically Art Deco depiction of Europa became replaced by depictions that were much more whimsical.

Vally Wieselthier was born in Vienna in 1895 to a prosperous Jewish family.[15] Her father was an attorney of the imperial court, which enabled the family to have a lifestyle of luxury, complete with servants and nannies. She studied at the Kunstgewerbeschule (College of Applied Arts) with the well-known ceramist Michael Powolny from 1914 to 1920. The school, founded by Josef Hoffmann and Koloman Moser, was much associated with the Wiener Werkstätte (Vienna Workshops), which dominated the culture of Vienna at this time. Wieselthier spent her student days during World War I in relative security as a student. Thereafter, she became a leading member of the Wiener Werkstätte and one of the leading ceramic artists in Europe. In 1922 she opened her own studio, where she employed a number of people. She exhibited several works in the Austrian Pavilion at the 1925 Exposition Internationale des Arts Décoratifs in Paris. Encouraged by the popularity of the Paris exposition, the Federation of American Artists organized "The International Exhibition of Ceramic Art," which opened at the Metropolitan Museum of Art in 1928. This traveling exhibition spread knowledge of the current trends in European ceramics to other American cities.

Although Christian citizens in Vienna had, for the most part, accepted longtime Jewish residents, attitudes changed with the influx of approximately twenty-five thousand Orthodox Jews after the outbreak of World War I. Many of the refugees suffered in abject poverty because Christian factory owners refused to hire them. According to Ruth Beckermann, "While the Germans were condemning the Jews in the east to forced labor, the Austrians were condemning them to forced unemployment."[16] The rise of anti-Semitism led to the first major Nazi attack on Jews in Vienna with the destruction of the well-known Café Produktenbörse in December 1929.

Wieselthier was invited to visit Cleveland when the Metropolitan's exhibition traveled to the Cleveland Museum in February and March 1929. The exhibition's Viennese ceramics had the greatest impact, especially the five examples by Wieselthier. Guy Cowan attended the exhibition with one of his best students from the Cleveland School of Art, the young Viktor Schreckengost. It is also likely that while staying in Cleveland, Wieselthier first became acquainted with Waylande Gregory, the head designer at the Cowan Pottery. Wieselthier brought a distinct Viennese style and sophistication to America. Marianne Horman has described her:

> She undoubtedly personified the modern woman that emerged in the 20's, self-confident, provocative and fashionably styled. She dressed in the style of the modern woman, wore her hair short, preferred figure-hugging, low-cut dresses, bold hats, striking jewelry and was always seen with a cigarette, the symbol of the emancipated woman.[17]

Wieselthier's last stay in Vienna was from September 9 to November 23, 1931. The growing anti-Semitism in Europe may have become too much for her, and she established residence in New York. There was also a growing

Figure 11 Valerie ("Vally") Wieselthier, *Europa and the Bull*, 1938. Glazed earthenware. H. 25". (Private collection; photo, Randl Bye.)

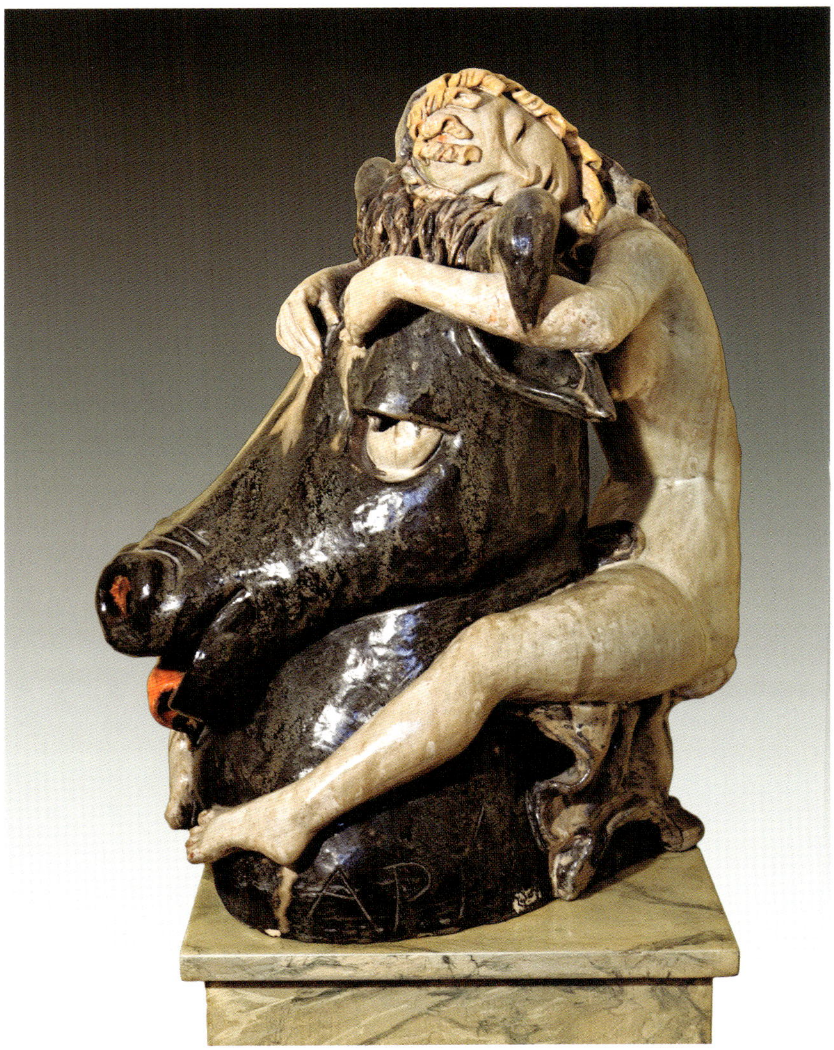

antagonism to modern styles, as evidenced by the Wiener Werkstätte's bankruptcy and closure in 1932. In the same year, she took up residence with her fiancé, Paul Lester Wiener, sharing an apartment at 509 Madison Avenue; after they separated, she moved to 301 East 38th Street. Wieselthier's *Europa and the Bull* (1938; fig. 11) is markedly different from the others discussed. Her Europa, blond and nude, straddles the bull's shoulders and rests her head on his, with her arms wrapped around his horns and her toes curled around the front of his neck. Only the bust of the bull is included, with his eyes rolling up in ecstasy and his red tongue hanging out. The scene conveys her affection and innocence and the bull's carnal desire.

Wieselthier would later visit Gregory and his wife in Warren, New Jersey, a short commute. When she developed cancer, Wieselthier began to spend more time at the Gregorys' home. She worked at her studio at 24 East 8th Street during the week but was afraid to take much medication then, fearing that the drugs would interfere with her responsibilities or with the quality of her work; she was financially stressed and needed to keep working. So on long weekends, she would take morphine while staying with the Gregorys, and could relax. She was very interested in Gregory's monumental ceramic sculpture and she could speak in German to his wife, Yolande, who was Austrian-Hungarian. The Gregorys would take her to see points of local interest and to local events. In time, she became quite familiar with people in the area. She died in 1945.

Viktor Schreckengost

The Viennese ceramics that were part of the international ceramics exhibition in 1929, particularly the work of Michael Powolny, had great impact in Cleveland and were inspirational for Ohio-born Viktor Schreckengost.[18] He had been attending the Cleveland Institute of Art, studying ceramics with Guy Cowan. Another important and influential art teacher was Julius Mihalik, who encouraged him to study in Vienna. Mihalik, who had taught at the Kunstgewerbeschule, wrote a letter of introduction, and Schreckengost arrived there in 1930 for a year's study with Powolny, Josef Hoffmann, and others. (Powolny had also been Wieselthier's principal ceramics professor.) The humorous quality of Schreckengost's later ceramic sculpture is a testament to the Viennese legacy, but he went beyond his training to develop his own focus and personal style. On Easter break he traveled to Russia; he may have been witness to the social and political changes that were beginning to rock Vienna and elsewhere in Europe.

Schreckengost returned to Cleveland and began working part-time at the Cowan Pottery—in fact, he was the very last artist to be hired there. Although the enterprise had filed for bankruptcy late in 1929, it was allowed to operate, maintain some of its staff, and fill orders until it closed its doors permanently in 1931. It was during these troubled times that Schreckengost created the Jazz Punchbowl, a limited edition that was originally designed for Eleanor Roosevelt. These bowls, decorated with skyscrapers, cocktail shakers, and ocean liners, are the most popular works to come from the pottery and may be among the most significant ceramic vessels produced in America.

During the later 1930s Schreckengost produced a series of ceramic sculptures depicting African-Americans. One of the best in the series is *The Abduction* (fig. 12), which shows a nude Europa reclining languidly on the back of the bull. No water or dolphins support the bull, however, as in the Europas by Gregory (see fig. 10) and Manship. Rather, the figures are accompanied by two African-American winged putti representing love. This Europa wears a bandanna, a common accessory among southern African-American and Caribbean women. Schreckengost modeled this Europa in the likeness of American entertainer Josephine Baker, who created an international sensation when she performed her *Danse sauvage* wearing only a string of artificial bananas and whom the artist had seen perform in Paris.[19] As the situation in Europe was becoming increasingly worse, it seems significant that Europa was becoming Americanized. This was one of three sculptures by Schreckengost to be awarded first prize in ceramic sculpture in the Syracuse Museum of Art's 1938 Ceramic National Exhibition.

Figure 12 Viktor Schreckengost, *The Abduction*, 1938. Glazed earthenware. H. 14". (Viktor Schreckengost Foundation.)

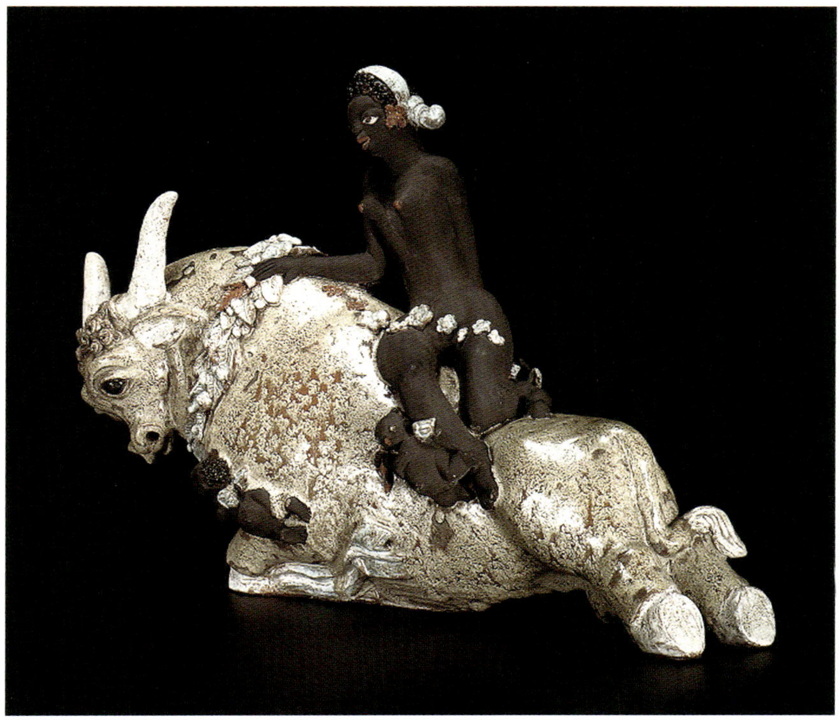

Russell Aitken

The humorous but sophisticated influence of Viennese wit is even more evident in the work of Russell Aitken, who had also been a student of ceramics of Guy Cowan at the Cleveland Institute of Art. After graduating in 1931 he found that the Cowan Pottery had closed, but he was able to decorate some of the remaining Cowan blanks, which were later fired at the Cleveland Institute. In October 1932 Aitken arrived in Vienna to study with Michael Powolny at the Kunstgewerbeschule. The Wiener Werkstätte had already closed by his arrival, and he must have seen an even more drastic scene in Vienna than had Schreckengost. Abruptly concluding his studies at the Kunstgewerbeschule as early as January 15, 1933, for no apparent reason, Aitken never received a certificate for his studies there. Nevertheless, he traveled to Germany and Hungary. Being an accomplished athlete and sportsman, he rode horseback, went canoeing, and even joined the Corps Hilaritas Wien, a Viennese dueling society. Social and political issues were generally not part of his work.

After he returned from Europe, Aitken joined a group of others and set up a cooperative venture in Cleveland known as the Pottery Workshop. Aitken created work there and gave instruction in pottery making. He moved on after a couple of years to White Sulphur Springs, West Virginia, where he continued to give instruction in ceramics. By 1935 he had opened a studio in New York City. His ceramic sculptures were represented in the Harry Salpeter, Sloane, Ferargil, and Brownell-Lamberton galleries, also in New York City. His best-known work, *Futility of a Well-Ordered Life* (1935), was based on Salvador Dalí's *Portrait of Gala with Two Lamb Chops Balanced on Her Shoulder* (1933), now in the Fundación Gala Salvador Dalí, Figueras, Spain. *Futility of a Well-Ordered Life* entered the collection of the Museum of Modern Art, and although the museum played only a tangential role in the development of ceramic sculpture in the thirties, it made a rare exception in acquiring this work.

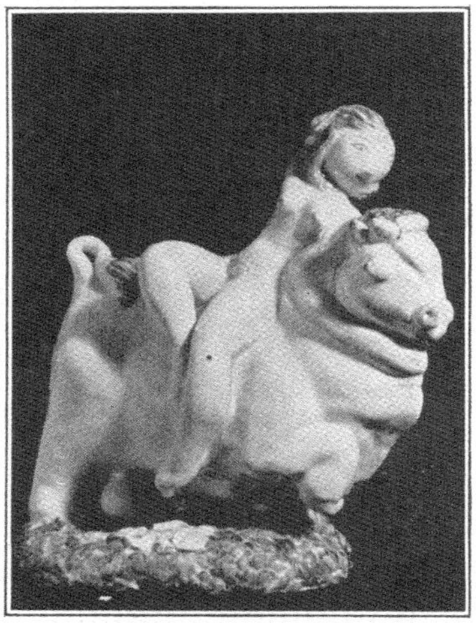

AITKEN'S *EUROPA*
The jollity and ogling is typical.

Figure 13 Russell Aitken, *Europa*, ca. 1935, glazed earthenware. Dimensions unrecorded. (Collection unknown; photo from *Time*, October 18, 1937.)

Aitken's first *Europa* (fig. 13) was described in *Time* magazine as a "jolly maiden atop a jolly, ogling bull."[20] It was included in the Cleveland Museum's 1935 "May Show," where it received a special award, and also won an honorable mention at the Syracuse Museum of Art's 1936 Ceramic National. Its current location is unknown.

Aitken's later *Wyoming Europa* (see fig. 1) is a more detailed and complex work. Like Schreckengost's *The Abduction*, this Europa-as-cowgirl is concertedly Americanized. Aitken created several works relating to the American West and was particularly interested in the sculpture of Frederic Remington. When he was a student at the Cleveland Institute of Art, Aitken was close friends with Hughlette Wheeler, a teacher of sculpture who had been a cowboy, and Aitken himself had worked as a cowboy in Arizona and Montana.[21] He ceased creating ceramics in the 1940s, and his ceramic work is rare today. He was a well-known sportsman and philanthropist and put together significant collections of decoys and tribal art.

Charles Yerkes Dusenbury

Charles "Chuck" Yerkes Dusenbury was born in Detroit in 1915 and graduated from Highland Park High School. He attended the Art Academy at Cranbrook for no credit during 1935–1936 as a scholarship student. In those years he studied drawing and painting with Zoltan Sepeshy. In 1937 he turned to sculpture and studied modeling with Marshall Fredericks, which gave him some grounding in sculpture before entering more advanced classes with Carl Milles in 1937 and 1938. From 1938 to 1939 he assisted Milles in his studio as a "technician." In 1940 he became interested in ceramic sculpture and studied ceramics with Maija Grotell, the year he created his version of Europa and donated it to Cranbrook, where he would later teach in 1946 and 1947. He married Cranbrook's master weaver, Marianne Strengell, in September 1940; they had two children but were divorced in 1948. Dusenbury was one of six people associated with Cranbrook who formed the Saarinen-Swanson Group, which produced a collection of modern furnishings for the middle class that debuted at Hudson's Department Store in Detroit. Dusenbury died in 1999 in Tryon, North Carolina.

Dusenbury's *Europa and the Bull* (fig. 14) is one of the most amusing. As a student at the Cranbrook Art Academy, Dusenbury had studied sculpture with Carl Milles in 1937–1938 and then studied ceramics with Grotell.[22] Although Grotell was a vessel maker, she allowed Dusenbury to create a ceramic takeoff on Milles's *Europa* (see fig. 6), two versions of which were on display at Cranbrook. Dusenbury's ungainly female sticks out her tongue at the seemingly bewildered bull. Although the awkward pose of

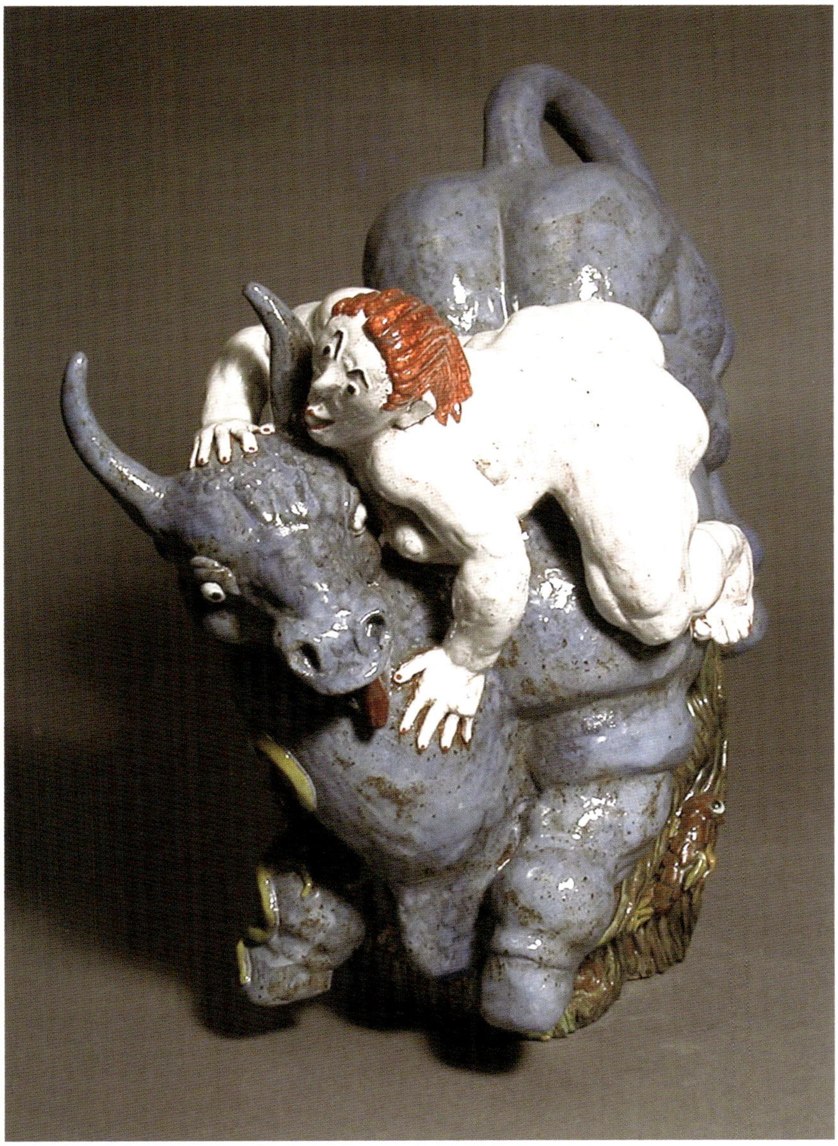

Figure 14 Charles ("Chuck") Yerkes Dusenbury, *Europa and the Bull*, 1940. Glazed earthenware. H. 14". (Cranbrook Art Museum, gift of the artist.)

both figures relates to Milles's original design, the comic nature relates more to the Europas by Cleveland artists Viktor Schreckengost and Russell Aitken.

The entry of the United States into World War II at the end of 1941 seemed to bring comical interpretations of Europa to an end. So we have come full circle with Europa, from Milles's version to Dusenbury's version, both of which are at Cranbrook today. For each of the artists discussed in this article, their expressions of Europa were, and remain, major examples of their artistry in their careers.

ACKNOWLEDGMENTS The author would like to acknowledge those who helped obtain photography for this article, including photographer Randl Bye, Barbara Perry, Gene Schreckengost, Chris Kende, Nadine ("Dina") Bluemel, Gregory Wittkopp, and Christine McNulty of Cranbrook; Leslie Cade of the Cleveland Museum of Art; Lauren Hansgen of the Cowan Pottery Museum, Rocky River Public Library; and Karen Convertino, Everson Museum of Art.

1. The seminal study that started a revival of interest in such artists as Aitken, Waylande Gregory, Viktor Schreckengost, and Vally Wieselthier is Ross Anderson and Barbara Perry, *The Diversions of Keramos, American Clay Sculpture, 1925–1950*, exh. cat. (Syracuse, N.Y.: Everson Museum of Art, 1983).

2. Francis Sejersted, *The Age of Social Democracy: Norway and Sweden in the Twentieth Century* (Princeton, N.J.: Princeton University Press, 2011).

3. For Milles, the most significant study is Meyric R. Rogers, *Carl Milles: An Interpretation of His Work* (New Haven, Conn.: Yale University Press, 1940). See also Ann Jonsson, "D'un mythe à l'autre: L' 'Europe' de Carl Milles et sa symbolique en Suède," in *D'Europe à l'Europe, II: Mythe et identité du XIXe siècle à nos jours. Actes du colloque tenu à Caen*, Caesarodunum 33, ed. Rémy Poignault, Françoise Lecocq, and Odile Wattel de Croizant (Tours: Centre de Recherches A. Piganiol, 1999), pp. 157–62.

4. Joan Marter, "Sculpture and Painting," in *Design in America: The Cranbrook Vision, 1925–1950*, exh. cat. (New York: Abrams, in association with the Detroit Institute of Arts and the Metropolitan Museum of Art, 1984), p. 243.

5. Ibid., pp. 245–46.

6. For Manship, see Edwin Murtha, *Paul Manship* (New York: Macmillan, 1957); and Harry Rand, *Paul Manship*, exh. cat. (Washington, D.C.: Published for the National Museum of American Art by the Smithsonian Institution Press, 1989).

7. Curator Lauren Hansgen of the Cowan Pottery Museum in Rocky River, Ohio, discussed the connection between Cowan and Manship in an email to the author, October 21, 2015. Manship was a guest speaker at the Exposition of Art in Trade held at Macy's department store in New York City, May 2–7, 1927. Cowan Pottery had the largest ceramic exhibit. See Mark Bassett and Victoria Naumann, *Cowan Pottery and the Cleveland School* (Atglen, Pa.: Schiffer Publishing, 1997), pp. 118–19.

8. For an excellent study of Guy Cowan, see Bassett and Naumann, *Cowan Pottery*.

9. For Gregory, see Thomas C. Folk, *Waylande Gregory: Art Deco Ceramics and the Atomic Impulse*, exh. cat. (Richmond, Va.: University of Richmond Museums, 2013).

10. Letter from Richard Raesman to George Booth, January 18, 1933, regarding Gregory's contract with Cranbrook, Cranbrook Archives. Hood and Saarinen apparently remained friends after the international architectural design competition for the *Chicago Tribune*'s new headquarters, which carried a $50,000 prize. More than 260 entries were submitted. Hood won the competition, and Saarinen's design came in second place. I would like to thank Leslie Edwards, Cranbrook's head archivist, for sharing pertinent correspondence regarding Waylande Gregory.

11. "Conversation with Maija Grotell at Cranbrook," May 24, 1968, in Jeff Schlanger and Toshiko Takaezu, *Maija Grotell: Works Which Grow from Belief* (Goffstown, N.H.: Studio Potter Books, 1996), p. 63.

12. Gregory's *Europa and the Bull* was displayed at the Everson Museum's Ceramic National exhibition in 1938. Although the Everson Museum has dated this work to 1937, a photograph shows it in an exhibition at the Montclair Museum, "Ceramic Sculpture by Waylande Gregory," in March 1934. The photograph is reproduced in Folk, *Waylande Gregory*, p. 60.

13. Waylande Gregory, cited in Anna W. Olmsted, "Art Chat: Gregory's Europa New Accession at the Museum," *Post-Standard* (Syracuse, N.Y.), December 29, 1941. Gregory Archive, Cowan Pottery Museum, Rocky River Public Library, Ohio.

14. Waylande Gregory, "Living with Art," *Daily Home News* (New Brunswick, N.J.), October 23, 1966. Waylande Gregory Archive, Cowan Pottery Museum, Rocky River Public Library, Ohio.

15. For Wieselthier, see Marianne Hussl-Hörmann, *Vally Wieselthier, 1895–1945: Vienna, Paris, New York: Ceramics, Sculpture, Design of the Twenties and Thirties* (Vienna, Austria: Böhlau, 1999). After Wieselthier died, Gregory took plaster impressions of her hands that are now in a private collection. St. Etienne Gallery in New York City mounted a memorial retrospective exhibition three years after her death, in 1948. Although the opening date is unrecorded, it closed on January 27, 1948.

16. Ruth Beckermann, *Die Mazzesinsel: Juden in der Wiener Leopoldstadt, 1918–1938* (Vienna, Austria: Löcker, 1984), pp. 16ff.

17. Ibid., p. 213.

18. For Schreckengost, see Henry Adams, *Viktor Schreckengost and 20th-Century Design*, exh. cat. (Cleveland, Ohio: Cleveland Museum of Art, 2000), and Adams, *Viktor Schreckengost: An American Da Vinci* (Windsor, Conn.: Tide Mark, 2006).

19. Adams, *Schreckengost and Design*, p. 58.

20. "Ceramics," *Time*, October 18, 1937, Cleveland Museum Archives.

21. Aitken's friendship with Hughlette Wheeler and his interest in Frederic Remington are discussed in Anderson and Perry, *Diversions of Keramos*, p. 78. Aitken even collected some of Remington's bronze sculptures.

22. I would like to thank Leslie Edwards for sharing this information. See also Martin Eidelberg, "Ceramics," in *Design in America*, pp. 222–23.

Figure 1 Edith Heath (1911–2005) at wheel, undated. (Courtesy, Berkeley Environmental Design Archives.)

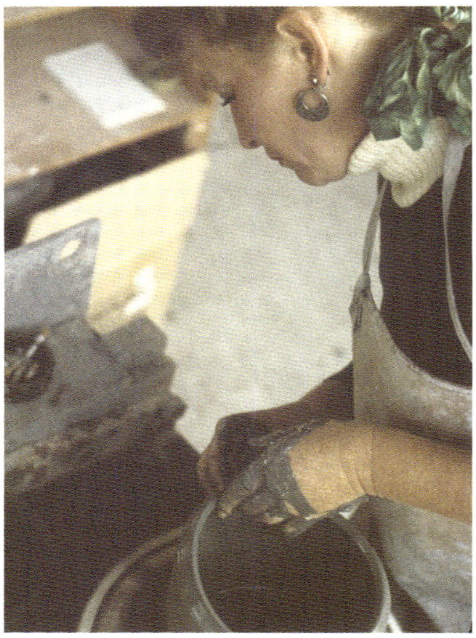

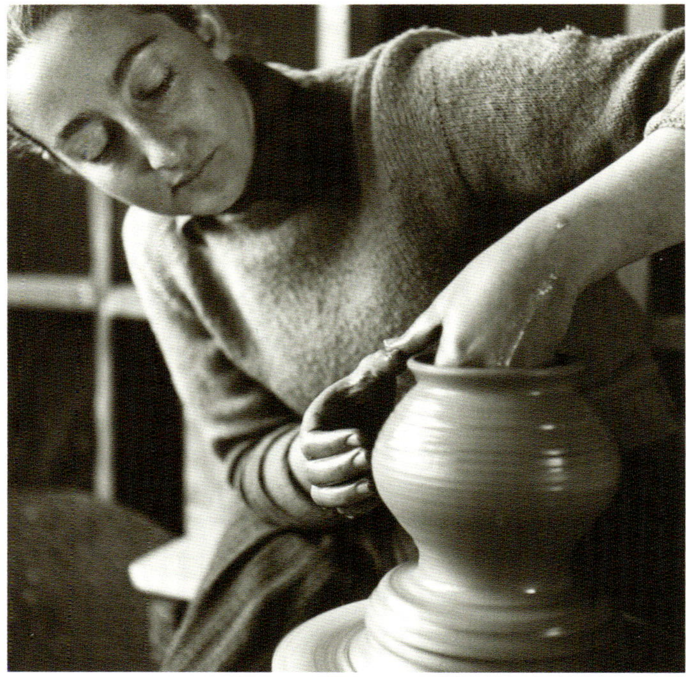

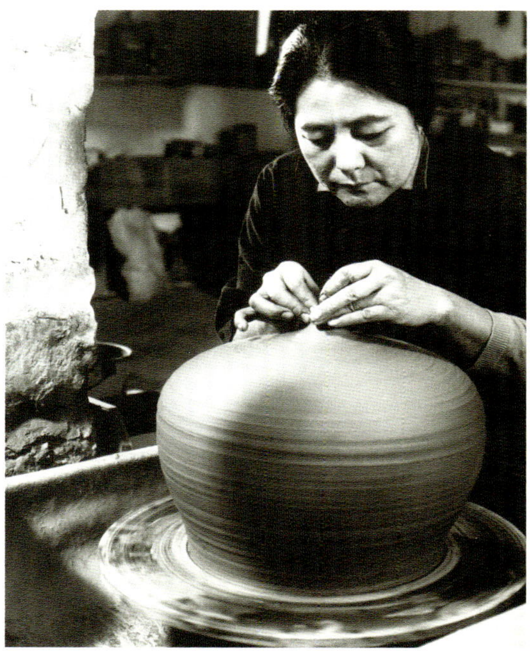

Figure 2 Karen Karnes (1925–2016) at wheel, ca. 1952–1954. (Courtesy, UNC State Archives Raleigh; photo, Edward Dupuy.)

Figure 3 Toshiko Takaezu (1922–2011), ca. 1960. (Courtesy, American Craft Council.)

Ezra Shales

Throwing the Potter's Wheel (and Women) Back into Modernism: Reconsidering Edith Heath, Karen Karnes, and Toshiko Takaezu as Canonical Figures

▼ IF GEORGE OHR (1857–1918) and Peter Voulkos (1924–2002) tend to be praised as artists—and not mere potters—because they "broke free" of conventional craftsmanship on the potter's wheel by casting off function as a normative condition and rupturing classical form, why is it that no woman has ever inherited Ohr's moniker, the "mad potter"? This deeply American narrative, a heroic liberation from European and Asian traditions, begs the question whether the "woman problem"—the lack of women in the studio craft canon—might be productively reconsidered by putting it in closer relation to the "wheel problem" and the question of what it means to be a pioneer in ceramics.

Generally, radical artistic license and aberrant tooling have been characterized as masculine discourses, whereas the maintenance of decorum is regarded as a feminine tendency and as effeminizing for men as the decorative. Three generations of critics have deployed the alliterative keywords *violation*, *violence*, and *volcanic* to position Voulkos as a centrifugal masculine force at the center of the mid-century revolution who transformed studio pottery into art. It is fine to claim Voulkos waged a battle against decorative art and traditional wheel-thrown art pottery, but it is time to suggest the ways in which a woman's work made on a wheel was equally controversial and to outline ways that his female colleagues can be integrated into this narrative of modernist disruption.

In the 1950s and 1960s, successful women who were working in clay received such accolades as "delicate," "lyrical," and "intimate," historically coded terms that facilitated the segregation and demotion of women's contributions to modernism and visual arts in general, as art historians such as Linda Nochlin, Cheryl Buckley, and Pat Kirkham have eloquently demonstrated.[1] If, after 1969, this narrative broadened to include the first women ceramists who were acclaimed as radically uninhibited, they worked largely in a Pop and declaratively "fine art" mode, such as Patty Warashina (then Bauer) and Viola Frey. While Voulkos both accelerated and defiled the potter's wheel in a Dionysian avant-garde explosion, did women's work in clay constitute an alternative Apollonian tradition, or is there some less schematic way to include them in art history?

This essay specifically reevaluates the late 1940s to late 1960s work of Karen Karnes (1925–2016), Toshiko Takaezu (1922–2011), and Edith Heath (1911–2005) (figs. 1–3), and challenges where existing scholarship locates them in the margins of revolutionary ceramic art. Instead of positioning their contributions within the dominant art historical narrative that serves

Voulkos's legacy, can we contextualize them and their use of the wheel in new ways? Each of these three women exhibited nationally, and their pottery gained visibility through their successful participation in the Syracuse Museum of Art's medium-specific national competition ("Ceramic National") begun in 1932; each show traveled to numerous far-flung venues, including not just metropolitan areas such as New York City but also many small ones, until 1972. Yet these artists have never been written about critically as a threesome because even within ceramics they were considered to inhabit different modes of production—one in art, one in craft, and one in design. If their names were afloat in the ceramics scene, none crossed over to the fine-arts world as Voulkos's did. He accomplished that momentarily, when he had a solo exhibition at the Museum of Modern Art in 1960.

While many outside the ceramics field remain unaware of Voulkos even today, in Rose Slivka's semi-erotic description of his muscular athleticism, through Christopher Knight's reviews in the *Los Angeles Times* in the 1990s, and on to recent scholarship, a hagiography has developed within the subgenre of ceramic art history. Critical appraisals have celebrated his virility and established the story that ceramics needed to depart from the traditional uses of the wheel in order to become sculpture. Indeed, a 2015 exhibition at the Yale University Art Gallery, "The Ceramic Presence in Modern Art," reiterates Voulkos's centrality and includes an essay on George Ohr as his primary antecedent, despite his rediscovery occurring at a later date. Yale reaffirms Rose Slivka's 1964 view on how and when clay turns into modern art—by spurning function and tradition—and positions a few women as mere adjuncts, none as Voulkos's equal.

A simple Google image search of "Voulkos potter's wheel" brings up evidence of manly potting across several decades, proving that he never abandoned the wheel as a theatrical stage to be bare-chested. Yet few women have been photographed working the wheel in such dramatic and combative terms. The one reference in the corpus of scholarship that does contextualize Voulkos in reference to the power of a woman artist is a comment by Frances Senska, Voulkos's teacher at Montana State: "Sometimes, when you watch Pete throw, you can see Marguerite Wildenhain."[2] With this inversion in mind, it is worth wondering whether the technique of the potter's wheel is valuable as a place to revisit the assignation of gender roles, and reconsider where and when manual transgression comes into play. If ceramics was a workspace to attack decorum—get muddy and existentialist—it was still a place where women were generally depicted maintaining propriety. The uncanny was possible to attain in functional pottery, but not equally accessed: products by Heath, Karnes, and Takaezu might be valuable to suggest where, when, and how American women did break new ground, and to see the wheel in a new light. And a longer view of the wheel might generate a more eccentrically shaped history that is less reductively modernist and less of a reaffirmation of current art-market values, a major pitfall of much art historical enterprise.

By highlighting the distinct ways that each of these three women used the potter's wheel, I hope to suggest an art historical framework conducive to evaluating both tableware and the more self-consciously fine-art ceramics. I propose that women working in clay might have left "quietly subversive" legacies worth adding to our canon, especially if craft and design are to be considered as representative of a zeitgeist as fine art. By resituating ceramic artistic production within the antipodes of license and decorum, instead of the avant-garde and its opposites, the academic and conservative, a more synthetic history that goes beyond the trope of individual genius or radical insurgency will emerge. Were women working in clay as reserved and conservative as scholarship suggests, or did they generate alternative modernities?

When they are revered today, Edith Heath is cast as California "good design" worthy of display in any museum of modern art, Karen Karnes is celebrated as a virtuosic and humble production potter, and Toshiko Takaezu is praised for achieving a Zen grace in ceramic sculpture, but this essay will set them into orbit around the potter's wheel in order to see both the tool and their work in a common cultural context. Such a Venn diagram of art, craft, and design will emphasize their disciplinary differences in an attempt to reveal the breadth of women's endeavors in ceramics. In 1969–1970's *Objects: USA* catalog and touring exhibition no work by male artists (even Rudy Staffel or Richard Devore) is described as "delicate," whereas both Karnes and Takaezu are characterized in this manner. Might we see these potters in a more complex light and offer their work a more nuanced reading today, perhaps as each testing the boundaries of gentility?

Photographic portraits of these three women aid an appreciation of potting as quietly content labor, a busying of hands that updates Vermeer's *Lacemaker* in the suggestion of a total absence of thinking or intellectual labor (fig. 4). Rapt attention to the hump of clay on their wheel diminishes a sense

Figure 4 Johannes Vermeer (1632–1675), *The Lacemaker*, 1669/1670. Oil on canvas. 8" x 9". (Courtesy, Musée du Louvre, Paris, France; photo, Bridgeman Images.)

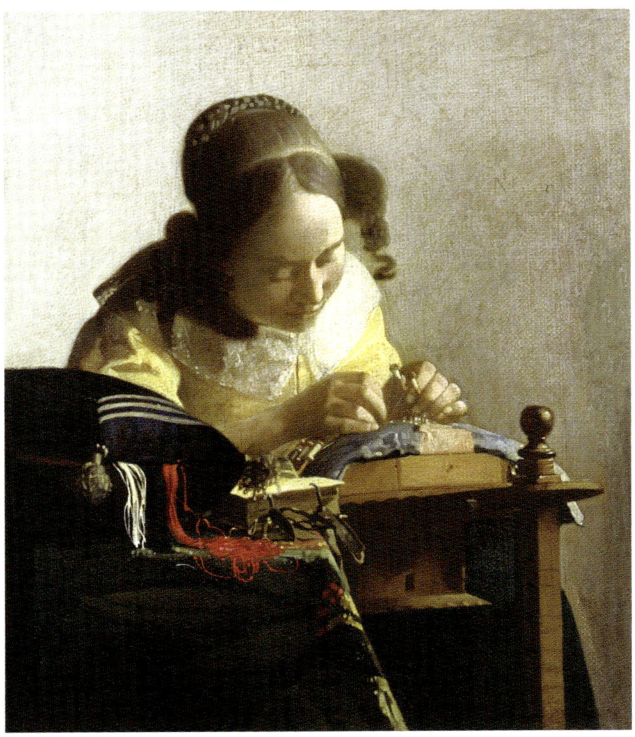

of the activity's athleticism. Surely, Heath, Karnes, and Takaezu appear to be centered, nonthreatening, and peaceful. The lack of eye contact with the potters in these images seems to reenact social codes of feminine modesty; these are not women who offer brazen glares. Peter Dormer used Vermeer's image to question whether "the essential quality of craft activity [is] the pleasure of losing oneself in a practice activity."[3] Certainly the wheelwork seems to subdue or at least displace an individual's autonomy. The image displeased Dormer for its alignment of handicraft with cultural conservation, but was one he sanctioned to claim that craft is a peculiarly domestic affair. M. C. Richards would probably have approved of these images of potting as a representation of Mihaly Csikszentmihalyi's notion of achieving "flow" and harmony in work.[4] There is no singular interpretation; one can see these images of handicraft both as an existential challenge in which individuals demonstrate self-reliance and as modernization vigorously contravened. We like to think of potting as hazy, romantic primitivism, a mythical lifeway, but the tool is mechanical as well. Sustaining the perspectives of both Dormer and Richards is my aim: Can we go beyond appreciating these women potters as modern variants on Vermeer's *Lacemaker*? Did they occupy such assigned gender roles, or did they make unruly forms and depart from conventions despite portraiture that suggests otherwise?

The Wheel as Machine

To think of wheel-thrown pottery as disruptive is to move in the opposite direction from regarding the activity as spiritually "centering," to use the title of Richards's wildly influential book of that name, *Centering in Pottery, Poetry, and the Person* (1964). Richards's view—that the wheel is a therapeutically transformative tool, a means to engage in Zen-like individual meditation—is now commonplace. Yet the potter's wheel did not hold only one meaning in the early and mid-twentieth century; it was a symbolic tool used to serve numerous interests. For Voulkos it was about speeding life up, exploring life out of balance; for Richards the wheel is often seen as a place of refuge and solace, where the world slows down.

Problematically, the term *potter's wheel* is remarkably vague and refers to numerous types of tools. Some are rudimentary kick wheels, others variable-speed electrified motors. In his hyperbolic praise for Voulkos as an innovator, Andrew Perchuk has written, somewhat misleadingly, "the archetypal potter's tool—the wheel—was largely unknown for much of the twentieth century, and throwing on the wheel became widespread only in the late 1940s."[5] Yet I argue that the identity of the wheel was not so clear-cut nor its florescence so acute, either in the 1940s or the 1970s. Voulkos was part of a wave of enthusiasm for electrifying the wheel, which remains the most plentiful type today, but this was not the "archetypal" wheel. Foot-powered wheels enjoyed resurgence in the decade after *Centering* was published. In the 1970s, to use a treadle or a kick wheel was a considered countercultural move off the grid—toward some presumed more historical and vernacular type. But many diligent art students in the 1930s used treadle wheels, and several classrooms and studios had wheels in motion when Voulkos was spinning his, though few were as powerfully motorized. Even if Richards centered on a kick wheel, and convinced many readers that it was the perfect metaphor to slow down, the majority of her readers in the 1970s probably used electric wheels. If "centering" was Richards's image to help her readers achieve balance in a manner opposed to modern industrial society, the wheel remained fundamentally mechanical—about standardized rotation—and a strange metaphor to identify with organic life.

How did we come to talk about the potter's wheel as if it were an ancient and unchanging tool? Perhaps through the use of nineteenth-century metaphors that set the device almost in opposition to the machine. Jiggering plates with profiles and casting in plaster were eighteenth-century techniques that cut into production achieved using the wheel, and steam power and the scale of production grew exponentially in ceramic manufacturing in the nineteenth century. In Robert Browning's wildly popular poem *Rabbi Ben Ezra* (1864) the reader is asked to "note that Potter's wheel, / That metaphor! and feel / Why time spins fast. . . ." Indeed, Browning urges his reader to reflect on "God, whose wheel the pitcher shaped" in contrast to "machinery just meant / To give thy soul its bent." Similarly, the sense of the wheel as a source of universal cultural continuity is the tone of Henry Wadsworth Longfellow's *Kéramos* (1878), which surveys an immense temporal span of cultures with the potter's wheel as a constant life force—and concludes, as Browning's poem does, that only "Potter and clay endure." *Kéramos* naturalizes the wheel by setting the solitary potter under the shade of a hawthorn tree, a seductive and alluring vision, but alas—this solitary condition was likely first realized in the twentieth century. Potters at such idyllic pastoral encampments as Black Mountain College and communes like Marguerite Wildenhain's Pond Farm in northern California turned Longfellow's romantic poetry into a playbook.

Meanwhile, throwing pottery on a wheel in industrial production endured into the twentieth century, far longer than many expected. But factory conditions were never solo affairs for the potter. The "great wheels" used in ceramic manufacturing in the eighteenth century, such as at Josiah Spode's, were dependent on divided labor: one operative was dedicated to produce power and another manipulated clay. In the mid- and late nineteenth century, the factory wheels gained new energy sources, as belts harnessed water, and then steam power. In small factories, such as Colorado Springs's Van Briggle Pottery, professional throwers like Harry Bangs squeezed out hundreds of pots a day on the wheel, rows of pitchers that were standardized by the regularity of the clay as much as the belt-driven steady speed and the use of measuring devices like calipers mounted directly to the wheel. Throwing remained economically viable much longer in industry than we assume. The phenomenon of the potter working independently as a start-to-finish production facility is a twentieth-century invention; even more recent is the association of a potter's wheel with anti-industrial use.

The mechanical purpose of the wheel is to achieve a consistent speed to facilitate controlled form-giving. Ideally, the head, or upper wheel, on which the clay sits revolves at 60–100 rotations per minute. Oscillation or eccentric movement of wheels results in disfigured pottery. Wheels that are foot-propelled relay this momentum to the platform above, but there are many kinds. In double-wheels or kick wheels, a flywheel set on a bearing sustains momentum and transfers the centrifugal rotation via an axle to the head.

To challenge the notion of an archetypal potter's wheel, a closer look at mid-century Californian knowledge exchange is instructive, and educator Susan Harnly Peterson's oral history is especially enlightening. Hired in 1952 to teach at the University of Southern California, she was horrified to find that Glen Lukens, her predecessor, had treadle sewing machines serving as pottery wheels. She immediately sought a budget to replace them with kick wheels.[6] Those conditions were subpar compared with the facilities at Whittier Union High School, in suburban Los Angeles, where her classmate from Alfred University, Joan Jockwig Pearson (later Watkins), was teaching at the time. There were also divergences in the design of the treadle wheel: a downward-moving pedal was common in the adapted sewing machines, while a horizontal bar lever was championed by many followers of Bernard Leach. This question of which wheel, the kick or the treadle, was superior would seesaw again in the 1990s, when the treadle grew in desirability for its perceived historical role in folk cultures. The irony is that the treadle wheel that so many Leachians used was actually one improved in the nineteenth-century Stoke-on-Trent factories, not a vernacular cultural product.

In order to understand the reinvention of the wheel by ceramic artists in the 1940s, it is worth noting that Peterson recalled attending Mills College in 1946 where F. Carlton Ball "had built a potter's wheel from a picture in a German book of a treadle wheel."[7] Ball and Glen Lukens, his teacher, constructed wheels from Singer sewing machine treadles, utilizing bearings from Ford Model A automobiles. Their efforts at the University of Southern California's School of Architecture integrated pottery wheels into a broader university education.[8] Peterson had learned to throw pottery on a wheel when at Alfred, the New York State College of Ceramics, which prided itself on an encyclopedic curriculum of design, glaze, kiln, and clay technology.[9] At Alfred in the first decade of the twentieth century, Adelaide Alsop Robineau had been one of the first women to study the wheel and gain notoriety as breaking a gender barrier. It was only in the 1940s that women were using the wheel in sizable numbers in schools. In the 1940s, other role models included Marguerite Wildenhain, who, upon her arrival in the United States, built her own kick wheels. Gertrud Natzler had her kick wheel shipped from Vienna when she came to America in 1940, and attracted mature students such as Laura Andreson.[10]

In 2004 Peterson recalled with pride being an educator and innovator in wheel technology and how Peter Voulkos and Paul Soldner came to visit her when they were new to Los Angeles. They admired the variable-speed electric wheel that she and her husband, Jack Peterson, had built. Jack, who had also attended Alfred in ceramic engineering, held a position at Gladding, McBean, one of the largest firms on the West Coast specializing in architectural cladding.

> We talked about wheels, we talked about the kilns . . . . [M]y husband had had Master Motor Company, which made gear reduction motors, hand-make a gear reduction motor for a low potter's wheel that you could put hundreds of pounds of clay on. That's just exactly what Voulkos needed. So he got the same motors and he built his own wheels with pipe legs and plywood.[11]

The earliest patent for a pedal-regulated variable-speed electric wheel dates to 1937, but the advancement took two decades to implement and was not commonplace until the 1970s.[12] Electric wheels had been demonstrated at the Golden Gate International Exposition in 1939–1940, but were not affordable or widespread for two decades. When use of the electrified potter's wheel expanded with the rise in art programs, the variable-speed foot pedal control became a new convention, one that would prove second nature to a society already accustomed to accelerating their automobiles with a gas pedal.

Peterson's memories of the four hundred small potteries around Los Angeles in the 1940s are also helpful in order to gain perspective on the diversity of approaches and numerous small firms—most of which press-molded and jiggered ware but some of which did use the wheel. To pay closer attention to the historical context, one should note that in department-store demonstrations, the potter's wheel sold artistic goods and affirmed American independence from Europe since the 1920s. From 1915 to 1925, the Fulper Pottery's John Kunsman was a regular sight manning a potter's wheel at Bamberger's department store in Newark, New Jersey, using the treadle-powered type of wheel that Peterson considered unpalatable.[13] Advertisements in the 1920s and 1930s attest that several American companies sold art pottery as an "inexpensive rare gift," emphasizing that "every piece [was] shaped by hand"—on the wheel—and "not moulded by machine." This encouraged the misrepresentation of the potter's wheel as a tool that was not mechanical. The Fulper Pottery's emblem was a round vignette that merged the profile of a wheel, vase, and thrower, and unified these elements with speed lines that radiated from the clay to the worker's smock. Such a celebration of handiwork was also undertaken in the retail world on the West Coast, especially in stores such as Gump's in San Francisco, where the pottery of Marguerite Wildenhain and Edith Heath was first sold in the United States, and which even sponsored an award in the Syracuse Museum's annual ceramic national exhibition for "best designed pottery suitable for mass production" with the understanding that wheel-thrown pottery was still suitable for research and development for the mass-production of molded wares.[14]

The potter's wheel was also a spectacle in the 1939 New York World's Fair, when "Hand Thrown Vases" were sold at the pavilion sponsored jointly by labor unions and modern companies, among them the Homer Laughlin China Company (fig. 5). The display featured "modern" ways of making pottery—jiggering it and firing it in a continuous tunnel kiln, for example—but also spotlighted skill on the potter's wheel. Judging from advertisements, fairly sedate monochrome vases were peddled alongside historicist

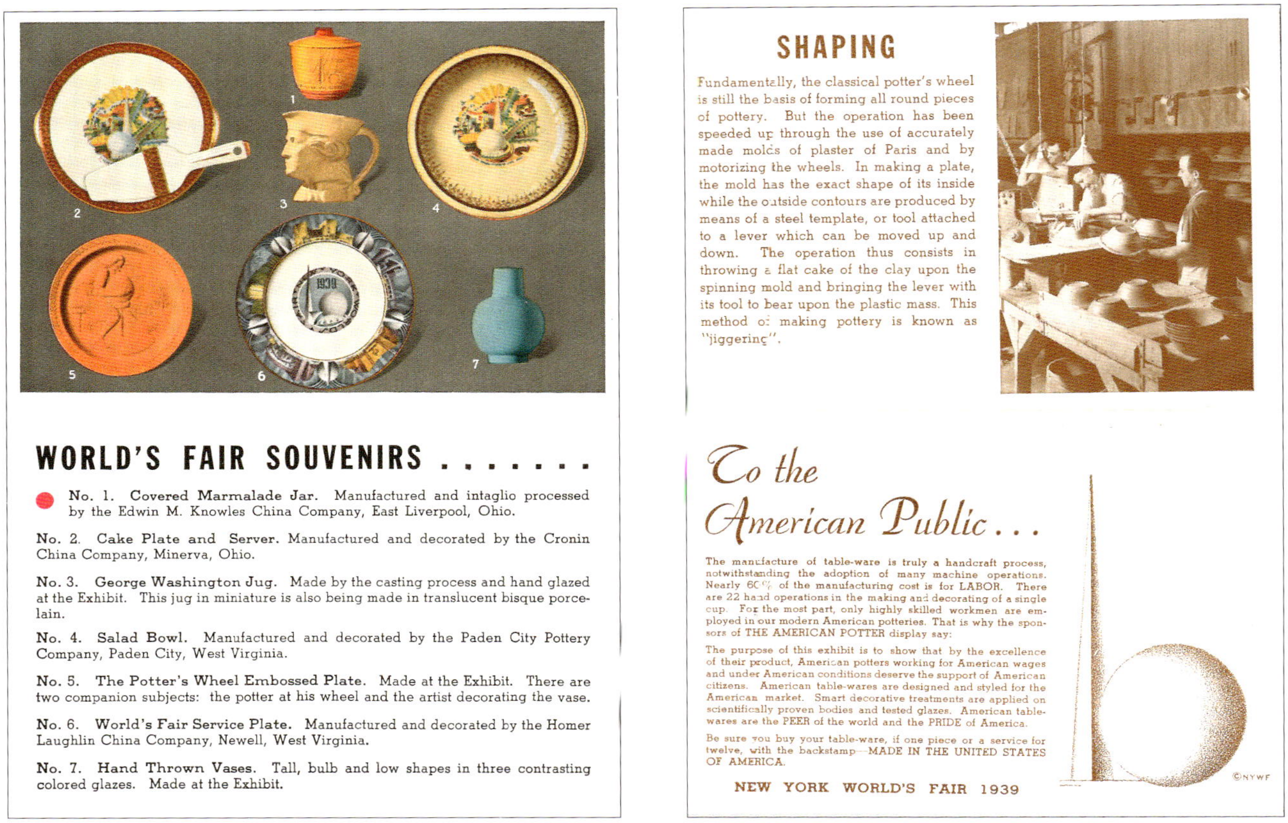

WORLD'S FAIR SOUVENIRS . . . .

No. 1. **Covered Marmalade Jar.** Manufactured and intaglio processed by the Edwin M. Knowles China Company, East Liverpool, Ohio.

No. 2. **Cake Plate and Server.** Manufactured and decorated by the Cronin China Company, Minerva, Ohio.

No. 3. **George Washington Jug.** Made by the casting process and hand glazed at the Exhibit. This jug in miniature is also being made in translucent bisque porcelain.

No. 4. **Salad Bowl.** Manufactured and decorated by the Paden City Pottery Company, Paden City, West Virginia.

No. 5. **The Potter's Wheel Embossed Plate.** Made at the Exhibit. There are two companion subjects: the potter at his wheel and the artist decorating the vase.

No. 6. **World's Fair Service Plate.** Manufactured and decorated by the Homer Laughlin China Company, Newell, West Virginia.

No. 7. **Hand Thrown Vases.** Tall, bulb and low shapes in three contrasting colored glazes. Made at the Exhibit.

SHAPING

Fundamentally, the classical potter's wheel is still the basis of forming all round pieces of pottery. But the operation has been speeded up through the use of accurately made molds of plaster of Paris and by motorizing the wheels. In making a plate, the mold has the exact shape of its inside while the outside contours are produced by means of a steel template, or tool attached to a lever which can be moved up and down. The operation thus consists in throwing a flat cake of the clay upon the spinning mold and bringing the lever with its tool to bear upon the plastic mass. This method of making pottery is known as "jiggering".

To the American Public . . .

The manufacture of table-ware is truly a handcraft process, notwithstanding the adoption of many machine operations. Nearly 60% of the manufacturing cost is for LABOR. There are 22 hand operations in the making and decorating of a single cup. For the most part, only highly skilled workmen are employed in our modern American potteries. That is why the sponsors of THE AMERICAN POTTER display say:

The purpose of this exhibit is to show that by the excellence of their product, American potters working for American wages and under American conditions deserve the support of American citizens. American table-wares are designed and styled for the American market. Smart decorative treatments are applied on scientifically proven bodies and tested glazes. American table-wares are the PEER of the world and the PRIDE of America.

Be sure you buy your table-ware, if one piece or a service for twelve, with the backstamp—MADE IN THE UNITED STATES OF AMERICA.

NEW YORK WORLD'S FAIR 1939

©NYWF

Figure 5 Brochure, New York World's Fair, 1939. (Author's collection.)

knickknacks, but all were celebrated as "truly a handcraft process," even jiggering. In addition to being an exhibitor at the World's Fair, Californian F. Carlton Ball demonstrated his potter's wheel alongside Maria Martinez making coiled ware, both situated within a spectacle purporting to envision the "World of Tomorrow."[15]

To add one more perspective to the discourse of ceramic technology, in the first half of the twentieth century archaeologists argued over the development of the wheel in ancient America and Europe and its role as a measure of civilization. Otis Mason and other ethnographers speculated that pottery was historically women's work, made by hand. The potter's wheel was seen as a device that sped up society, leading to industrialization and standardization, and was overtly conceptualized as a mechanical innovation. Amateur potter and antiquarian Henry Chapman Mercer, founder of the Moravian Pottery in Doylestown, Pennsylvania, himself distinguished between using a wheel and making pottery "by hand"—that is, without a wheel—as if the revolutions of the tool reverberated cataclysmically, neatly cleaving temporal phases of human culture. In 1954 archaeologist V. Gordon Childe celebrated the potter's wheel as a radical innovation that the Sumerians devised at the same time as the chariot wheel, noting that "a professional can shape in ten minutes a vessel that might take a housewife ten hours to build by hand."[16] Just as the accuracy of Childe's interpretation has been challenged, so too should we note his risible gendering of preindustrial and mechanical labor. Childe seems to have conceived of the "housewife" potter as either an indigenous "primitive" or a modern amateur. The modernization of pottery with the wheel was conceived as a masculine application of technology.

A saner voice in these discussions concerning the history of technology was that of anthropologist George M. Foster at the University of California, Berkeley, who challenged both the scholarship regarding the gendering of the wheel and ceramic production in general. Foster paid attention to the real complexity of pottery, accounting for trimming pottery after pots are thrown—which also used wheels. "Much ware can neither be classified as wheel-made nor hand-made, since various combinations of the two methods are used," Foster concluded, dismissing assumptions about gendered as well as technological determinism.[17] Foster's rebuttal of stereotypes might very well have had something to do with his location in the Bay Area, where Wildenhain, Heath, Laura Andreson, and Beatrice Wood were all visible as exhibiting potters.

To claim that the potter's wheel signified either slow or fast production at mid-century is obviously a risky generalization, one as fraught as ascribing technological invention to one gender or another. By the time Richards's *Centering* celebrated the potter's wheel as a respite from industrialized mass culture, both popular and elite dispositions might have looked on the flywheel as a mechanical contrivance not worthy of respect as a machine; such interpretations depended on subjective opinion. In June 1957 at the First Annual National Conference of American Craftsmen held at the Asilomar conference facility in Pacific Grove, California, when Elena Netherby (a founder of the Mills College Ceramic Guild whose wheel-thrown pottery received citations of praise at the Syracuse Museum's ceramic nationals in 1949, 1956, and 1957) sat beside Peter Voulkos, it is probable that he knew her as a woman who ran a contracting business building mid-century suburban ranch homes around San Francisco. She also had the distinction of filing a patent on a kick wheel, design #no. 61,363; Voulkos might have known that, too. He even might have used a Netherby wheel when he demonstrated for Laura Andreson's students at UCLA or at Scripps.[18] And he was also probably aware that Jayne Van Alstyne, one of his teachers at Bozeman and one of Peterson's colleagues at Alfred University, was interested in improving the kick wheel, and drafted blueprints for a "Dandy Potter's Wheel" as carefully as she drew sketches for General Motors, where she worked as an industrial designer. More famous are the electric and kick wheels of Paul Soldner and Alfred University's Ted Randall, which developed into lucrative businesses. Van Alstyne's blueprint for a wheel shows that women such as herself and Peterson saw both the electric wheel and the kick wheel as living technologies, and ones that they had mastered but might still be worth improving. To be certain, the ambition to live as an artisanal potter in the shadow of the atomic bomb seems anything but rational, but it is a complex decision worth savoring in terms of these nuances of gender and the technological landscape.

Three Wheels Turning: Heath, Karnes, and Takaezu

The following three profiles expand on the complex ways that we conceive of clay in relationship to the technology of the wheel by looking more closely at both artistic work and lives. The work of Edith Heath importantly recenters the role of the wheel in design, mass production, and invention. Instead of seeing Heath ware as factory production that was boldly modern, Heath herself deserves to be regarded as locating handicraft in production and incorporating the wheel as a modern, even contemporary tool, from the 1940s through the 1970s. Her designs still are faithfully made, a remarkable seven-decade run. Her low-slung handle continues to seduce hipsters; it purportedly satisfied Heath's desire for a handle that enabled her to accommodate both vices of drinking coffee and smoking a cigarette in one hand—a pose of daring informality (figs. 6, 7). This freed up her other hand to direct her factory.

Karen Karnes, recently reconsidered in the book *A Chosen Path* (2010), dedicated her life to pottery in conjunction with the wheel but also labored to challenge simple definitions of virtuosity and technique in ceramics, shifting out of her comfort zone repeatedly to experiment and expand her oeuvre.[19] Her coil-built stools are difficult to integrate adequately into her tableware and reflect her ability to think "beyond" the wheel in a manner as untraditional as Voulkos or any other twentieth-century ceramist (fig. 8). Like Voulkos, Karnes deserves to be interpreted as revealing the process of unlearning technique as much as the mastery of the wheel.

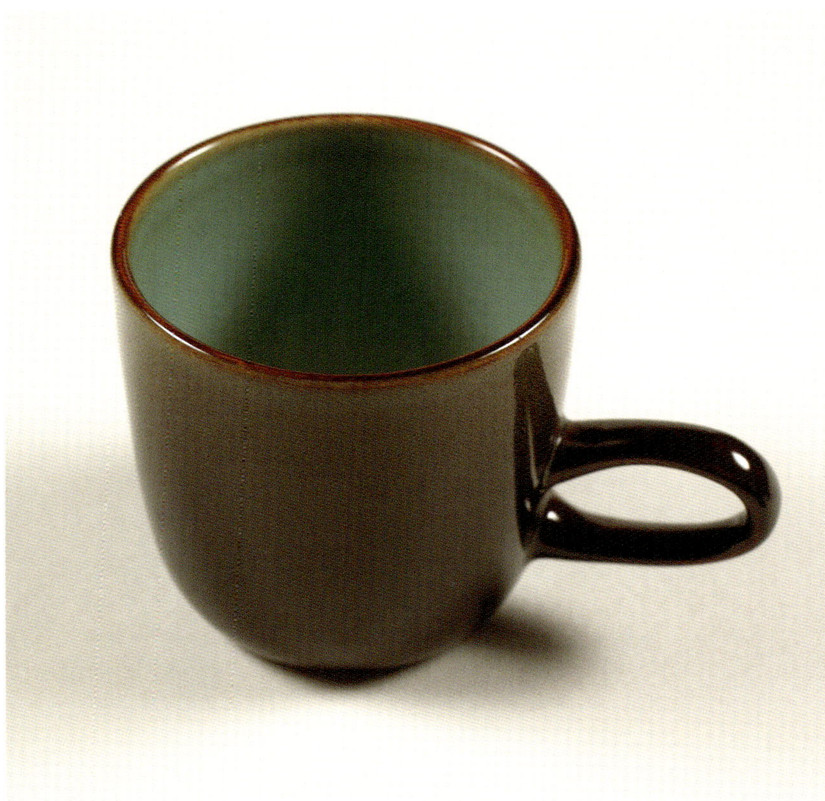

Figure 6 Edith Heath (1911–2005), *Coupe* (model cup), 1948. Glazed stoneware. H. 5". (Courtesy, Berkeley Environmental Design Archives.)

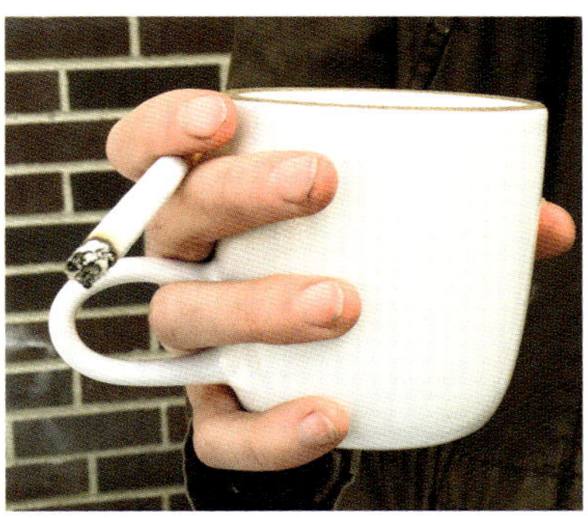

Figure 7 Demonstration of how the cup illustrated in fig. 6 can be held while smoking. (Photo, Caitlin Brown.)

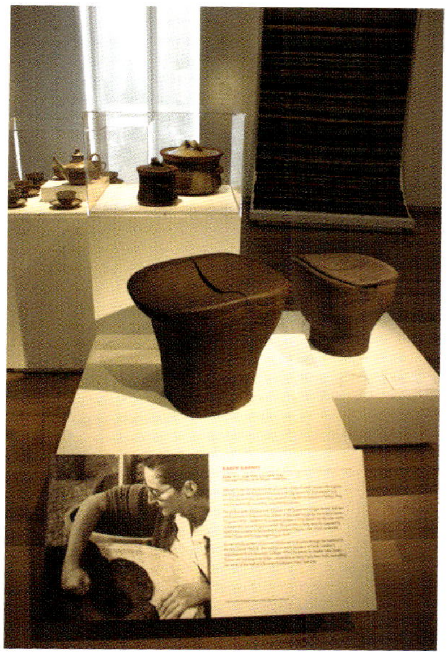

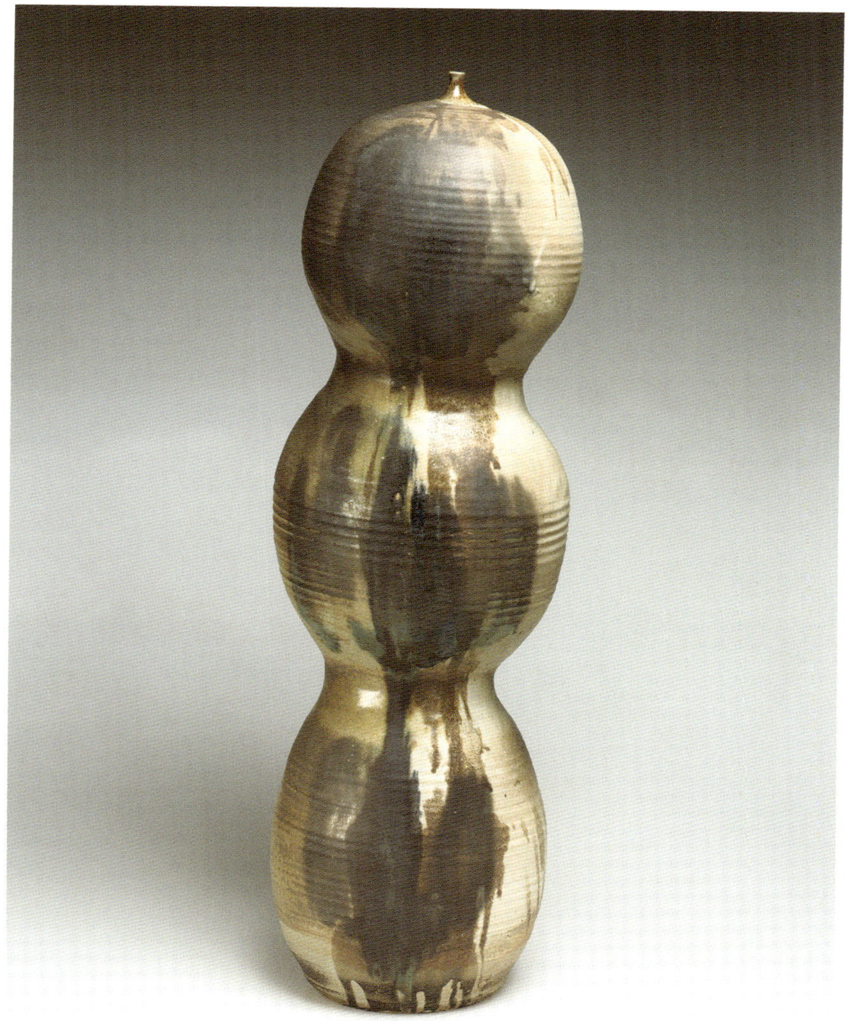

Figure 8 Karen Karnes (1925–2016), *Stools*, 1960–1967. Stoneware. Each 18" x 21¾" x 16¾". (Courtesy, Museum of Arts and Design, New York; photo, Ezra Shales.)

Figure 9 Toshiko Takaezu (1922–2011), *Tamarind*, ca. 1960. Stoneware. H. 35". (Collection of Peter Russo; photo, Brian Oglesbee.)

Of these three potters, Toshiko Takaezu was closest to Voulkos in friendship. Like him, she was trained as a potter first then decisively shifted her allegiance to the self-conscious production of fine art. Her work after 1960 used the wheel but disassociated thrown forms from functions, and she became famous for closing her vessels in on themselves (fig. 9). If any contemporary of Voulkos deserves to be seen as sympathetic to his upending of the wheel, it is Takaezu, but, curiously, she has never been seen as "unmaking" pottery or as veering toward his allusions to destruction. Rather, she has been interpreted in a most narrow vein, exoticized as a Hawaiian-Japanese-American and considered a landscape painter of sorts.

EDITH HEATH'S WHEEL: Central to Mass Production

Edith Kierzner Heath first gained exposure to ceramics at the Art Institute of Chicago in the mid-1930s, and by the end of the 1940s her tableware was considered to embody the new postwar California lifestyle. Her mix-and-match monochromatic forms, initially in a palette of four colors (sage green, blue, sand, and shining sand), celebrated coastal associations and continued the idea of a "distinctive informal dinner service" cultivated in the 1930s by the likes of Russel Wright (1904–1976) and Frederick Hurten Rhead (1880–1942). That she ran her own company for four decades and continued to throw prototypes on the wheel for all those years make Heath a remarkable pioneer of modern craft, and shows her to have been hands-on beyond the range of most other designers and entrepreneurs. She is said to have converted a treadle sewing machine into a potter's wheel in order to maintain production once she graduated from art school, and then, in 1946, she used the pottery studio at the legendary boutique department store Gump's to keep up with orders; she and her husband, Brian, opened their Sausalito factory the next year. Such anecdotes reveal the broader economic context in which studio pottery grew. Heath exhibited both locally, at the Legion of Honor and Merchandise Mart, and nationally, in Syracuse Museum annuals from 1946 through 1956. Unlike Karnes and Takaezu, who worked alone, Heath achieved all this through incorporation, joining forces with her husband and building a loyal and close family of skilled employees. For Heath, the kick wheel was not an escape from mechanization but a spring that fed production and the score of hands who were her team.[20]

The contexts in which Edith Heath exhibited reflect the choices that were open to her and which she cultivated: an ideal of art in everyday life and a belief in scaling production to need and want. She participated in exhibitions of interior design as well as the medium-specific shows organized by the Syracuse Museum. At Syracuse's 11th Ceramic National Exhibition, in 1946, Heath showed bowls and a plate ranging in price from $15 to $35. Objects made by numerous peers using the wheel were also exhibited that year, among them Maija Grotell's five large vases, priced considerably higher at $175–$275, and Laura Andreson's $100 cookie jar. At the 12th Ceramic National Exhibition, in 1947, Heath exhibited a wine bottle and three cups, an eleven-piece tea set (the teapot with a silver handle by Franz Bergman was featured in LACMA's recent "California Modern" exhibition), a vase and a turquoise matte bowl, all hand thrown. By the years of the 18th and 19th Ceramic National Exhibitions, 1954 and 1956, Heath was included in a special category, "Artists in Industry Exhibit," alongside Glidden's ram-pressed tableware and Viktor Schreckengost's "Primitive" line for Salem China. By 1959 Heath's production output, nearing a quarter million pieces a year, was made in a factory that she designed with Robert Marquis and Claude Stoller; a model of enlightened entrepreneurship, it provided each worker with a window. The factory is a wonder today for still running as it was envisioned then.

In between these benchmarks, a 1949 article in the *New Yorker* by Sheila Hibben praised the arrival of California tableware made by individuals and some medium-sized firms that employed a dozen to one hundred people, and singled out Heath. "She has managed to retain an astonishing feeling of spontaneity and freedom in her forms," Hibben gushed, claiming the quality and design superior to the eccentricities of Eva Zeisel's "Town and Country" made by Castleton.[21] The number of outlets for wheel-thrown ceramics at the time in Manhattan suggests the aftermath of World War II was a high-water mark for craft. One could shop for handicraft pottery at many small shops: New Design on 75th Street; Rabun Studios on Madison and 68th Street; Cardel on 58th Street; Jacques Seligmann, Rena Rosenthal, and Van Diemen-Lilienfeld Galleries on 57th Street; America House on 52nd Street; Americraft on 51st Street; and Cavallon Olds on 11th Street. Larger boutiques in such department stores as B. Altman & Co. and Wanamaker's also carried these wares. Southern Highlanders maintained a store on 50th Street that carried Jugtown pottery from rural North Carolina, but Heath had buyers from Marshall Field & Co. of Chicago, Neiman-Marcus (later without the hyphen) of Dallas, and Bullock's of Los Angeles. Her craftsmanship was not a touristic novelty but an aesthetic with large geographic distribution to a discerning constituency. Hibben's article was followed three months later by a classified advertisement that suggests the stature of Heath's company and the currency of her name: the rental of "the most completely equipped home in Fort Lauderdale" had Robsjohn Gibbings furniture and a "Kitchen completely equipped with everything electrical.

Figure 10 Edith Heath to Edgar Kaufmann jr., April 10, 1947. (Courtesy, Berkeley Environmental Design Archives.)

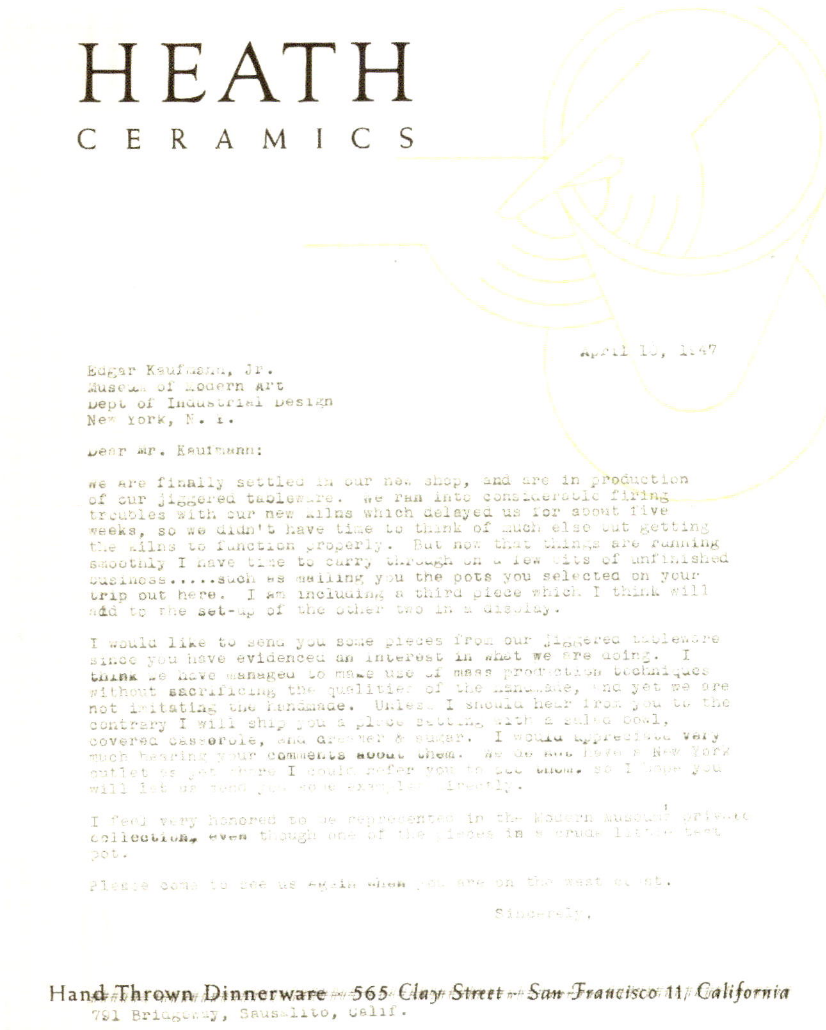

HEATH
CERAMICS

April 10, 1947

Edgar Kaufmann, Jr.
Museum of Modern Art
Dept of Industrial Design
New York, N. Y.

Dear Mr. Kaufmann:

We are finally settled in our new shop, and are in production of our jiggered tableware. We ran into considerable firing troubles with our new kilns which delayed us for about five weeks, so we didn't have time to think of much else but getting the kilns to function properly. But now that things are running smoothly I have time to carry through on a few bits of unfinished business.....such as mailing you the pots you selected on your trip out here. I am including a third piece which I think will add to the set-up of the other two in a display.

I would like to send you some pieces from our jiggered tableware since you have evidenced an interest in what we are doing. I think we have managed to make use of mass production techniques without sacrificing the qualities of the handmade, and yet we are not imitating the handmade. Unless I should hear from you to the contrary I will ship you a place setting with a salad bowl, covered casserole, and creamer & sugar. I would appreciate very much hearing your comments about them. We do not have a New York outlet as yet where I could refer you to see them, so I hope you will let us send you some examples directly.

I feel very honored to be represented in the Modern Museum's private collection, even though one of the pieces is a crude little test pot.

Please come to see us again when you are on the west coast.

Sincerely,

Hand Thrown Dinnerware – 565 Clay Street – San Francisco 11, California
791 Bridgeway, Sausalito, Calif.

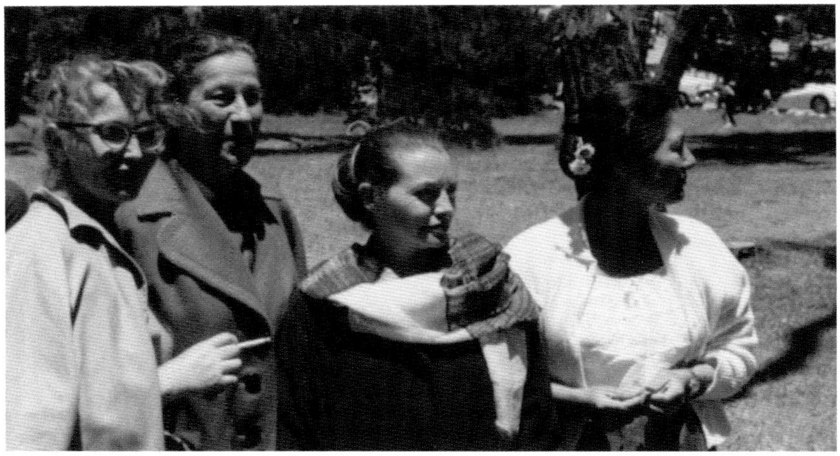

Figure 11 Edith Heath, Marguerite Wildenhain, Lenore Tawney, and Toshiko Takaezu, June 1957. (Courtesy, American Craft Council.) The four artists are pictured at the First Annual National Conference of American Craftsmen, sponsored by the American Craftsmen's Council and held at the Asilomar Conference Grounds, in Pacific Grove, California, June 12–14, 1957.

Edith Heath dinnerware. Blenko glassware. No expense spared."[22] Heath had become a known luxury brand signifying comfort in just a few years.

Correspondence in 1947 between Edith Heath and Edgar Kaufmann jr. (his preferred spelling), the curator at the Museum of Modern Art, and in 1949 with Edgar's mother, Liliane (the buyer for Vendome Gifts, an exclusive shop in Kaufmann's department store in Pittsburgh, owned by her husband), suggests that two years before Hibben's glowing notice in the *New Yorker* the curator had seen value in Heath's "Hand Thrown Tableware," as it was identified on her letterhead (fig. 10). Not only was Kaufmann jr. collecting her prototypes and adding them to the museum's collection, he seems to have alerted his mother to them as well.

Heath Ceramics stationery depicts a pair of stylized hands throwing a cylindrical jar, an image that lingered despite the fact that the company sold jiggered ware after relocating to Sausalito in 1947. The line drawing was an apt means to describe the company's research and development as well as good branding. In 1964–1965, when Heath made prototypes for Wedgwood, she made them first on the wheel. It was not a technology to leave behind or graduate from, and remained in use as a tool in prototype design by firms such as Arabia in Helsinki, Royal Copenhagen, and several Staffordshire pot works for much of the twentieth century. Production might require jiggering and molding, but the wheel was still a place to feel one's way into a form.

Despite Edith Heath's acumen at the wheel and the success of her company, it is worth noting that Rob Forbes, founder of Design Within Reach and a graduate of Alfred University who was a thrower of wares for a decade, recently asked, "How is it that I wasn't aware of Heath Ceramics, given that the company has been cranking out pottery since the 1940s in my own stomping grounds of Northern California?"[23] Although Edith Heath was invited to the First Annual National Conference of American Craftsmen in 1957 and hobnobbed with renowned potters such as Marguerite Wildenhain and Takaezu (fig. 11), her work was rarely valued as "handicraft" by the time Forbes trained in the 1970s—most potters regarded it as "design." What counted as wheel-thrown work in art academies and craft fairs after 1970 was work that was strictly solo performance. Heath and her work were not taught to several generations of M. C. Richards acolytes, who were interested primarily in centering clay and self. As self-expression grew to be the prized value of wheel-thrown pottery, a general appreciation of the wheel as a tool to manipulate precision withered both pedagogically and aesthetically.

KAREN KARNES: Learning and Unlearning Technique

Although Karen Karnes was cast as a sculptor in the revisionist 2011 exhibition "Crafting Modernism," originated by the Museum of Art and Design (MAD), New York, she has always described herself as a potter. Her work has always been firmly functional and titled as such. Back in 1951, before the art market heated up to today's level, her peers Leza McVey and Peter Voulkos also called their work pottery and exhibited it as such. Understanding that Karnes's and McVey's radical hand-built clay forms were categorized as "pottery" and not "sculpture" in the 16th Ceramic National Exhibition of 1951 is difficult for many to accept and contradicts the prevailing twenty-first-century desire for definitive categorization. The Karnes and McVey pieces were classified as pottery alongside the much more traditional wheel-thrown work of Peter Voulkos. The same year, a vase by Karnes won a national prize sponsored by Lord & Taylor (fig. 12). A decade later Karnes won a Palma d'Oro in 1963 at the Milan Triennale for hand-built garden stools, and, in 1969–1970's "Objects: USA" exhibition she represented herself with these stools made with a coil technique, though she again elected to call herself a "production potter."

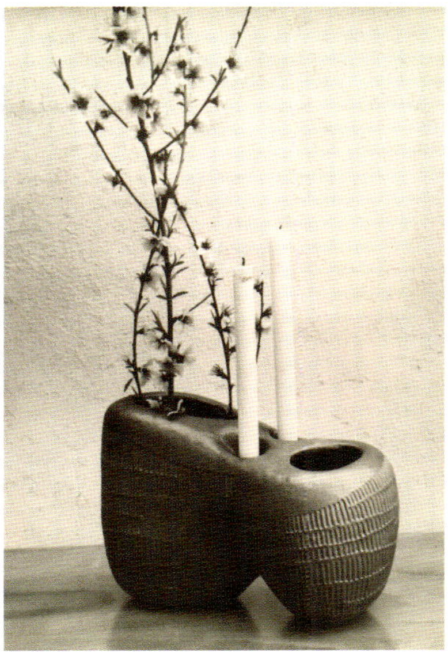

Figure 12 Karen Karnes (1925–2016), *Double Vase with Candlestick Holders*, 1951. Stoneware. H. 9½", W. 13⅞". (Courtesy, Everson Museum of Art.) The illustration is from Karnes's submission to the 16th Ceramic National.

The relabeling of *Double Vase* in MAD's 2011 "Crafting Modernism" show now negates the intended function of the piece, which originally was titled *Double Vase with Candlestick Holders* (1951). Photographic records in the Everson Museum show it in use, holding branches in one reservoir and a tall white candle in each of two others. In contrast to locating the work on a white pedestal in 2011, these archival images articulate Karnes's intentions to set table: etiquette and social concerns are essential. What at first might appear to be surrealist biomorphic indentations on the pot's top are explained away by the original title; they are for holding candles. The organization of the two cavities also becomes more clear: one sphere is intended to contain *wabi-sabi* clippings of weeds and the other kept as an access path to replenish water. These functional rationalizations discourage perceiving the vase as an abstract form. She made the vase in Sesto Fiorentino, Italy, while accompanying her second husband, David Weinrib, on his grant-funded travel. A cosmopolitan New Yorker, she knew museum collections well, and so one can credit the complexity of *Double Vase with Candlestick Holders* to an appreciation of what would have been labeled "primitive pottery" by her childhood haunts, the Metropolitan Museum of Art and the Museum of Natural History. However, the compelling surface pattern of rhythmic incisions remains mysterious but impossible to label as derivative of any specific historical model. The incisions resemble both body scarification and linoleum block cuttings. In comparison, the Voulkos wheel-thrown pottery on display alongside Karnes's piece was a "popcorn bowl" and a "decorative rice bottle"—a mélange of heartland Americana with Anglo-Japanesque folksiness.

Karnes purportedly had no experience with clay until she met Weinrib, despite having studied at Music & Art High School in Manhattan, where Lee Rosen, founder of the ceramic firm Design-Technics, taught pottery, and then at Brooklyn College with Serge Chermayeff. Meeting Weinrib while taking an evening course at the Newark School of Industrial Art, Karnes decided to flee her role as a suburban housewife and elope with Weinrib to live as a bohemian artist. She usually credits her first interest and work in clay to when she and Weinrib spent a summer living in a tent in East Stroudsburg, Pennsylvania, making ceramic lamp bases for Design-Technics. Weinrib connected to the firm through his colleague from Alfred, Nancy Wickham (later Boyd), who won a prize at the 11th Ceramic National Exhibition in 1946 for her $4.25 Design-Technics breakfast set. If Karnes did find pleasure and skill in hand-built clay forms that summer, when she finally enrolled at Alfred University she gained exposure to the wheel as a discipline. Herbert Cohen, a fellow New Yorker who won a scholarship from the Henry Street Settlement to study at Alfred at age sixteen, remembered watching her trim the foot of a pot in one of their classes and leaving it too thick—but was prevented from interfering by his shyness and modesty.[24] This memory is important in order to locate Karnes as a skilled painter and designer who then pushed herself in ceramics to become a good potter. She came to the material with a sophistication that would prevent her from being seduced by an ideal of virtuosic technique but did focus on the wheel as a skill worth pursuing. As befits a mature student who had confidence, Karnes could exploit elementary techniques.

Although she was more skilled on the wheel after her year at Alfred University and subsequent years throwing pottery at Black Mountain College, in the late 1950s Karnes gravitated to press-molded forms and also began to coil-build large freestanding sections of bird baths and planters. In an interview with Dido Smith in *Craft Horizons*, she described this decision as one made alongside her physiological transformation when she became pregnant.[25] She was profiled as a beatnik homesteading in a cooperative community, Gate Hill, on the Hudson River thirty-five miles north of New York City, asserting her freedom from mainstream society while remaining dependent on all of the urban forces of capitalism bustling downstream. She sold her casseroles on Park Avenue at Bonniers and her garden furniture to weekenders. The Swedish emporium favored unique and handmade pottery, shaggy *flossa* weavings, and tastefully sleek teak furniture by designers such as Jens Risom. Karnes's teacups were tender and refined as well as texturally gritty in that context. She was not making mugs to swig from crudely but cups to sip from in genteel repose, to slow the world down. When Karnes situated herself just an hour outside the city, her admirers and patrons drove up the new Palisades Parkway to visit "the Land," as Karnes, M. C. Richards, and experimental filmmaker Stan VanDerBeek called it. To put this socioeconomic dynamic in context, during the 1950s many in Gate Hill were dependent on urban commerce. Karnes's neighbor John Cage was under contract with Jack Lenor Larsen, the weaver-entrepreneur, to design advertisements and Larsen Incorporated's Christmas cards. Gate Hill's many musicians performed in and around Manhattan.

Karnes began to make stools and chairs, usually in pairs, about 1957 or 1959. These heavy and squat ceramic forms at first glance resemble obstacles rather than seating. Like elephant-foot wastebaskets popular with big-game hunters, they evoke the unexpected even if they are decorative furnishings (fig. 13). This is not what Harold Rosenberg meant when he coined the phrase "the anxious object," but the term is well suited. Part of the disturbing nature of Karnes's stools and chairs is that in their scale and shape they call to mind associations with toilets. They are also precarious, narrower at the base and flaring outward. The color and girth of the brown coils increase our visual associations with human evacuation. Is the stool another type of stool? The anatomy of excrement and anthropomorphic forms of sanitary appliances seems playful.

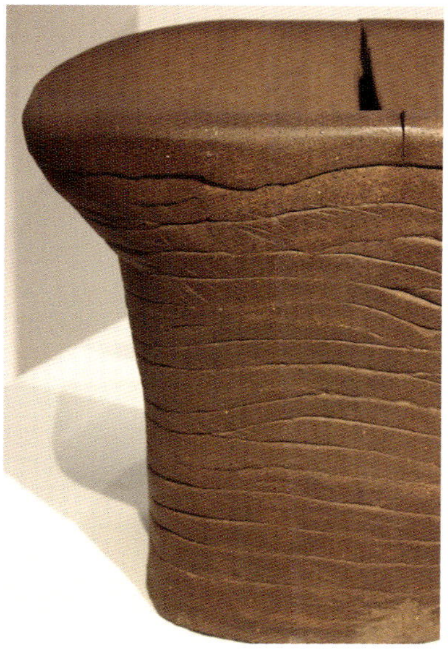

Figure 13 Detail of one of the stools illustrated in fig. 8. (Photo, Ezra Shales.)

As a building method, the coil is a fundamental manipulation of pliant stuff like dough or mud, to the degree that it is almost a preschooler's exercise. However, Karnes elevates the coil by exaggerating and emphasizing its visibility. Her ability to preserve the plastic and viscous quality of the clay is both elegant and extremely difficult to achieve. The novice's method of coiling clay becomes reinvented and complex in Karnes's hands. Whereas the coiled sides rise in anatomical strata like geological intestines, the seat is smooth and taut like a piecrust. There is a contrast within the structure. On the one hand, the bucket seat seems to slope gently in a curvature that matches the human body, but on the other hand this seat is sliced open dramatically. The deliberate gap permits rainwater to drain out and yet also lends intentional mystery to the interior. By touching this topography, an admiring finger circumscribes the ceramic form to understand the process, after which the swelling shape seems less unnerving. One can sense the potter working in a centripetal manner around the hollow center and moving eccentrically to sustain an outward thrust. Karnes balanced making the walls lean out and keeping the clay from collapsing.

They might seem as if they are made to be outside, liberated from a white cube, where they might connect to rocks and leaves and resemble tree stumps, in both textural and literal manners, but Karnes made her stools and tables to go inside as well. The genre of the garden stool flourished in the postwar attitude and erased the distinction between patio and open living room, exterior and interior. Sliding glass doors and expanses of plate glass transformed mid-century domesticity and opened up new possibilities for informal living. Karnes's birdbaths and planters also occupied these new liminal spaces of the "open floor plan" and migrated with the seasons; her stools fit this repertoire even if they only grudgingly welcome sitters.

Karnes made about a dozen stools and chairs, mostly pairs but one trio, and they were in important private collections. The most accessible to the public was a single seat in the Noguchi Museum in Queens, New York (a venue devoted to the artist and architect Isamu Noguchi), where for years it was the one work of sculpture displayed that the artist that himself had

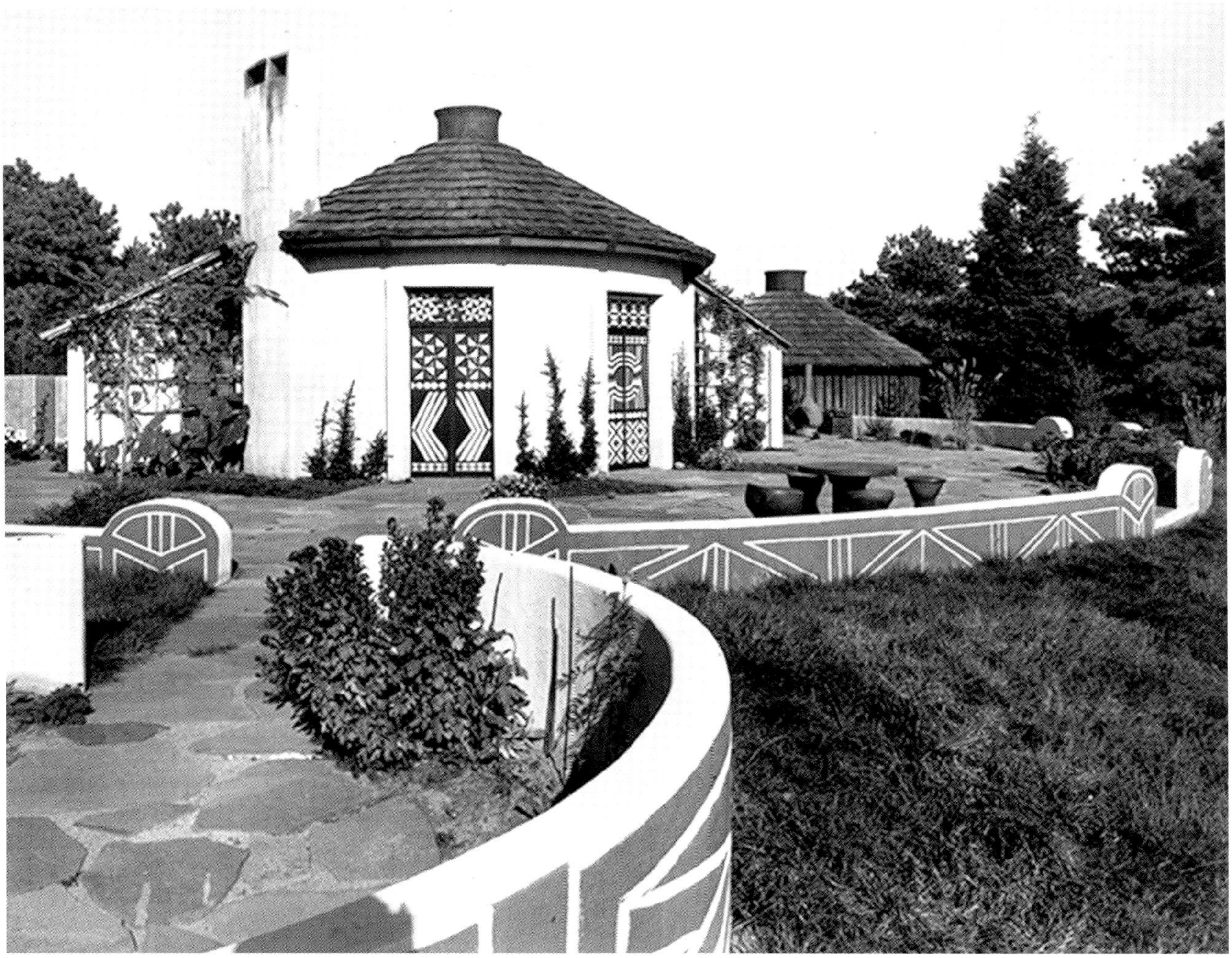

Figure 14 Karen Karnes seating at Jack Larsen's Round House. (Press photograph stamped and labeled *New York Daily News,* January 9, 1967; photographer unknown.)

not made. Learning of Karnes's work from Noguchi, Jack Lenor Larsen also commissioned work from Karnes, and when his African-inspired Round House opened in 1963 and received critical attention in dozens of national newspapers and periodicals, many noted that Karnes had contributed a fireplace, a pair of garden stools, sinks and tiles to the bathroom, light switch plates, and finials on the roof (figs. 14, 15). Locating Karnes's stools in their appropriate environment is a challenge. There is no single place but rather myriad material forms, social acts, and emotional atmospheres; in the Round House the four stools could be considered tribal and "primitive" as well as delicate and down-to-earth. The seat looked austere and biomorphic among Noguchi's sculpture. Karnes's stools ended up in her own home alongside seating by Eames and Aalto, icons of modernism, and also beside overstuffed wing chairs upholstered in chintz. Does their roughness air reservations about submitting to authority or the predicament of conformity, or does it aspire to resemble an organic fungus? They sit quietly, repudiating formality.

The stools seem to have willpower; they are aggressive social objects. "Karen's chair sits you, you don't sit on it," noted Zeb Schachtel, one of

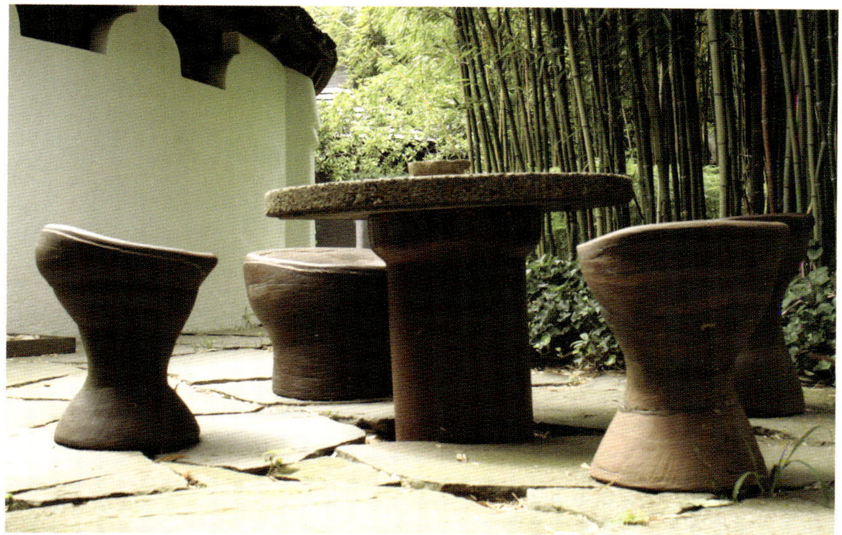

Figure 15 Karen Karnes seating at Jack Larsen's Round House, 2015. (Photo, Ezra Shales.)

her oldest friends and a customer since the 1960s, who had three ceramic seats on the porch of her beach house. That Karnes embarked on making her seating just before her mother bought her an Eames Lounge Chair, a recently debuted design at the time, is suggestive. As in the Eames chair, one sits on a slope. Both spin upward on smaller bases. The chairs exude physicality, which looms as an integral part of their social function. Looking at them one can envision the diminutive Karnes heaving them up, almost equal to her own weight, as she moved the wet clay into form and into the kiln all by herself.

Whereas the Eames chair spun on its base, Karnes's coiling compels the viewer to circumambulate, inverting the idea of centering and oscillation as social actions. Her large planters and birdbaths comprise alternative miniature intentional communities, little worlds complete in themselves. The stools, begun after she was left by Weinrib and had to manage as a single mother, seem resilient and antisocial even if they also demand people to complete them. Do the stools materialize companionship or its absence? One set of stools especially seems to visualize the ideal of coupling and also of cellular division—mitosis. Karnes noted that she needed Weinrib's help to flip the clamshell-like halves of the largest planters onto pillows before joining them together; the stools posed a physical challenge. Her inquiry into making ceramic seating emerged specifically during this moment before or during their separation. She made large seating in suites of two and three after her marriage to Weinrib disintegrated in 1956, and ceased when she began her next long-term relationship, with Anne Stannard, another ceramic artist.

It is probably difficult for young artists today to understand why Karnes refrained from calling her ceramic work "sculpture," and why she continued to throw casseroles on the wheel time and again over more than five decades. Her work could be seen as art, craft, or design today. Her garden stools at the Milan Triennale in 1964 were gathered around a hearth, alongside a hammock and a chair made by Eva Zeisel (figs. 16, 17). She was acknowledged with a Palma d'Oro, but there was very little interpretative analysis of the work. Were

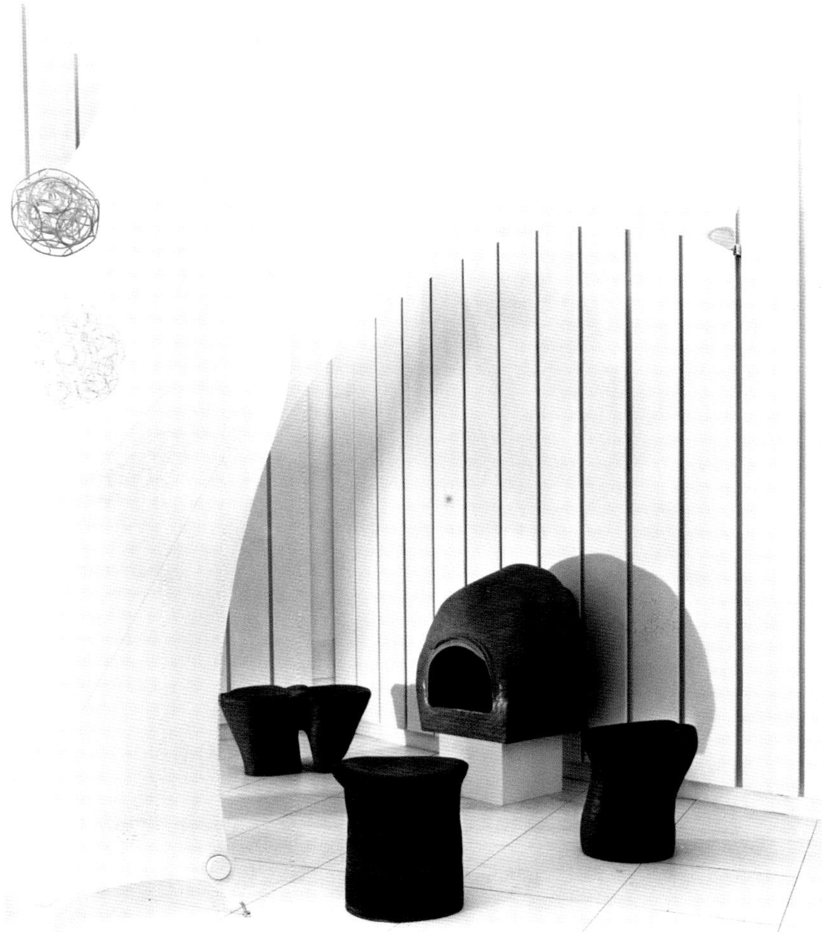

Figure 16 Karen Karnes (1925–2016), ceramic seating and hearth, Milan Triennale, 1964. (Archive photo courtesy Biblioteca del Progetto.)

the stools a declaration of unlearning skill, in that they exploited coiling as a visual process, or were they merely picturesque? At the time of his exhibition at the Museum of Modern Art in 1960, Voulkos's work was acclaimed as "free ceramic sculpture" with an "interest in natural textures," terms seemingly applicable to Karnes's seats. Her stools have never before been regarded as overtly antidomestic in the same way, and this is the first appraisal to suggest they have elements of the uncanny. Even when the Karnes stools toured the nation in "Objects: USA" in 1969, they were singled out as "one or two of a kind," not plumbed for meaning beyond their obvious function. In relation to the praise lavished on Karnes as a virtuoso on the potter's wheel, these stools were incongruous, obstacles more than examples.

TOSHIKO TAKAEZU: Explicitly Thrown, Implicitly Fractured

When Toshiko Takaezu arrived at her mature style in the early 1960s, she signaled the shift by declaring that she made "ceramic forms" on the wheel, no longer mere bowls. She now squarely situated her pottery, once categorized by implied function, as art, using a new terminology. Her decision to reject the word "pottery" along with keeping the wheel as her central tool is one way to understand her proximity to Voulkos, her friend.[26] Although she is famous for making closed forms on the wheel that boldly refute function, Takaezu was not the first to arrive at that innovative strategy. Her gesture of closing her pots into spheres was not a violation of convention and her work has never been read as subversive. She was a teacher, quietly yet willfully driving the young elite of Princeton University to question their values and ponder her work's validity. To pursue a more complete interpretation of Takaezu's work and suggest alternative readings, this profile will emphasize her penchant for adding a rattle to the interior of her closed forms, an act that

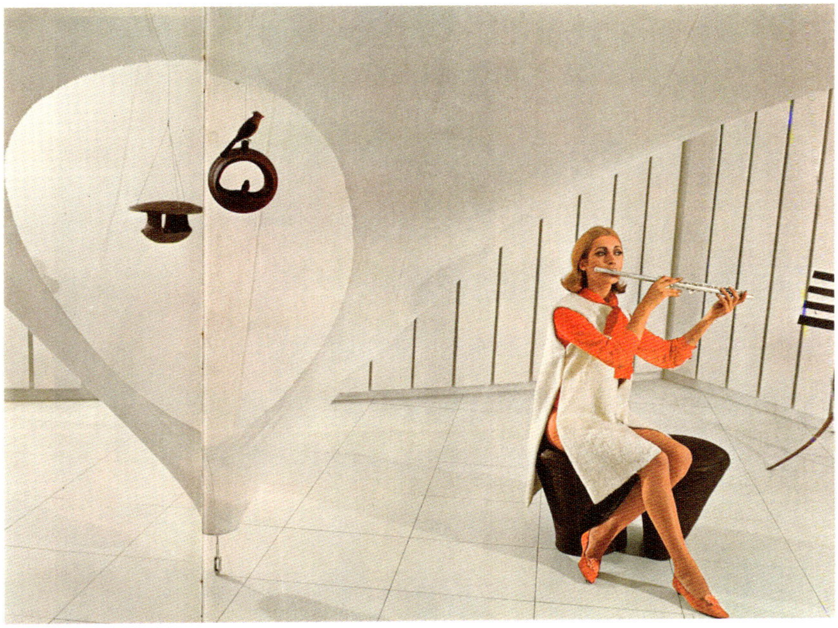

Figure 17 Karen Karnes (1925–2016), *Double Seat* (under model) and *Bird Houses*, Milan Triennale, 1964. Illustration in *Saturday Evening Post*, October 17, 1964.

requires evaluating the pottery's sound and touch and weighing these engagements with sensual knowledge as equal in priority to sight. It is through full sensorial engagement that Takaezu's work ceases to be merely beautiful and inches closer to conveying more complex and contradictory messages. The sense of modernist rupture and fragmentation that is literal in Voulkos is also present in some of Takaezu's works as a crack or fissure, but not the enduring mark. Her illusionistic layering of glazes and concentric form suggest completion, not disintegration.

Like Karnes at Alfred, Takaezu was a mature student by the time she landed at Cranbrook Academy of Art. Her journey to Cranbrook was sponsored by the Hawaiian department store, McInerny. Although her studies with Maija Grotell at Cranbrook constitute a visible legacy in her work through the 1950s, before that she studied at the Hawaii School of Art with Claude Horan, who had graduated from Ohio State University. Upon Takaezu's graduation from Cranbrook in 1954, the *Los Angeles Times* ran a profile of her as a "Japanese Potter" and explained her pathway from the Potters' Guild in Honolulu, where "she made commercial pottery by semi-mass-production methods" until, "[d]issatisfied, she began working in sculpture."[27]

Takaezu began to call her work "art" instead of "pottery" between 1962 and 1965, the 22nd and 23rd Ceramic National Exhibitions; she announced that she was making abstractions, and that to make a work of art abstract was necessarily to depart from function. In her first appearance at the Ceramic National Exhibition, in 1962, Takaezu exhibited a large, dark purplish-black bowl; and in her second attendance at the prestigious show, in 1965, she presented an eight-inch-high pumpkin-shaped clay object titled *Stoneware Form*. Although the 1962 bowl had a lively glaze treatment, it was still very much a unified skin with a limited tonal range; the 1965 "form," however, was mostly dipped in a dark glaze so as to leave one rectangular panel on which there were two dark brushstrokes, a painting on a pot that was reminiscent of Robert Motherwell's monochromatic gestural forms in which splattering preserved the original sense of viscosity. Both of these works have coarse throwing lines and accentuate Takaezu's tool of choice. Like her teacher Maija Grotell, she tended to work against firm clay, and the stoneware she favored in the 1950s and 1960s never hid the joinery of her assembled units but retained the facture of deep, rippling grooves from the wheel's rotational collision with her imprint.

The power of the word *form* can be traced to two distinct influences on Takaezu: Lenore Tawney's 1963 exhibition "Woven Forms," held at the Museum of Contemporary Craft, and the 1947 book *The Search for Form* by her teacher Eliel Saarinen. Even though his work was never free of classical tendencies, Saarinen's thesis was that intuition and "personal creative experience," not historical forms, should be at the core of an education. He begins by describing his successful effort to gain freedom from "the dope" of architectural history and to see for himself, a shift he grounded as both personally and historically necessary for the modern artist to be decisively modern. His argument is by turns learned and intuitive. For instance, his explanation that students should not study the past too much was predicated on the pseudo-scientific and down-to-earth observation that "we cannot live physically on food digested by others."[28] Grotell and Marianne Strengell, Takaezu's immediate influences at Cranbrook in the ceramics and weaving departments, instilled a sense of discipline, and the student work left from those years was about a search for pattern and rhythm but within predetermined types of forms: Takaezu made *flossa*, traditional Scandinavian rugs, and worked within the boundaries of what constituted a vase. Her friend Alice Parrott, a weaver, reminisced that Cranbrook emphasized a "placemat neatness" in student work.[29] Saarinen's emphasis on intuition must therefore be understood as a celebration of freedom within certain constraints, ones that can be interpreted as tradition but also as decorum.

The aesthetic belief in the value of a placemat, vase, and candlestick was retained at Cranbrook until the 1960s, when a generational shift began to accept off-loom weaving and more raw ceramic gestural constructions as appropriate student outcomes. However, analyzing pottery in terms of form already was part of the Cranbrook education. Saarinen's book tackled "form migration" from East to West and vice versa. Chinese art is always Chinese art, Saarinen opined, but "sculpture can express ideas in the West that reflect those of the Chinese."[30] Takaezu's aptitude for conducting formal analysis and seeing form as separate from cultural context can be noted in her own descriptions of her works from the 1950s that she called "bottles": "Perhaps these [two-spouted] forms were inspired by pre-Columbian pottery in the Cranbrook Collections. But I certainly did not copy them." She continued, "I do not make a tea pot and say 'this is for use.' Form is my first concern and often the pieces have a later use."[31]

Describing one's pottery as a "form" was a self-conscious and slightly pretentious way to name it abstract art. The novel strategic jargon arose in exhibition checklists, including Voulkos's and those of many others, as the 1960s went on. The gesture caused as many divisions and as much consternation as it gave its users a sense of liberation; the field of studio production, especially in clay, was immediately sundered. Warren Mackenzie famously wrote a tirade in response to Rose Slivka's "New Ceramic Presence" in *Craft Horizons* in which he sarcastically lamented, "In the future, we will give up any attempt to make functional ceramics as an expressive form."[32] To be free of function suggested ontological differentiation, not merely terminological change. The appellation was a way to assert status and new aesthetic ideals in 1963, when Takaezu opted to use it, but was not yet fueled by marketplace ambitions as would occur in later decades.

Even if Takaezu's cue for this name game came from Lenore Tawney, they should be seen as fellow travelers—while noting that Tawney's was the more forceful and more audacious personality. After making tapestries both pictorial and abstract, Tawney arrived at the decision to call her individual works "Woven Forms," as well as to use that for the 1963 exhibition title once she had resolved to make a formal reorientation away from the loom—her weavings literally departed from the weaver's tool.[33] While Takaezu is credited with making an innovative step in closing her pots into spheres, her turn away from function was not a rejection of the wheel. The wheel remained her primary tool. She might have seen her teacher Claude Horan make such a form, and, if not, she certainly admired firsthand Joseph Hysong's *Narrow-necked Bottle* (1961), a closed vessel that was on the cover of the Syracuse national catalog in which she made her debut. The formal decision to close ceramic vessels antedates the rhetorical strategy to name—or not name, as it were—a piece of pottery a *Form*. The friendship between Lenore Tawney and Toshiko Takaezu has long been recognized in publications because of the numerous photographs of their interactions. For instance, in 1957, at the First Annual National Conference of American Craftsmen in Asilomar, Tawney and Takaezu were photographed together—alongside Heath and Wildenhain (see fig. 11). The snapshot is suggestive of a halcyon moment before Slivka's essay supposedly separated the fields of art, craft, and design as divergent in status and hierarchy as well as intent.

Whereas Tawney declared that she was an artist and not merely a weaver, Takaezu maintained a stranger position whereby she made art but taught pottery. The wheel was a discipline she maintained, forbidding students to save any work for the first stretch of a semester in order to focus on the process instead of products. Teaching first at the Cleveland Institute of Art and then at Princeton a few days a week, she avoided a full-time position in academia even as she spent an enormous amount of time traveling and conducting workshops. This latter activity is important in terms of situating Takaezu in the fraught relationship between art and craft. The pace of workshops is steady and seemingly averse to the ideal of a focused and slow practice, but Takaezu and many others seemed energized by the process of conducting demonstrations, a peculiar type of event distinct to the milieu of craft. These short-term residencies of one or two weeks, still a defining feature of the field, were captured in photographic essays that showed Takaezu transforming her process into discrete steps, another measure of how the tool of the wheel defined her practice and persona. Whereas Voulkos was an entertaining instructor, a performer with bravura, Takaezu and Karnes were known for their rigor and quiet intensity. Tawney and most others who defined themselves as artists declined to do "demos" as visitors to schools or museums, and the division remains a professional boundary marker to this day.

Although Takaezu's forms were defined in *Objects: USA* as "delicate," "lyrical," and "intimate," the catalog also noted a peculiar feature: her pots rattled.[34] Her inclusion of clay within the interiors resulted, intentionally, in clinking sherds. In 1969 both the pots and the rattling were considered a "blending of the East and West." The problematic exoticization of Takaezu as a foreigner was a constant in her life; museums in the past decade continue to refer to her as a "Japanese potter," even though she was born in Hawaii. Her older brothers had been enlisted in the U.S. military but when she went to Cranbrook, and ever after, she was interpreted as having a significant aesthetic relationship to Asia. Although ancient indigenous American potters also made rattling pots—sometimes inside tripod legs—*Objects: USA* interpreted the sounds as Hawaiian waves "breaking upon silent shores." Takaezu navigated sexism and both anti-Japanese sentiment and the romanticization of Japan—with quiet acceptance. She considered herself as American as Peter Voulkos.

Historically, one mode of testing pottery was the clarity of its ring, to sense if there were cracks and to gauge the density of the clay body. Takaezu undermined such a sense of purity: sound in Takaezu's pottery is a quietly articulated threat, not spoken by the author to their patrons but implanted in the pot as a performative gesture. The mortal sound is anything but oceanic or becalming. The rattle affords a premonition of breakage. It runs counters to the placid grace and contemplative seduction of the throwing lines and glazes. If Takaezu's ridged throwing lines are the structure—the warp over which she runs glazes as a weft—then the rattle runs counter to this sense of integration by calling attention to itself as a fragment that defines the interior, a space to which we will never gain access except by breaking into it.

Whereas Voulkos's cuts into clay literalize modernist rupture, Takaezu's are made with great subtlety. To compare pottery to photography, Roland Barthes offered readers two terms, *studium* and *punctum*, to suggest that such culturally loaded artifacts operate on at least two levels.[35] When we read the pot as a pot, or a photograph as a document of a person or place, *studium* denotes our cultural, linguistic, and political apprehension of facts. *Punctum* denotes the personal touch that establishes a direct relationship with the object or person. If Takaezu established an attractive, seductive artifact that suggested it was a completed gesture in its visible throwing marks, she also introduced trauma, so that her pottery is not solely about integration and raising a bowl into a closed sphere on the wheel. If a splash of pink and gray glaze on a Takaezu "form" suggests sky meeting water, then this beauty is given an undertow, or a specific gradient of sand underfoot by the noisy pebble that rolls unseen in our hands but out of our grasp.

This interpretation focuses on Takaezu's smaller works. Her Star series, however, which she began at age seventy-five, is a dramatic expansion in size—and the crowning achievement of her life's work (fig. 18). The monumentality of her late works is incredible to reckon; centering twenty to thirty pounds of clay in her seventies and eighties is not a minor achievement. Stacking thrown units, even with the aid of college students, required Takaezu to stand on chairs and ladders above the wheel. The grandeur of these large forms as a field is compelling and immersive. They populate a room not with a human presence but as geological protrusions. Takaezu's visual illusions of layered glazes are tempting and yet often illusionistic. On her small plates she implied the intersection of two planes and then made us reach out for confirmation only to discover the perceived glaze demarcations were intangible. On her large Star works, the sense of touch becomes all the more imperative yet just as inconclusive. Takaezu's ceramic monuments appear so triumphant that they might contradict the hypothesis just put forward, that the sound and space of the interior of her forms were an emotional counterpoint. However, these large works can also be assessed as stacked thrown forms that contain within them a cavity the approximate size of the artist's own body. These ceramic forms, named after constellations (which themselves are mythical forms of illusionary bodies), remain unsettling and ambiguous and decidedly part of classical Western culture. When visitors walk among the Stars, self-concealment and the substitution of clay forms for human bodies become a complex game of abstraction and anthropomorphic suggestion. The surfaces of the stars are tactile and illusionistic—we perceive optical lines and horizons in the glaze, but our fingers do not confirm they are truly present.

Conclusion

There are three obvious reasons to value Heath, Karnes, and Takaezu. The physical objects they made have endured and remain avidly collected despite their modest critical acclaim and despite a lack of interpretation about the work. Second, their work documents the breadth of ways that the potter's wheel was a disruptive technology in the twentieth century. Wheel-throwing was a process that crossed the categories of art, craft, and design. The wheel shows these to be culturally constructed terms, not incompatible disciplines. Third, they stand as under-recognized women artists who succeeded in spite of a system conspiring against them. In 2016 the Museum of Arts and Design opens a new retrospective, "Voulkos: The Breakthrough Years," and

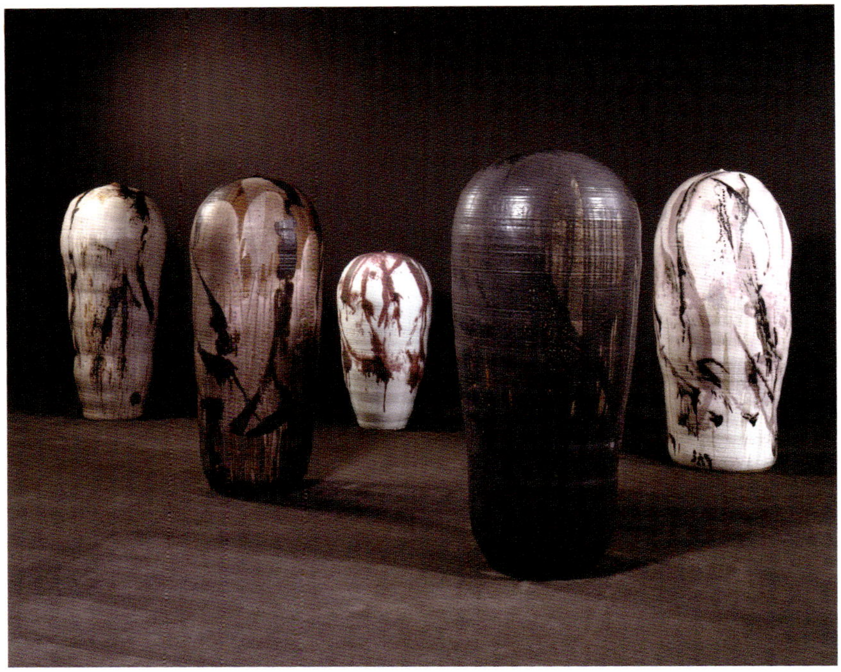

Figure 18 Toshiko Takaezu (1922–2011), *Nommo*, *Unas*, *Nephythys*, *Po Tolo (Dark Companion)*, and *Isis (Sirius)* from the Star series (fourteen works), 1999–2000. Glazed stoneware. H. ranging from 47" to 66". (Collection of the Racine Art Museum; gift of the artist; photo, Michael Tropea.)

it will be intriguing to see if the exhibition reflects waves of feminism since Rose Slivka's 1964 deification. Advance promotion states that in "defying mid-century craft dictums of proper technique and form, Voulkos completely reinvented his medium," reiterating Slivka's critical appraisal even more stridently than did Yale's 2015 exhibition "The Ceramic Presence in Modern Art," which merely borrowed her essay's title. While Voulkos might have made interesting art, to indulge in hagiographic celebration ad nauseam reveals a lack of imagination; worse, museums' capability for criticism is curtailed by Voulkos's lion's share of the art market. A Voulkos is still not worth as much as a pot by Pablo Picasso or Lucio Fontana or Jackson Pollock, but it remains a categorical leap beyond the work of any women artists.

M. C. Richards tried to articulate the sense of precarious balance in "centering," "which will enable us to step along that thread [of life] feeling it not as a thread but as a sphere."[36] Her goal was not the perfection of an artifact. "What I care about, it is the person," wrote Richards. "This is the living vessel: person. This is what matters."[37] Here, we can sense that Richards's centering and Voulkos's violation were not necessarily opposites. Perhaps after sipping from a Heath cup, sitting on a Karnes garden stool, or listening to a Takaezu pot rattle, it is we who experience metaphorical human bodies as fragile porcelain dolls. If we move away from canons, marketplace frameworks, and especially away from either a purely verbal or a purely literal reading of violated pottery to a more nuanced and multisensorial appreciation for what the wheel meant in ceramics at mid-century, the abundance of diverse modes of self-realization is breathtaking. The wheel was a way to break tradition and also sustain convention, "both/and," not "either/or." Through a real reverence for the complexity of the tool and these artists, we can begin to understand how it was both a habit of working and a way of thinking. It also became a way of unmaking, and undoing skill. Most important, by remembering the oscillation of this larger cast of individuals instead of one sole hero as a focus, we can approach the complexity of these women's work, where classicism and order intersect with trauma and the subtle threat of chaos. License and decorum might appear as antipodes, but perhaps self-invention requires revolutions in both extremes.

ACKNOWLEDGMENTS I thank Peter Russo and Sequoia Miller for their close reading of this manuscript and their insightful comments. Miller and John Stuart Gordon generously invited me to present a version of this essay at a Yale University Art Gallery symposium. Jenni Sorkin, Ulysses G. Dietz, Edward S. Cooke, Elisabeth Agro, and Sarah Warren were wonderfully responsive listeners; the comments of each improved the text. My research on Karen Karnes has benefited enormously from the generosity of Mark Shapiro, and Jay Stewart richly expanded my awareness of Edith Heath's legacy. I am grateful for my interviews with Herbert Cohen, Karen Karnes, David Weinrib, and Zeb Schachtel. My inquiry into the wheel began with stimulating discussions with the engaging Freddy Fredrickson in 2007, and my understanding of it has been refined through dialogue with Kevin Millward. For the significance of "quietly subversive" pots I thank Cheryl Buckley's contribution to the exhibition catalog *Women Designers in the USA, 1900–2000* (New York: Bard Graduate Center and Yale University Press, 2000).

1. Lee Nordness, *Objects: USA* (New York: Viking Press, 1970), pp. 61, 81; Linda Nochlin, "Why Have There Been No Great Women Artists?," *Art News* 69 (January 1971): 22–39; Cheryl Buckley, *Potters and Paintresses: Women Designers in the Pottery Industry, 1870–1955* (London: Women's Press, 1990); Pat Kirkham, ed., *Women Designers in the USA, 1900–2000: Diversity and Difference*, exh. cat. (New Haven, Conn.: Yale University Press, in association with Bard Graduate Center for Studies in the Decorative Arts, 2000).

2. Rose Slivka, *Peter Voulkos: A Dialogue with Clay* (New York: New York Graphic Society, 1978), p. 12; Andrew J. Perchuk, "Time in Mid-Twentieth Century Ceramics," *American Art* 21, no. 1 (Spring 2007): 11.

3. Peter Dormer, "The Ideal World of Vermeer's Little Lacemaker," in *Design after Modernism: Beyond the Object*, edited by John Thackara (London: Thames and Hudson, 1988), p. 144.

4. Mary Caroline Richards, *Centering in Pottery, Poetry, and the Person* (Middletown, Conn.: Wesleyan University Press, 1964), p. 4.

5. Perchuk, "Time in Mid-Twentieth Century Ceramics," p. 11.

6. Oral history interview with Susan Peterson, March 1, 2004, Archives of American Art, Smithsonian Institution. Elaine Levin, "Millard Sheets and the Impetus for the Ceramic Revolution of the 1950s," in Christy Johnson, *Common Ground: Ceramics in Southern California, 1945–1975* (Pomona, Calif.: American Museum of Ceramic Art, 2012), p. 14.

7. Oral history interview with Susan Peterson.

8. Billie P. Sessions, "Ceramics in Academia: A Site of Unique Circumstances," in *Common Ground*, p. 72 n. 28.

9. Cheryl Buckley, "Subject of History? Anna Wetherill Olmsted and the Ceramic National Exhibitions in 1930s USA," *Art History* 28, no. 4 (September 2005): 517.

10. Laura Andreson, interview with Elaine Levin, June 1974, in Levin, "Millard Sheets and the Impetus for the Ceramic Revolution of the 1950s," p. 17.

11. Oral history interview with Susan Peterson.

12. Patent no. US 2174061 A, available online at https://www.google.com/patents

/US2174061?dq=variable+speed+potter%27s+wheel&hl=en&sa=X&ei=1VllVLWtF4HC ASq3YC4Cg&ved=0CCYQ6AEwAQ (accessed November 11, 2014).

13. Ezra Shales, *Made in Newark: Cultivating Industrial Arts and Civic Identity in the Progressive Era* (New Brunswick, N.J.: Rivergate Books, an imprint of Rutgers University Press, 2010), p. 259.

14. *Twelfth National Ceramic Exhibition* (Syracuse, N.Y.: Syracuse Museum of Fine Arts; Onondaga Pottery Company, 1947), p. 11.

15. Oral history interview with Susan Peterson.

16. V. Gordon Childe, "Rotary Motion," in *A History of Technology*, vol. 1, *From Early Times to Fall of Ancient Empires*, edited by Charles Singer, E. J. Holmyard, and A. R. Hall (Oxford, Eng.: Clarendon Press, 1954), p. 204.

17. George M. Foster, "The Potter's Wheel: An Analysis of Idea and Artifact in Invention," *Southwestern Journal of Anthropology* 15, no. 2 (Summer 1959): 100. "Perhaps the most common argument against the evolutionary sequence hypothesis is that, since women do not use the wheel, they could not have invented it. How could a turntable customarily used by women, argues van Gennep, develop into a wheel used by men? This type of argument is based on the early assumption, now known to be only partially true in primitive societies, that hand-made pottery is alone the work of women. In many places, we now know, men using hand techniques are equally important in pottery manufacture" (p. 116).

18. Dr. Billie Sessions credits Carlton Ball with introducing the "Netherby wheel" in 1944 at the Mills College Ceramic Guild, although this seems rather improbable, considering its name, so the history and contributions of Elena Netherby remain worthy of future research. See Sessions, "Ceramics in Academia," p. 72 n. 28.

19. Mark Shapiro, ed., *A Chosen Path: The Ceramic Art of Karen Karnes*, exh. cat. (Chapel Hill: University of North Carolina Press, 2010).

20. Amos Klausner, *Heath Ceramics: The Complexity of Simplicity* (San Francisco, Calif.: Chronicle Books, 2006).

21. Sheila Hibben, "On and Off the Avenue: About the House," *New Yorker* (September 17, 1949): 86.

22. Classifieds, *New Yorker* (December 17, 1949): 92.

23. Rob Forbes, in Klausner, *Heath Ceramics*, p. 25.

24. Herbert Cohen, in conversation with the author, May 29, 2014. To benchmark Karnes's skill in retrospect is difficult. It is important to note that when she returned from Italy, she shipped her wheel back across the Atlantic, having formed an emotional (or financial) attachment to it.

25. Dido Smith, "Karen Karnes," *Craft Horizons* 15, no. 5 (May–June 1958): 10–14; Jody Clowes, "The Woman behind the Pot," in Shapiro, ed., *A Chosen Path*, 27–45.

26. Voulkos profoundly admired both Takaezu and Karnes, followed their work, and was on a first-name basis with both. While he famously played the rascal for the camera and crowds, these women did not—such behavior by women would not have been socially sanctioned—yet they laughed at his coarse and colorful language.

27. Undated clipping, *Los Angeles Times*, Toshiko Takaezu Foundation Archive.

28. Eliel Saarinen. *The Search for Form in Art and Architecture* (New York: Reinhold, 1948), p. xvi.

29. Alice Parrott, ". . . The Right, Unhurried Place," *Craft Horizons* 20, no. 1 (January–February 1960): 20.

30. Saarinen, *Search for Form*, p. 176.

31. Undated clipping, *Los Angeles Times*, Toshiko Takaezu Foundation Archive.

32. Warren Mackenzie, "Letter to the Editor," *Craft Horizons* 5 (November/December 1961): 45.

33. Elissa Auther, *String, Felt, Thread: The Hierarchy of Art and Craft in American Art* (Minneapolis: University of Minnesota Press, 2010).

34. Nordness, *Objects: USA*, p. 81.

35. Roland Barthes, *Camera Lucida: Reflections on Photography*, translated by Richard Howard (New York: Hill and Wang, 1981).

36. Richards, *Centering*, p. 6.

37. Ibid., pp. 34–35.

Ellen Archie, Ronald W. Fuchs II, Jennifer Mass, and Erich Uffelman

"The most dangerous imitations": A Group of Spurious Chinese Export Porcelain Decorated with Fame and the American Eagle

▼ IN 1933 HOMER EATON KEYES, the first editor of *Antiques* magazine and an expert on Chinese export porcelain, warned that "the most dangerous imitations of Chinese Lowestoft are those achieved by removing the decoration from genuine old pieces or Oriental ware and substituting a rarer design that materially enhances their apparent value."[1] What Keyes was referring to was a growing number of pieces of fake Chinese export porcelain decorated with American emblems that were emerging on the antiques market in the 1920s and early 1930s. Among these was what Keyes described as "weird versions of the Cincinnati pattern"—pieces decorated with a figure of Fame holding an eagle with a shield on its breast emblazoned with initials (fig. 1).[2]

Approximately twenty pieces of these "weird versions" are known, all of which are tea and coffee wares.[3] They are found with both overglaze blue and gold and sepia-and-gold enamel decoration, and with four different border patterns: a gold band; a blue-and-gold intertwined chain and star; a blue intertwined chain and star; and no border at all.[4] Four different sets of initials—W, NEA, WSA, and DL—personalize the shields on several pieces, and a simple floral spray decorates the shield of one other, suggesting that there were at least five different services.[5] A possible indication that all is not what it seems with this group is the fact that, in some instances, the same initials are found on pieces with different borders.[6]

The pieces appeared on the market in the mid-1920s. Henry Francis du Pont, one of the leading collectors of Chinese export porcelain of the period, bought a "Lowestoft Tea Caddy and Cover, Order of the Cincinnati" from New York antiques dealers Ginsburg & Levy on December 11, 1925, for $350, and another "Cincinnati Lowestoft Tea Caddy, blue trellis dec. on top" on December 2, 1927, from dealer J. Grossman for $850 (fig. 2).[7] Similarly decorated pieces were acquired by other leading collectors, such as Helena Woolworth McCann and Euchlin and Louise Reeves.[8]

At first glance, these pieces seem to date to the late eighteenth or early nineteenth century, and relate to George Washington's export porcelain decorated with the Society of the Cincinnati insignia. The high price paid by du Pont and other knowledgeable collectors suggests that these works were considered genuine, and published as such by the 1950s.[9] But stylistic differences, physical evidence, and scientific evidence all bear out Keyes's warning that some, of the pieces in this group are, in his words, "FORGED DECORATION ON GENUINE CHINESE PORCELAIN / Hybrid armorial eagle with pseudo-Cincinnati emblem" (fig. 3).[10]

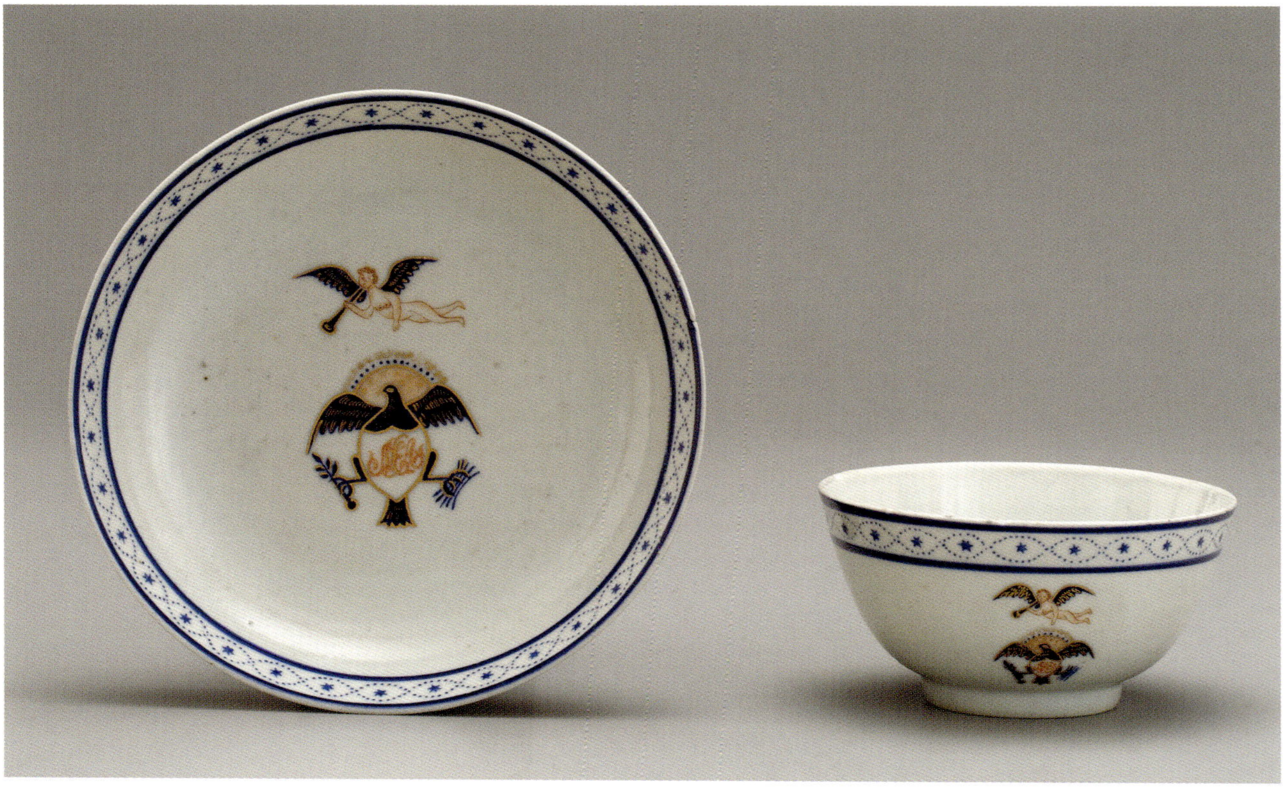

Figure 1 Teabowl and saucer, Jingdezhen, China, 1760–1775, decoration added in the United States ca. 1925. Porcelain. D. of saucer 5½". (Reeves Collection, Washington & Lee University, Gift of Richard and Catharine Hubbard; photo, Robert Hunter.)

Stylistic Evidence

The allegorical figure of Fame originates from the ancient Greek goddess Pheme (Fama, in the Roman pantheon), who personified positive renown and negative rumor and who was the messenger of Zeus. According to George Richardson's *Iconology; or, a Collection of Emblematical Figures*, published in 1779, Fame was "represented by the figure of a woman with large white wings at her shoulders, holding a trumpet. . . . [T]he wings denote the candour and velocity of Fame; the trumpet signifies that the voice of Fame resounds, like this instrument, and encourages men to imitate the virtuous."[11] This depiction was standard for Fame from at least the sixteenth century, when she became popular as a decorative motif on a range of objects, including ceramics.[12]

The eagle is an adaptation of the Great Seal of the United States, which was adopted by Congress in 1782 as the country's official emblem. The seal consists of a shield with thirteen red and white stripes, representing the original thirteen states, under a blue chief, representing Congress, borne on the breast of an American eagle.[13] Designed for use on official documents, the Great Seal also became a popular motif on objects ranging from furniture to textiles to ceramics. It proved a popular decorative motif for Chinese export porcelain, and on many pieces it was common for the stripes and chief of the shield to be replaced with the initials of the original owner.[14]

The figure of Fame and the eagle that decorate the pieces under study are similar to those found on what is arguably the most famous Chinese export porcelain made for the American market: George Washington's

Figure 2 Tea caddy, Jingdezhen, China, 1760–1775, decoration added in the United States ca. 1925. Porcelain. H. 5½". (Courtesy, Winterthur Museum, Bequest of Henry Francis du Pont, 1963.716a,b.)

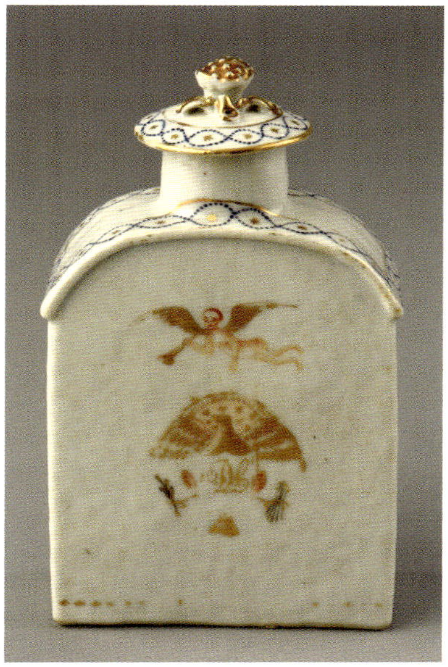

Figure 3 Illustration from Homer Eaton Keyes, "Imitations of Chinese Lowestoft," *Antiques* 24, no. 4 (October 1933): 143.

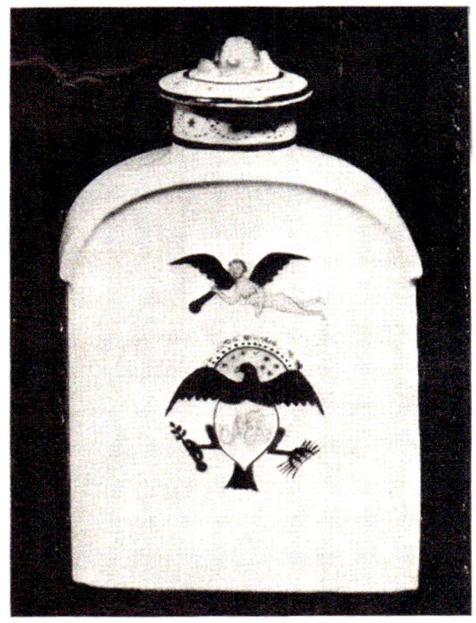
Fig. 8 — FORGED DECORATION ON GENUINE CHINESE PORCELAIN
Hybrid armorial eagle with pseudo-Cincinnati emblem

Cincinnati service. Commissioned in 1784 by Samuel Shaw, the first American merchant to go to China, the original service consisted of 302 pieces, each of which was decorated with an underglaze blue border and a figure of Fame holding the badge of the Society of the Cincinnati: an eagle with a shield on its breast containing a vignette of Cincinnatus, the ancient Roman hero after whom the society is named (fig. 4).[15]

The Society of Cincinnati, an organization of Revolutionary War officers founded in 1783, was named after Lucius Quinctius Cincinnatus, a Roman farmer turned military leader who rescued Rome and then relinquished his power to return to his farm. Cincinnatus was seen by many as the model of a selfless patriot, and George Washington, a farmer who had led American troops to victory, served two terms as president, and then retired to return to Mount Vernon, was seen by many as a modern-day Cincinnatus.

Figure 4 Salad bowl, Jingdezhen, China, 1784, decorated in Guangzhou (Canton), China. Porcelain. W. 9¼". (Reeves Collection, Washington & Lee University, Gift of the Edward H. Thompson Family.)

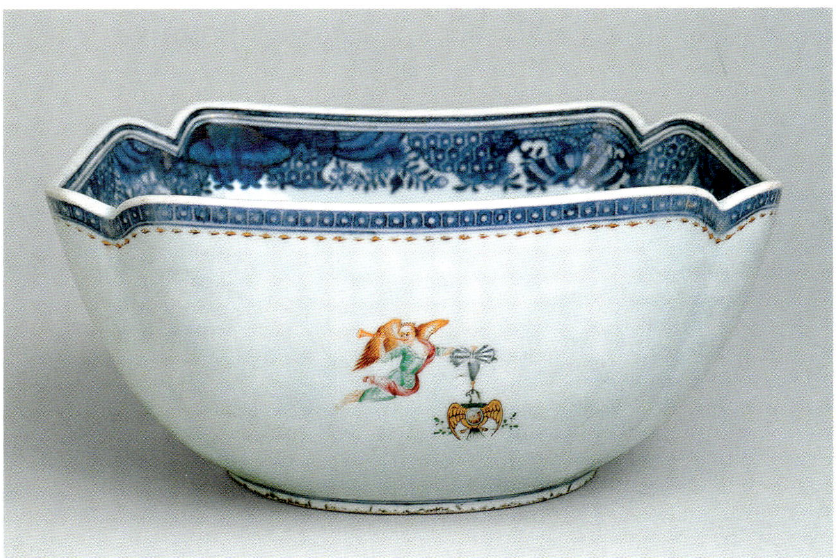

Washington, the first president of the society, bought the service in 1786 and used it at Mount Vernon and the president's mansions in New York and Philadelphia. Conscious that his actions were carefully watched and that he was creating the model for the office of the U.S. presidency, Washington may have used the Cincinnati service to reinforce his identity as a modest, selfless leader in a new, democratic society who would serve his country without expectation of reward and then relinquish power back to the people.[16]

Washington's Cincinnati service is one of the most prized examples of American-market export porcelain, and it has been avidly studied, published, and collected since the mid-nineteenth century.[17] Any piece from the service was considered one of the crowning glories of any great Chinese export collection, and was priced accordingly; Henry Francis du Pont paid $71,550 for sixty-nine pieces from the service in 1927 and 1928.[18]

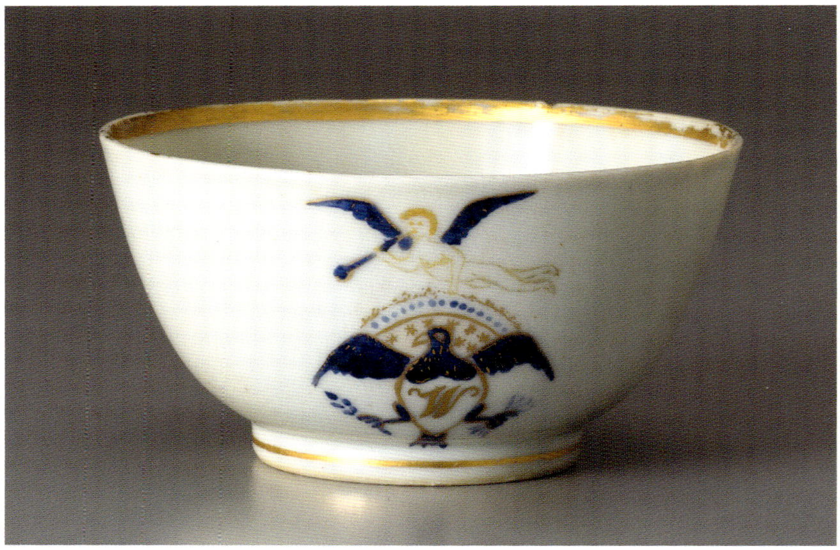

Figure 5 Teabowl, Jingdezhen, China, 1800–1820, decoration added in the United States ca. 1925. Porcelain. H. 1⅞". (Reeves Collection, Washington & Lee University, Gift of Euchlin and Louise Herreshoff Reeves; photo, Robert Hunter.)

The juxtaposition of Fame and the eagle on the pieces in the group under study was almost certainly meant to suggest some connection with the Society of Cincinnati, and several of the pieces have the initial W on the shield, further suggesting a connection to George Washington (fig. 5).[19]

Despite the similarity of the decoration to pieces of American-market export porcelain known to be genuine, elements of the design on the pieces in this study raise questions about their authenticity. The most questionable element is the nude figure of Fame. While Fame was occasionally depicted in the nude in the eighteenth and nineteenth centuries, she more commonly was shown at least partially clothed. Rees's *Cyclopedia* recorded in 1810 that "the common representation of Fame exhibits her in a flying attitude, sounding a trumpet . . . with a flowing robe."[20]

Depictions of a naked Fame would have been especially unusual in the United States, where depictions of nudity in art were often frowned on. While the growing popularity of neoclassical design made figures of nude or partially draped women more common in the eighteenth and nineteenth centuries, they were generally not seen as suitable for American audiences. Regarding a collection of casts of Greek and Roman statues, in 1802 the United States minister to France Robert R. Livingston wrote, "though these statues are viewed by the most delicate women here without a blush, yet the modesty of our country women renders a covering necessary."[21]

Artists, craftsmen, and merchants were aware of the difficulty of marketing paintings or objects with nude figures to American consumers. In 1809 artist John Vanderlyn warned one of his clients that he might not want Vanderlyn's painting of a nude because, "I am not sure the subject will please so as to be desirous of possessing it for it is perhaps not chaste enough for American eyes, at least to be displayed in the dwelling of any private individual to either the company of a parlor or drawing room." When James Monroe ordered new furniture for the White House from France, his agent Joseph Russell wrote saying they had "great difficulty in getting *Pendules* [clocks] without *Nudities*, and were . . . *forced* to take the two models we have bought on that account."[22]

Thus, it seems unlikely that an eighteenth- or early-nineteenth-century American consumer or merchant would have commissioned porcelain decorated with such a blatant depiction of nudity. And while one piece or set could be seen as an anomaly, to have twenty pieces, representing up to five sets, is less plausible.

Some of the borders that appear on pieces in the group are also curious. The most common border consists of two intertwined, meandering lines enclosing stars—sometimes executed in all blue enamel, sometimes in blue enamel with gold stars (see fig. 1). While stylistically the border conforms to the neoclassical style popular in the late eighteenth and early nineteenth centuries, a search of collections and the literature has found no similar border on genuine pieces.[23]

Physical Evidence

The execution of the decoration also raises questions: the application of the blue enamel seems less carefully painted than what is typically found on genuine Chinese export, and the enamel is thick in some places and thin in others. As Keyes described,

> Seldom, however, can the modern imitator give a wholly convincing air to his handiwork. Something will be wrong about his touch, or in the forms he outlines. The swift, sure, calligraphic stroke of the Chinese artist will be wanting, and his amusing distortions of human and animal figures will be bunglingly translated. Such disqualifications will usually be felt, even if not fully recognized by the sensitive eye; but the subtlest admonition to consultation and comparison thus conveyed should never be ignored.[24]

Even more damning is the fact that several pieces in the group bear evidence of earlier decoration that has been removed. When viewed in raking light, ghosts of scratched-off borders, medallions, and coats of arms are visible. For instance, ghosts of a swag-and-bamboo border appear underneath the intertwined chain and star border on the saucer illustrated in figure 1 (fig. 6).[25] Swag-and-bamboo borders date roughly between 1765 and 1775, suggesting the correct date for the porcelain and their original decoration.[26]

The teabowl illustrated in figure 1 shows residual evidence beneath the eagle of a floral spray, a popular motif in the late eighteenth and early nineteenth century.[27] Here, too, is clear evidence of scratching, suggesting that the original decoration was abraded off (fig. 7).

Another teabowl in the Reeves Collection (1967.1.565) shows faint traces of a circular medallion containing an initial, a style that was popular in the first decades of the nineteenth century. Similarly, a cream jug in the same collection (1967.1.379) shows faint traces of an urn-shaped shield with mantling, which most likely was a coat of arms or a shield bearing initials, a motif popular from the 1770s into the early nineteenth century.

It is possible that some of the original decoration, like the circular medallion with initials, was painted in gold. Gilded decoration, which was

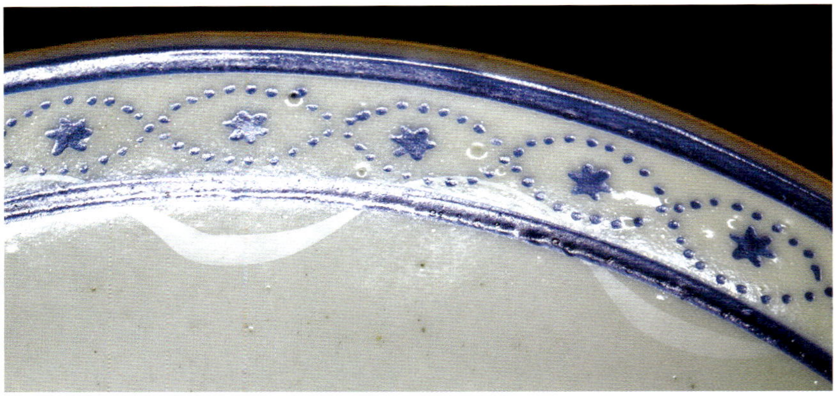

Figure 6 Detail of the saucer illustrated in fig. 1, photographed in raking light.

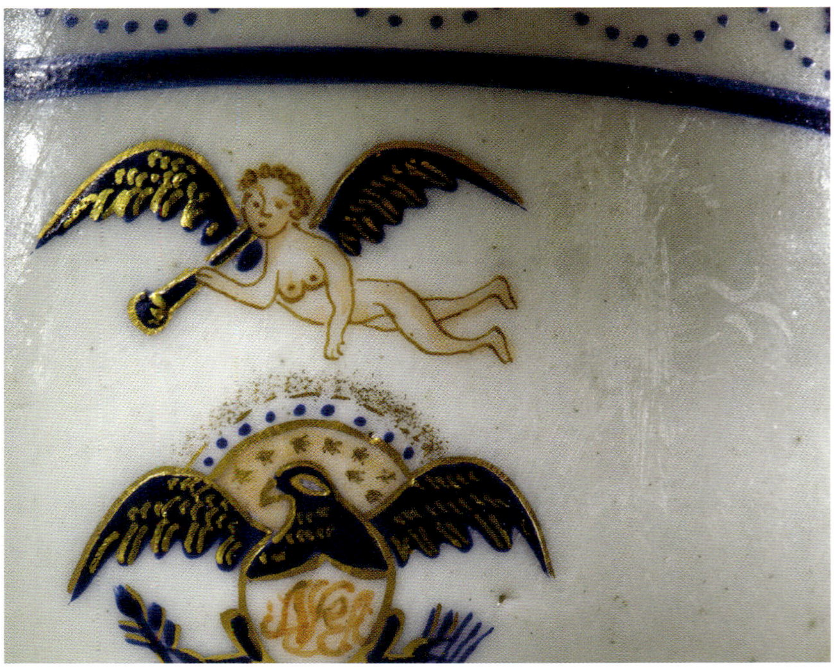

Figure 7 Detail of the teabowl illustrated in fig. 1, photographed in raking light.

fired at a low temperature and therefore fixed poorly to the surface, was especially susceptible to wear and so would have been easy to remove with acid or even just physical abrasion. Overglaze enamels would have been more difficult to remove than gilded decoration, but were still relatively easy to remove, probably with acid and abrasion, as is seen in the scratch marks on the teabowl illustrated in figures 1 and 7.

Chemical Evidence

The most convincing evidence that the objects under discussion are not genuine is the fact that the enamels used on a number of them do not match eighteenth- and early-nineteenth-century Chinese enamel technology. The enamels were analyzed using X-ray fluorescence spectroscopy, a nondestructive technique in which an incident X-ray beam is used to excite atoms in the overglaze enamels, glaze, and body. The X-rays emitted from all of these components can provide information about the colorants and fluxes used.[28]

Zinc, probably used as a fluxing agent, was found in high concentrations in the blue enamel on several of the pieces in the study.[29] The presence of zinc is strongly suggestive of twentieth-century manufacture. It is present in the enamels on a group of Chinese export porcelain, decorated with a scene of the signing of the Declaration of Independence, that is thought to have been made in China in the 1920s or 1930s, and is also present in the enamels on fake pieces of George Washington's Cincinnati service that were created in Europe or America in the 1970s by adding the figure of Fame and the Cincinnati eagle to genuine late-eighteenth-century Chinese export plates (fig. 8).[30]

A tin oxide opacifier also was inferred in the blue enamel on several pieces, rather than the typical lead arsenate opacifier typically found on genuine Chinese enamels of the eighteenth and nineteenth centuries.[31]

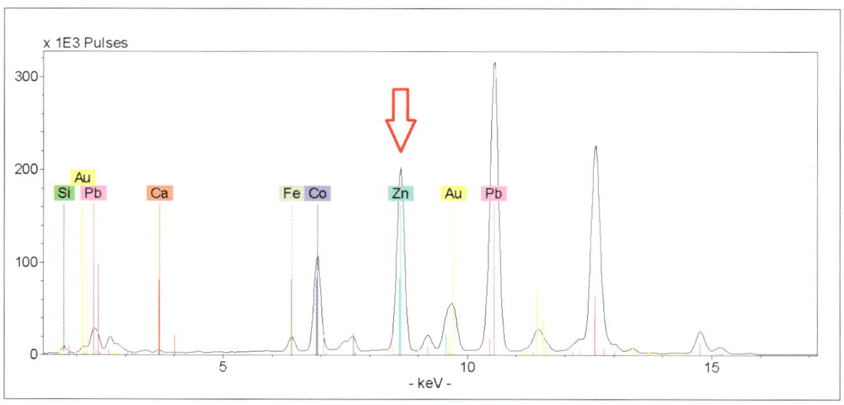

Figure 8 p-XRF analysis of the blue enamel on the eagle on the saucer illustrated in fig. 1. Note the peak representing zinc (Zn) shown at the arrow.

Another anomalous element is bismuth, which is found on several pieces in the gilding decorating Fame and the eagles. Bismuth was used as a flux to bind the gold to the glaze but has not been identified in gilded decoration on genuine late-eighteenth-century Chinese export porcelain, suggesting that it is a marker for later gilded decoration.[32] In fact, bismuth is found on the eagle and figure of Fame on one of the teabowls but is not found on the gilded rims and small floral sprays, suggesting that the latter probably date from the cups' original decorative scheme and that Fame and the eagle do not.[33]

The final, and perhaps most damning, anomalous element present on some of the pieces in this group is chromium, which is found in the sepia enamels used to paint the figure of Fame on several pieces.[34] Chromium was discovered in 1797 and is first known to have been used on ceramics in 1802 at Sèvres.[35] Its use in China is not known before the twentieth century, making it highly unlikely that the decoration could be period (fig. 9).[36]

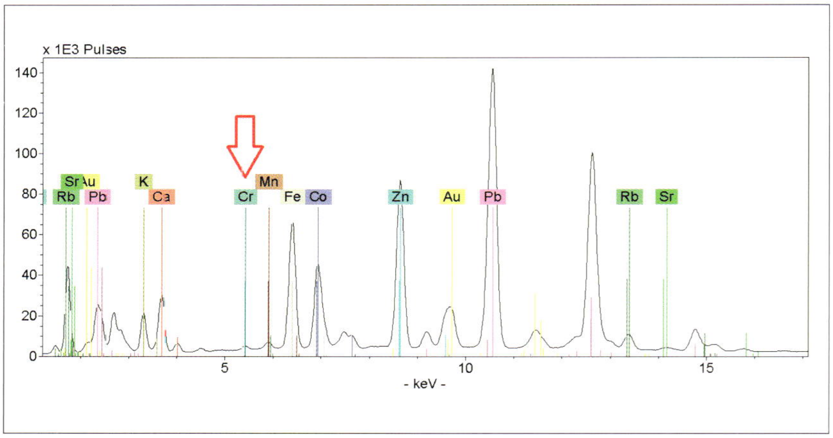

Figure 9 p-XRF analysis of the sepia enamel on the figure of Fame on the teabowl illustrated in fig. 1. Note the small but significant peak presenting chromium (Cr) shown at the arrow.

Although the chemical composition of the enamels on related wares virtually guarantees them to be later, several pieces from the group were found to have period-appropriate enamel compositions, consistent with late-eighteenth- and early-nineteenth-century Chinese export porcelain.[37] While this raises the possibility that these pieces could be genuine, their stylistic similarity to the others—with the incongruous nude figure of Fame, the similar style of painting, and the presence of initials that match those found on presumed fakes—suggests that these, like the others, bear later decoration on earlier porcelain.[38] Just because a new chemical composition was introduced does not mean that old ones did not continue to be used, so if the forger (or forgers) made the pieces over a period of time, they could likely have obtained different batches of enamels with different compositions.

Discussion and Conclusions

The combination of stylistic, physical, and chemical evidence shows that at least some, and most likely all, of the pieces decorated with the naked figure of Fame and an eagle are not late-eighteenth- or early-nineteenth-century originals but, rather, fakes made in the early twentieth century.

But why would such fakes have been made, and why in the first decades of the twentieth century? Studying these fakes not only helps us distinguish the genuine from the spurious, it also provides a picture of the collecting and decorating tastes of the 1920s and 1930s, the development of the field of ceramics study, and the growing skill and knowledge of ceramic scholars, collectors, dealers, and fakers.

Chinese export porcelain decorated with American symbols began being produced soon after merchants from the United States first began trading with China in 1784, and it continued to be made into the 1820s. While the pieces of specially commissioned porcelain with American symbols are relatively small in number, they have long been popular among collectors, as they combined patriotism, the perceived romance of the China trade, and the historic notion that Chinese porcelain was the finest ceramic ever made.

To early collectors, Chinese export porcelain was not just a foreign product but also rather intrinsically American. As R. T. Halsey, one of the founders of the American Wing at the Metropolitan Museum of Art, wrote: "Possibly no element which went into the making of the American home is more characteristic than the Chinese table-ware which formed part of the home cargoes of our early ships in the Oriental trade."[39] Pieces decorated with eagles and pieces that were similar to George Washington's Cincinnati service would be sure to find appeal among patriotic collectors.

As interest in Chinese export porcelain made for the American market grew in the first decades of the twentieth century, demand increased, supply remained static, and prices rose accordingly. Not surprisingly, in the 1920s manufacturers in Europe and China began making reproductions of historic pieces, and some unscrupulous artisans and dealers began to produce and sell fakes to take advantage of this growing market.

There were several instances of such fakes emerging on the market in the 1920s. In Keyes's article in *Antiques* he documented not just the "weird versions of the Cincinnati pattern" that are the subject of this study but also the tomb of George Washington (a plate "from which the original enamel flower sprigs had been removed to make way for a spread eagle perched on accouterments of war") and the arms of the State of New York.[40]

Keyes warned that while some of the modern fakes made in Europe and China were relatively easy to identify, it was far harder to identify forgeries "achieved by removing the decoration from genuine old pieces or Oriental ware and substituting a rarer design. This, obviously, is done with dishonest intent; and because the porcelain itself is beyond criticism, the forgery is often hard to detect and almost impossible to prove."[41]

Among these fakes was a group decorated with the arms of the State of New York, that was the subject of a court case and articles published in 1930 in *Time* magazine and the *New York Times*.[42] As the *Time* article explained: "Genuine plain old Lowestoft is used as a basis for the fake ware. To give the china greatly increased value to collectors, artists are employed to decorate the plain ware with early American motifs, such as Salem ships. The modern decorations, which are fired on the china, are said to be readily recognizable by experts."[43]

The collector and dealer Edward Crowninshield was responsible for discovering the forgery, according to *Time*: "A more careful examination convinced him that the service was of plain Lowestoft porcelain which had been skillfully decorated on top of the glaze, then sandpapered to give approximately the correct 'feel.'"[44]

The *Time* article went on to describe how officials "added immeasurably to the detective-story air of the whole business by producing the traditional sinister oriental, an expert Japanese repairer of antique porcelain who labors in a little art shop on Sixth Avenue, Manhattan and is known as 'Mr. Chicago.' Inspector Liese insisted last week that Mr. Chicago of Japan is the real perpetrator of the forged Lowestoft, but since china-painting is not yet a criminal offense, Mr. Chicago remains outside the law."[45]

The removal of old enamel and gilding, the preparation and application of new enamel decoration, and the firing of new enameling in a muffle kiln would have been relatively easy, and in fact was hardly a new phenomenon. The ceramics scholar Joseph Marryat wrote in 1850 that

> the old stocks in the Royal Manufactory at Sèvres, were put up to auction, and bought by certain individuals, who also collected all the soft ware they could find in the possession of other persons. The object of this proceeding for a long time remained a mystery, but at length the secret transpired, that the parties had discovered a process, which consisted in rubbing off the original pattern and glaze, and then colouring the ground with turquoise or any other colour, and adding paintings or medallions in imitation of the style of the old "pête tendre;" thus enhancing a hundred fold the value of the pieces.[46]

Adding new decoration to old porcelain would have been even easier in the 1920s, as the proliferation of amateur china painters in the late nineteenth and early twentieth centuries and of ceramics restorers, such as the nefarious "Mr. Chicago," meant that the tools, overglaze enamels, and talent needed to add decoration to old porcelain were readily available. That is almost certainly what some unethical but talented person did: took mismatched pieces of old porcelain, removed the original and probably worn enamel and gilded decoration with acid and abrasion, and then, using George Washington's Cincinnati porcelain as inspiration, decorated the now-blank porcelain with a new border and central decoration of Fame and an eagle. The result was something that had an air of authenticity about it and was certain to appeal to collectors looking for pieces that evoked America's glorious past.

Though the Chinese export objects decorated with a nude figure of Fame and an eagle did not gain the same level of notoriety as the pieces decorated with the arms of New York, they were recognized as fake by Homer Eaton Keyes, who included them in his 1933 article "Imitations of Chinese Lowestoft."[47] However, that news seems to have been forgotten by the 1950s, when the pieces began to be published as genuine. In 1956 John Goldsmith Phillips illustrated a coffeepot from the Helena Woolworth McCann Collection, describing it as "American market 1795–1800."[48] But close study of the design, physical evidence, and chemical composition of the group has revealed that they are, in fact, just what Homer Eaton Keyes warned about back in 1933: "most dangerous imitations."

1. Homer Eaton Keyes, "Imitations of Chinese Lowestoft," *Antiques* 24, no. 4 (October 1933): 143. Keyes used the term *Chinese Lowestoft* to refer to what is now known as Chinese export porcelain. The term arose in the nineteenth century, when early ceramics scholars mistakenly thought that Chinese porcelain was actually English porcelain made in the city of Lowestoft. Though the error was soon recognized, the terms *Oriental Lowestoft* and *Chinese Lowestoft* remained in use into the 1950s.

2. Ibid.

3. Examined as part of this study were a teabowl and saucer, another teabowl, and a helmet-shaped cream jug in the Reeves Collection, Washington & Lee University; three teabowls, a teapot, stand, oval dish, and two tea canisters in the Winterthur Museum; two teabowls and matching saucers in a private collection; and a teabowl and saucer in the Cluthe Collection. Known only through publications are a tea caddy, illustrated in ibid., and a coffee pot, illustrated in John Goldsmith Phillips, *China-Trade Porcelain* (Cambridge, Mass.: Harvard University Press, 1956), pl. 100.

4. One piece, a cream jug in the Reeves Collection (67.379), has no border. Four teabowls—one in the Reeves (67.1.565) and three in the Winterthur Museum (59.86.58–60)—have a gold band at the rim. Five pieces—a teabowl and saucer in the Cluthe Collection and a tea caddy, teapot, and stand at Winterthur (63.716a,b and 73.240a–c)—have a blue-and-gold intertwined chain and star border. Nine pieces—a teabowl and saucer in the Reeves Collection (16.1.1a,b), two teabowls and saucers in a private collection, and a tea caddy, oval dish, and circular dish at Winterthur (63.718a,b, 59.85.61, and 54.64)—have a blue intertwined chain and star border. The tea caddy illustrated in *Antiques* and the coffee pot illustrated in Phillips's *China-Trade Porcelain* have intertwined chain and star borders, but because the illustrations are black-and-white it is not clear whether they are blue or blue-and-gold.

5. The initial W is found on four teabowls—one in the Reeves Collection (67.1.565) and three in the Winterthur Museum (59.86.58–60). The initials NEA are found on eleven pieces—a teabowl and saucer and a cream jug at the Reeves (16.1.1a,b and 67.1.329); a circular plate, oval dish, and tea caddy at Winterthur (54.64, 59.85.61, and 73.718); three teabowls and saucers in the private collection; and the coffee pot illustrated in Phillips's *China-Trade Porcelain*. The initials WSA are found on a teapot and stand at Winterthur (73.240a–c). The initials DL are found on a tea caddy at Winterthur (63.716). The initials on the tea caddy illustrated in *Antiques* are illegible in the illustration. A saucer in the Cluthe Collection has a floral spray.

6. "NEA" appears on pieces that do not have a border: a cream jug at Reeves (67.1.329); on pieces with a blue intertwined chain and star border (a circular plate, oval dish, and tea caddy) at Winterthur (54.64, 59.85.61, and 73.718); and two teabowls and saucers in a private collection.

7. Henry Francis du Pont Day Books, Winterthur Registrar's Office, Winterthur Museum, Wilmington, Del.

8. Phillips, *China-Trade Porcelain*, pl. 100; Thomas Litzenburg, *Chinese Export Porcelain in the Reeves Center Collection* (London: Third Millennium Publishing, 2003), p. 143.

9. Phillips, *China-Trade Porcelain*, pl. 100.

10. Keyes, "Imitations of Chinese Lowestoft," p. 143.

11. George Richardson, *Iconology; or, a Collection of Emblematical Figures . . .* , 2 vols. (London: Printed by G. Scott, 1779), 2:16, pl. LVII.

12. Fame appears on pl. 122 of *The Ladies Amusement*, a collection of designs published circa 1760 that would be "extremely useful to the Porcelaine, and other Manufacturers." Robert Sayer, *The Ladies Amusement* (Newport, Eng.: Ceramic Book Co., 1966), p. 122 and title page. Figures of Fame are known on transfer-printed porcelain made by Worcester and Longton Hall. Simon Spero, *18th Century English Transfer-Printed Porcelain and Enamels* (Carmel, Calif.: Mulberry Press, 1991), p. 33.

13. U.S. Department of State, Bureau of Public Affairs, "The Great Seal of the United States," www.state.gov/documents/organization/27807.pdf (accessed November 15, 2015).

14. Jean McClure Mudge, *Chinese Export Porcelain for the American Trade, 1785–1835* (Newark: University of Delaware Press, 1981), pp. 167–69.

15. In addition to the service that was sold to George Washington, Shaw commissioned a service for himself and nine tea sets decorated with the society's eagle (but not the figure of

Fame). Susan Detweiler, *George Washington's Chinaware* (New York: Henry N. Abrams, 1982), pp. 81–97; Amanda Lange, *Chinese Export Art at Historic Deerfield* (Deerfield, Mass.: Historic Deerfield, 2005), pp. 250–53.

16. Detweiler, *George Washington's Chinaware*, pp. 81–97.

17. Benjamin Lossing, *Mount Vernon and Its Associations* (New York: C.A. Alvord, 1859), pp. 239–40; Alice Morse Earle, *China Collecting in America* (New York: Charles Scribner's Sons, 1892), pp. 229–48.

18. Thomas McCreedy Dealer File and Day Books, Winterthur Registrar's Office, Winterthur Museum, Wilmington, Del.

19. Teabowl from the Reeves Collection (1967.1.564); teabowls from the Winterthur Museum (1959.85.58–60).

20. Abraham Rees, ed., *The Cyclopaedia; or, Universal Dictionary of Arts, Sciences, and Literature*, 41 vols. (1810–24), 14:125.

21. Robert R. Livingston to Robert G. Livingston, December 23, 1803, quoted in E. McSherry Fowble, "Without a Blush: The Movement toward Acceptance of the Nude as an Art Form in America, 1800–1825," *Winterthur Portfolio* 9 (1974): 104.

22. John Vanderlyn to John Murray, June 12, 1809, quoted in Fowble, "Without a Blush," pp. 103–21, 111; Joseph Russell to James Monroe, May 25, 1818, quoted in Betty C. Monkman, *The White House: Its Historic Furnishings and the First Families* (New York: Abbeville Press, 2000), p. 62.

23. David Howard, *Chinese Armorial Porcelain* (London: Faber and Faber, 1974), pp. 645–69, 741–76, 971–92; David Howard, *Chinese Armorial Porcelain, Volume II* (Chippenham: Heirloom and Howard, 2003), pp. 473–509, 600–643; Jean Gordon Lee, *Philadelphians and the China Trade* (Philadelphia: Philadelphia Museum of Art, 1984), pp. 54–57.

24. Keyes, "Imitations of Chinese Lowestoft," p. 143.

25. Saucer in the Reeves Collection (16.1.1b).

26. Howard, *Chinese Armorial Porcelain*, p. 386; Howard, *Chinese Armorial Porcelain, Volume II*, p. 261.

27. Teabowl in the Reeves Collection (16.1.1a).

28. Ronald W. Fuchs and Jennifer L. Mass, "Deciphering *The Declaration of Independence* on Chinese Export Porcelain," in *American Ceramic Circle Journal* 15 (2008): 168–87.

29. Zinc was found in the blue-and-gold enamels on three teabowls with the initials NEA (Reeves Collection 16.1.1a and two in the private collection); three saucers with the initials NEA (Reeves Collection 16.1.1a and two in the private collection); a cream jug with the initials NEA (67.1.379); a circular dish with the initials NEA (Winterthur Museum 54.64); an oval dish with the initials NEA (Winterthur Museum 59.8561); and four teabowls with the initial W (Reeves Collection 1967.1.565, Winterthur Museum 59.85.58–60).

30. Fuchs and Mass, "Deciphering *The Declaration of Independence* on Chinese Export Porcelain," pp. 168–87, 177–78; E. Uffelman, R.W. Fuchs II, P. A. Hobbs, L. F. Sturdy, D. S. Bowman, and D.A.G. Barisas, "Technical Examination of Cultural Heritage Objects Associated with George Washington," in *Collaborative Endeavors in the Chemical Analysis of Art and Cultural Heritage Materials*, edited by P. L. Lang and R. A. Armitage (Washington, D.C.: American Chemical Society, 2012), pp. 251–83, 257.

31. Tin oxide was identified in the blue enamels on three teabowls with the initial W (Winterthur Museum 59.85.58–60); a circular dish with the initials NEA (Winterthur Museum 54.64); and an oval dish with the initials NEA (Winterthur Museum 59.85.61). Pam Vandiver, "The Eighteenth-Century Change in Technology and Style from the Famille-verte Palette to the Famille-rose Palette," in *Technology and Style: Proceedings of a Symposium on Ceramic History and Archaeology*, edited by W. D. Kingerly (Columbus, Ohio: American Ceramic Society, 1986), pp. 363–81, 377.

32. Aniko Bezur and Francesca Casadio, "The Analysis of Porcelain Using Handheld and Portable X-ray Fluorescence Spectrometers," in *Handheld XRF for Art and Archaeology*, edited by Aaron Shugar and Jennifer Mass (Leuven: Leuven University Press, 2012), pp. 249–311, 272.

33. Bismuth was found in the gold of the eagle but not the rims on three teabowls with the initial W (Winterthur Museum 59.85.58–60).

34. Chromium was identified in the sepia enamels on three teabowls with the initial W (Winterthur Museum 59.85.58–60) and a circular dish with the initials NEA (Winterthur Museum 54.64).

35. Tamara Préaud, *The Sèvres Porcelain Manufactory: Alexandre Brongniart and the Triumph of Art and Industry, 1800–1847* (New Haven: Yale University Press, 1997), p. 58.

36. Fuchs and Mass, "Deciphering *The Declaration of Independence* on Chinese Export Porcelain," pp. 177–78.

37. Pieces studied that had an enamel composition consistent with eighteenth- and nineteenth-century famille-rose enamels include two tea caddies (Winterthur 63.716 and 63.718); a teapot and stand (Winterthur 73.24a–c); and a teabowl and saucer in the Cluthe Collection.

38. Ten pieces studied are decorated with the initials NEA. One piece, a tea caddy, was decorated using an enamel technology consistent with eighteenth- and early-nineteenth-century Chinese export porcelain; the remaining pieces—three teabowls and saucers in the Reeves Collection (16.1.1) and two in the private collection; a cream jug at Reeves (67.1.379); a circular dish in the Winterthur Museum (54.64); and an oval dish in the Winterthur Museum (59.8561)—all showed anomalies in their enamel composition that suggests they are later-decorated.

39. R.T.H. Halsey and Elizabeth Tower, *The Homes of Our Ancestors* (New York: Doubleday, Doran and Co., 1925), p. 201.

40. Keyes, "Imitations of Chinese Lowestoft."

41. Ibid.

42. Ibid.; "Faked Lowestoft," *Time* 16, no. 18 (November 3, 1930): 65; "Woman Is Jailed in Chinaware Deal," *New York Times*, October 22, 1930, p. 18.

43. "Woman Is Jailed in Chinaware Deal."

44. "Faked Lowestoft," p. 143.

45. Ibid.

46. Joseph Marryat, *Collections Towards a History of Pottery and Porcelain, in the 15th, 16th, 17th, and 18th Centuries: With a Description of the Manufacture, a Glossary, and a List of Monograms* (London: J. Murray, 1850), pp. 200–201.

47. Keyes, "Imitations of Chinese Lowestoft."

48. Phillips, *China-Trade Porcelain*, p. 202.

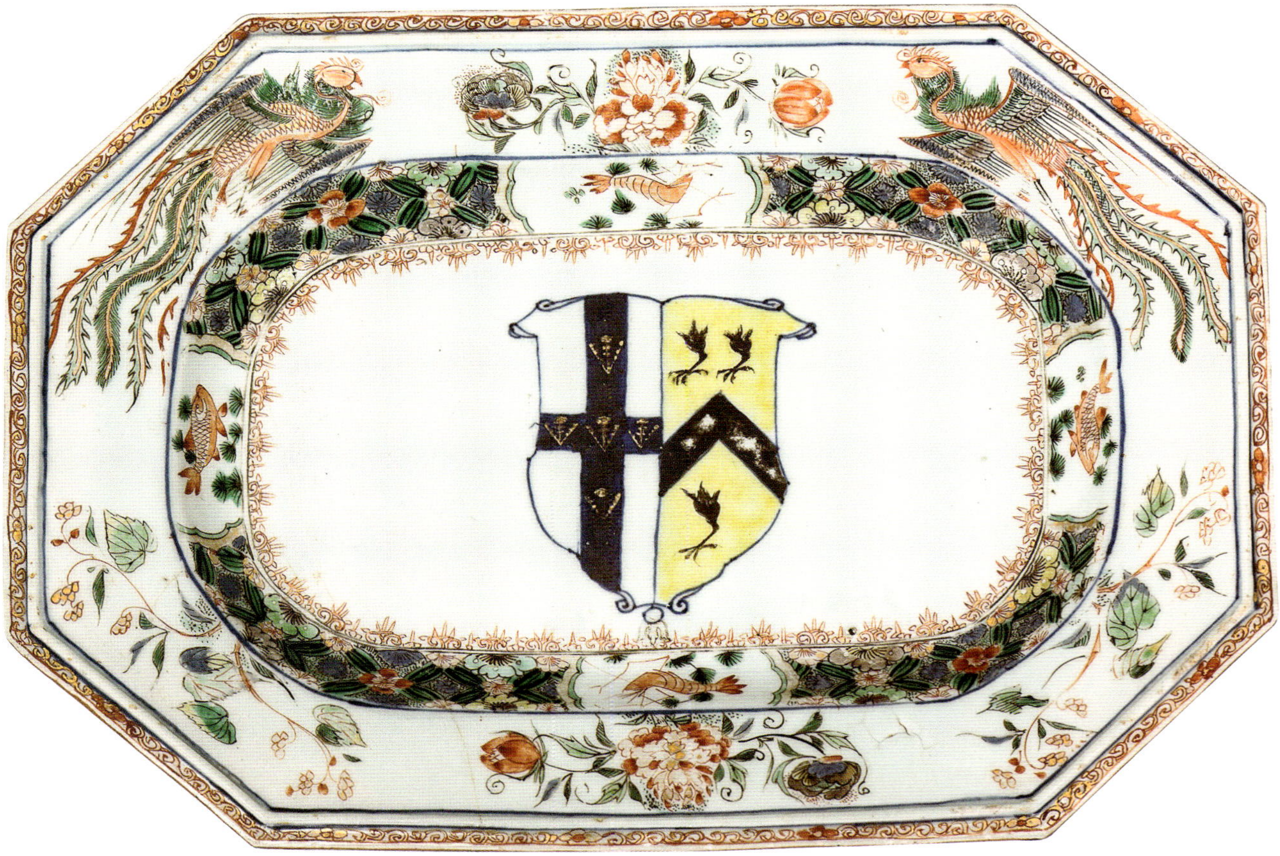

Figure 1 Dish (one of a pair), China, 1715. Porcelain. L. 12½". (Private collection; photo, Thomas M. Mueller.) Famille verte enamel decoration with the arms of Edward Harrison (1674–1732) and his wife, Frances Bray (1674–1752). This armorial dish shows no visible gold or silver application except for the five golden arrowheads on the cross on the left side of the armorial.

Shirley M. Mueller and Matthew Bunney

A Chinese Export Porcelain Mystery Solved Using Intensive Surface Analysis

▼ FOR SOME TIME, there has been an unanswered question regarding a specific Chinese export porcelain service. The 1715 armorial porcelains made in the famille verte style for Edward Harrison (1674–1732) and his wife, Frances Bray (1674–1752), appear to have two variations in the decoration.[1] On the dish illustrated in figure 1, the quadrants created by the blue enameled cross are blank and without color; the right side of the armorial has a brushed yellow ground. In the dish shown in figure 2 gold application is clearly visible in the quadrants of the cross as is the addition of silver on the right side.[2]

After Shirley Mueller studied errors made on Chinese export porcelain, she wondered whether the lack of gold and silver on the service pieces was a mistake of production.[3] That is, silver and gold were meant to be applied but the service was sent out before it was ready, possibly because of inadequate instruction on the part of the agent in charge of the arrangements (who is known as the supercargo) or possibly due to time constraints that prevented putting the finishing touches on some pieces. Others, however, maintain that the unembellished porcelains were handled significantly enough during their period of use so that the silver and gold were rubbed off.[4] After all, the metals were the last to be applied in the process of decoration and were either low-fired or not fired at all.

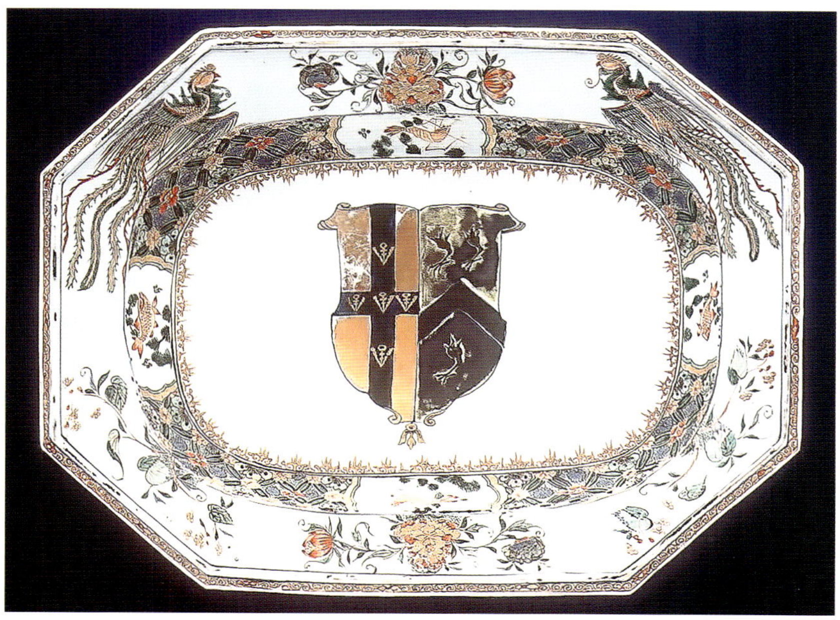

Figure 2 Dish (one of a pair), China, 1715. Porcelain with visible gold and silver application. L. 12". (Courtesy of the late Khalil Rizk of the Chinese Porcelain Company; photo, Chinese Porcelain Company.)

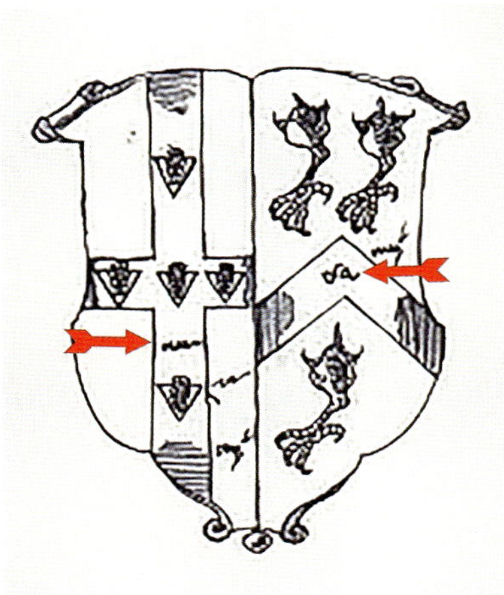

Figure 3 Armorial drawing of the first Harrison and Bray service ordered prior to 1715. Adapted from Clare Le Corbeiller and Alice Cooney Frelinghuysen, "Chinese Export Porcelain," *The Metropolitan Museum of Art Bulletin* 60, no. 3 (2003): 18.

Whether the cause was wear or mistake is especially relevant for the Harrison and Bray service. The original order, made before 1715, included instructions for colors to be painted on the cross (left) and inverted V (right) using heraldic language in the form of text abbreviations on the armorial itself. The order, however, was executed in underglaze blue and the instruction for colors were made part of the design as seen in figure 3. The "sa" (designated in fig. 3 by the red arrow on the right) stands for "sable," or black, and the barely legible "argt" (designated by the red arrow on the left) stands for "argent," or silver. The Chinese clearly did not understand the meaning of the text, which was written in English, and simply copied the design that they were given.[5]

Within a few years the service was reordered by the family in the famille verte palette, and it was hoped that color errors would be rectified. In many existing examples of the Harrison and Bray famille verte service, the cross decoration and inverted V are correctly executed (see fig. 2), but the absence of silver and gold on some of the armorials suggests that errors continued to be made, at least in part. The purpose of this paper is to make a definitive determination regarding this question.

The plain pair (see fig. 1) was examined with both the naked eye and an 8X loop, and neither silver nor gold was visible. The gold and silver on the adorned platters in figure 2, however, were easily seen with the naked eye.

To expand the investigation, the Ceramics Trace Model Study Method (CTMSM) was employed, though used in a manner different from its initial application.[6] As described by its originators, the technique was designed to distinguish Chinese porcelain hundreds of years old from more recent examples.[7] This is accomplished by focusing through the glaze of the ceramic piece in question using a high magnification (500X) digital Dino-lite microscope connected by a Universal Serial Bus (USB) to a laptop or desktop computer. The procedure is not destructive and consequently there is no residual damage to the examined ceramic.

In this study, however, the analysis was of the surface of the porcelain, not within the glaze. We call this adaptation of the CTMSM "Intensive Surface Analysis." Our purpose here was to determine whether residual silver or gold decoration existed on top of the glaze of Chinese export porcelain when it cannot be seen using the naked eye or an 8X loop.

Scanning the porcelains with and without silver and gold decoration with the hand-held microscope at up to 500X magnification produced a series of images that we recorded and compared. The photographs (figs. 4, 5) revealed that traces of both gold and silver adhered to the surface of the pair of plain pieces in the same areas in which the adorned porcelains show the metals.

Our findings indicate that gold and silver were originally applied to the two unadorned platters tested but apparently rubbed off over time. Whether the metals were overlaid in a faulty manner, which thereby contributed to easy surface abrasion cannot be determined with this technique. We can only say that the metals were present originally because vestiges of them remain. No remnants of gold or silver were found outside the areas decorated with these metals on the intact and the abraded platters.

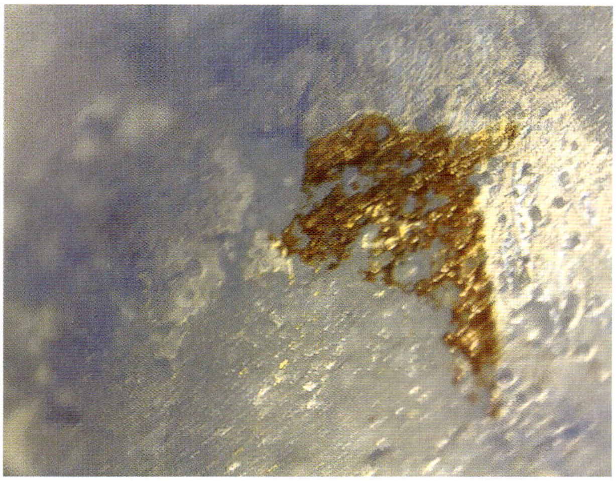

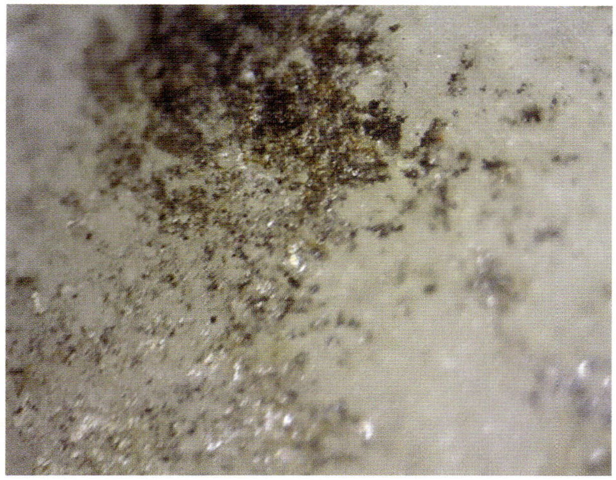

Figure 4 Vestiges of gold identified using Intensive Surface Analysis on the dish illustrated in fig. 1. (Photo, Matthew Bunney and Shirley M. Mueller.)

Figure 5 Vestiges of silver identified using Intensive Surface Analysis on the dish illustrated in fig. 1. (Photo, Matthew Bunney and Shirley M. Mueller)

The intact pieces of the famille verte service with the applied metal ornament apparently were not used. For their owners, the armorial Chinese export porcelain pieces must have been primarily decorative.

We conclude that the 1715 Harrison and Bray armorial service was embellished with silver and gold that wore off over time on the pieces that were used and that microscopic remnants of the precious metals are still visible in the appropriate part of the design using Intensive Surface Analysis. This technique is relatively easy to use, can provide important information where the human eye or 8X loop cannot, and is nondestructive. We believe it will become a method of choice for those who seriously examine not only Chinese export porcelain but other porcelain as well.

ACKNOWLEDGMENTS We are grateful to Angela Howard for providing her insights into the Harrison and Bray armorial service.

1. David Sanctuary Howard, *Chinese Armorial Porcelain I* (London: Faber and Faber, 1974), p. 187.

2. Shirley Maloney Mueller, "Surface Silver Decoration on Chinese Export Porcelain: A Survey," *Oriental Art* 47, no. 3 (2001): 56–64; Shirley Maloney Mueller, "Surface Silver Decoration on Chinese Export Porcelain: An Analytic Approach," *Oriental Art* 48, no. 4 (2002): 43–46.

3. Shirley Maloney Mueller, "Chinese Export Curiosities," *Oriental Art* 46, no. 1 (2000): 16–27.

4. In reference to a pair of famille verte plates (one of which is illustrated in fig. 1) that appeared in Christie's sale catalog #2404, January 25, 2011, lot 281, Angela Howard noted: "In the catalog the yellow does look somewhat suspiciously bright, so it might be interesting to test it. But there is no doubt that there is general wear all over. . . ." Private communication with the author, February 17, 2011.

5. Clare Le Corbeiller and Alice Cooney Frelinghuysen, "Chinese Export Porcelain," *The Metropolitan Museum of Art Bulletin* 60, no. 3 (2003): 18.

6. Zhou Yong and Zhou Qiang, *Survey of Ceramics Trace Model Study* (Guangzhou: Nanfang Daily Press, 2013); Zhou Yong and Zhou Qiang, *Trace Model Research and Material Evidence Authentication: Yuan Dynasty Underglaze Blue Porcelain* (Guangzhou: Nanfang Daily Press, 2014).

7. Zhou Yong and Zhou Qiang, *Survey of Ceramics Trace Model Study*; Zhou Yong and Zhou Qiang, *Trace Model Research and Material Evidence Authentication.*

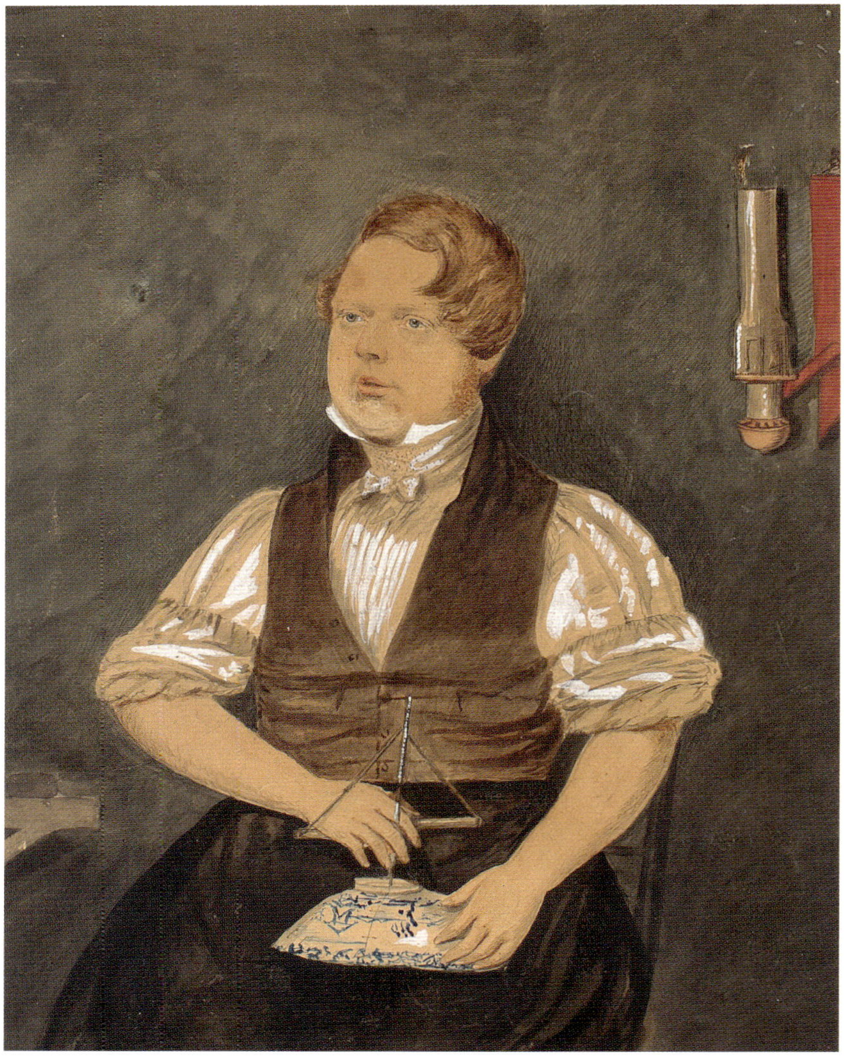

Figure 1 Artist unknown, England or America, ca. 1820-1830. Watercolor and gouache on paper. 9" x 6¾". (Private collection.) A rare period image of a china mender operating a bow drill to prepare a cracked bowl for a staple repair.

Angelika R. Kuettner

Simply Riveting: Broken and Mended Ceramics

▼ THE VICTIM OF A ROBBERY in Boston in 1773 placed an advertisement in a local paper offering a $10 reward to anyone who could help him regain his lost property. Listed between the engraved silver teaspoons and the fine "Cream-Pot of Pencil China" was "a pint blue & white China Mug that was crack'd and riveted."[1] Despite its less-than-perfect condition, the mug was clearly a valued object to be included among the others in the ad.

The practice of mending ceramic objects could well be as old as the pottery-making tradition itself. The British Museum boasts objects with repairs dating back at least to the sixth century B.C.[2] The collection of Chinese porcelains in Topkapı Palace, Istanbul, includes pieces with repairs made during the medieval period to objects centuries older.[3] And archaeological excavations in the Middle East reveal fragments with evidence of rivet repairs found in eleventh-century Islamic-world contexts.[4] Repaired ceramic objects are not just relics of ancient and medieval antiquity, however. They are found in colonial and federal America as well, and can be traced through newspapers, inventories, account books, letters, intact objects in museum collections, and archaeological finds.

Looking at mended ceramics with modern eyes, it is easy to overlook the context in which the repairs were made. How did those objects break? Who repaired them? What methods did they use? And ultimately were they preserved as family heirlooms or disposed of, only to be found in later archaeological excavations? This article explores these questions with a particular focus on the eighteenth and early nineteenth centuries in order to put broken and mended ceramics in historical context within the Anglo-American world (fig. 1).[5]

How Do Ceramics Break?

In the life of pottery and porcelain, breakage begins at the kiln; wasters are the ultimate proof, as any potter will lament. Sometimes there is evidence of repairs made to objects before they leave the kiln. It was probably in the leather-drying or air-drying stage that a piece of the rim of the plate illustrated in figure 2 was broken off and stuck back on or replaced by the Chinese potter before it was decorated and fired. A slip-decorated dish excavated in Kingsmill in Williamsburg, Virginia, reveals a similar repair. Perhaps made too thin in the initial object formation, the well of the dish cracked during the firing. The potter then used a different color slip to patch the dish before firing it a second time (fig. 3).

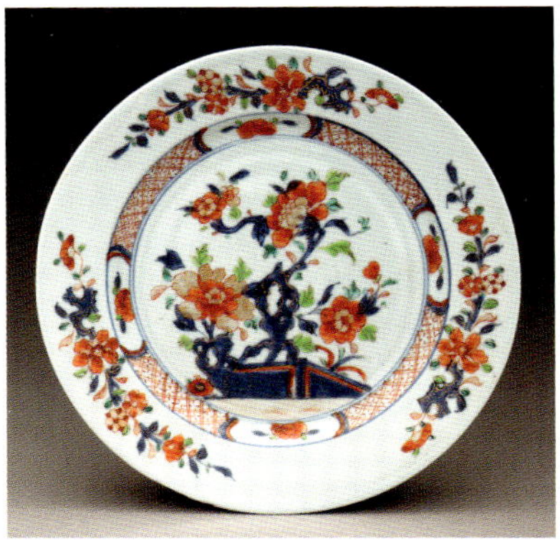

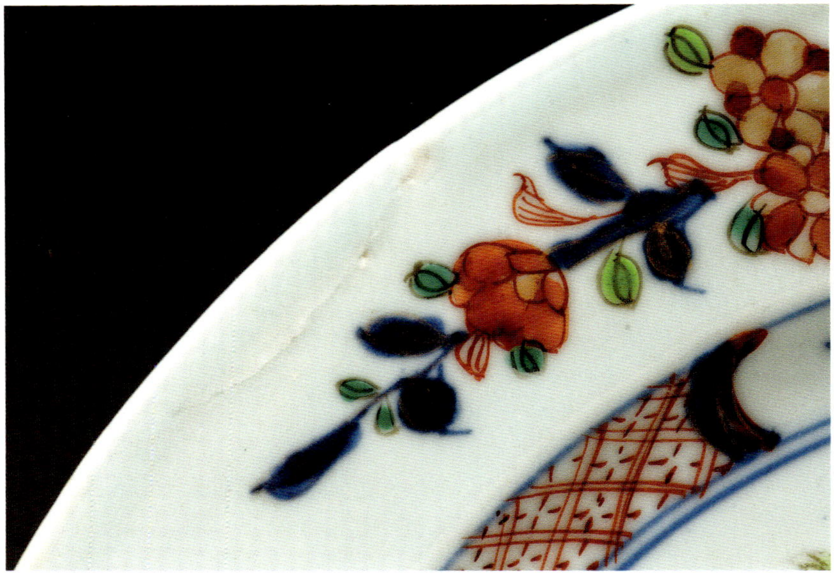

Figure 2 Plate, Jingdezhen, China, ca. 1735. Hard-paste porcelain. D. 9". (Colonial Williamsburg Foundation, Museum Purchase, 1968-6.) Note detail of rim showing repair before firing.

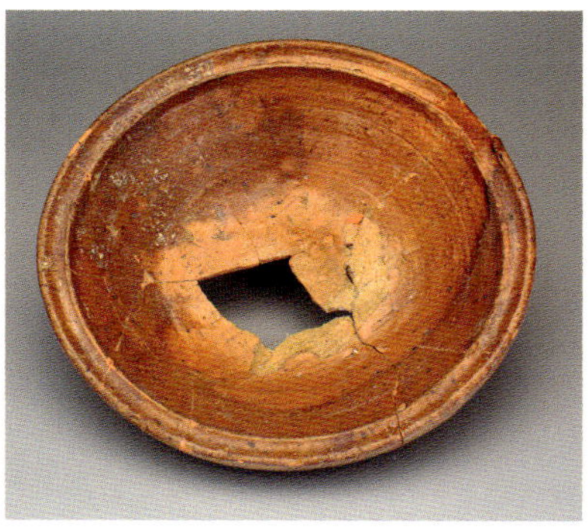

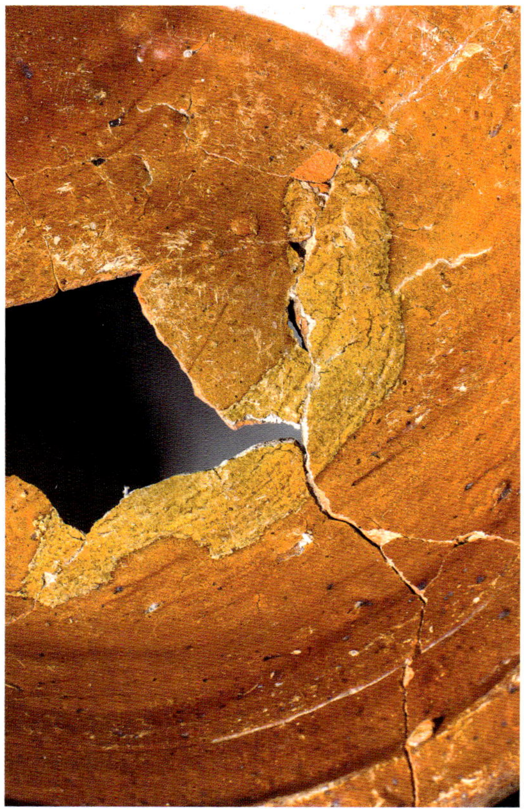

Figure 3 Milk pan, England, ca. 1730. Slip-decorated, lead-glazed earthenware. D. 14½". (Courtesy, Kingsmill Resort, Williamsburg, Virginia; photo, Robert Hunter.) Excavated in Kingsmill, Virginia.

If the ceramics were fortunate enough to survive the kiln, they would then have to survive packing and shipment. Poor packers, rough seas, or a combination of the two frequently brought frustration or sorrow to eighteenth-century consumers.[6] George Washington, for one, found his shipments of white salt-glazed stoneware "incomplete with two things broke."[7] Shipment was such a constant concern for Washington that he often ordered additional wares from his factors in England at the outset, anticipating breakage en route. He wrote one of his London factors, noting that, "The China came without any breakage, for which Reason I must counter order the addition to it desir[e]d in my last [letter]."[8] The arrival of an order of ceramics without at least a few broken dishes must have been an unexpected experience.

Although breakage seems to be more the exception than the rule overall, efforts were made by consumers to deter the destruction of their fragile purchases. In 1781, amid the height of battles during the American Revolution, Abigail Adams wrote from Boston to family acquaintance and fellow Bostonian James Lovell, who was residing in Philadelphia. Lovell had been instrumental in several successful naval skirmishes off the coast of Boston, and the outcome of those battles meant the possibility of securing shipment for some goods that Lovell had been holding for the Adamses. Aware of this, and understanding that goods traveling by sea rather than over land had a better chance of surviving intact, she wrote, "There may possibly be some opportunity opened by water for a safe conveyance of the articles about which you have already taken much trouble. If there should I would rather risk the Box of china that way than by land."[9]

Breakage generally occurs either through human error or an act of nature. An example of the latter was documented in 1739, when the *Virginia Gazette* recorded a storm in Philadelphia: "On Sunday Night, about the Hour of Eleven, there was a very severe clap of Thunder . . . Which Struck the House of Andrew Hamilton, Esq. . . . [and] broke some China."[10] Another account was recorded in the Marblehead, Massachusetts, home of the Reverend Barnard, which was struck by lightning that "came down the Side of the Chimney, tearing some of the Shingles of the Roof, and ran into the lower Room . . . it proceeded to a Closet, and broke some China Ware therein. . . . A bowl which stood in the Closet had a Hole through, the [size] of a Knitting-needle, occasioned by the lightning."[11] In 1760 the *South-Carolina Gazette* published an eyewitness account of an earthquake in Brussels where, late one evening in January, the door bells of the city's inhabitants started ringing "of themselves," and the "china and glass ware, by striking together, were dashed to pieces."[12]

Military actions can wreak havoc with ceramics. The *Pennsylvania Gazette* reported in 1745 that off the coast of Boston Harbor a naval engagement resulted in breaking all of one sailor's "Glass and China Ware."[13] And, of course, carelessness frequently resulted in broken ceramics. Satirist Jonathan Swift wrote in his tongue-in-cheek publication *Directions to Servants*, "When you carry a parcel of China plates, if you chance to fall, as it is a frequent misfortune, your excuse must be, that a dog ran across you in the hall, that a chamber-maid accidentally pushed the door against you, that a mop stood across the entry, and tript you up," and the excuses go on.[14] But the important point to note for this thesis is that the occurrence of ceramic breakage was a "frequent misfortune," the truth of which Swift eloquently expands in his spoof.

A cache of broken but nearly complete teawares excavated from the

Garden Terrace outside the tea room at Thomas Jefferson's Monticello may be the result of a cover-up by household servants when an entire serving tray was dropped.[15] Many eighteenth- and nineteenth-century prints depict

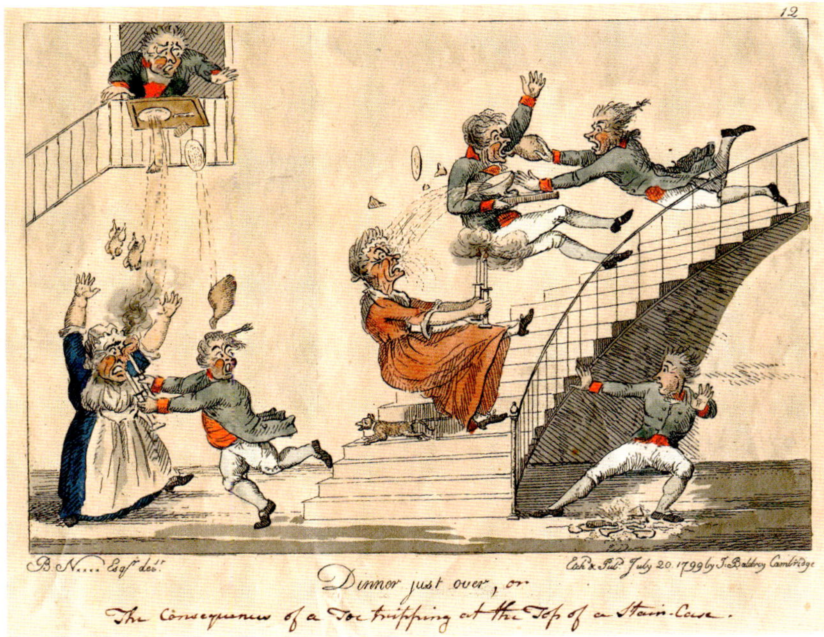

Figure 4 *Dinner just over, or The Consequences of a Toe tripping at the Top of a Stair-Case*, engraved and published by J. Baldrey, Cambridge, England, July 20, 1799. Black-and-white etching with period hand color. (Colonial Williamsburg Foundation, Museum Purchase, 1957-30.)

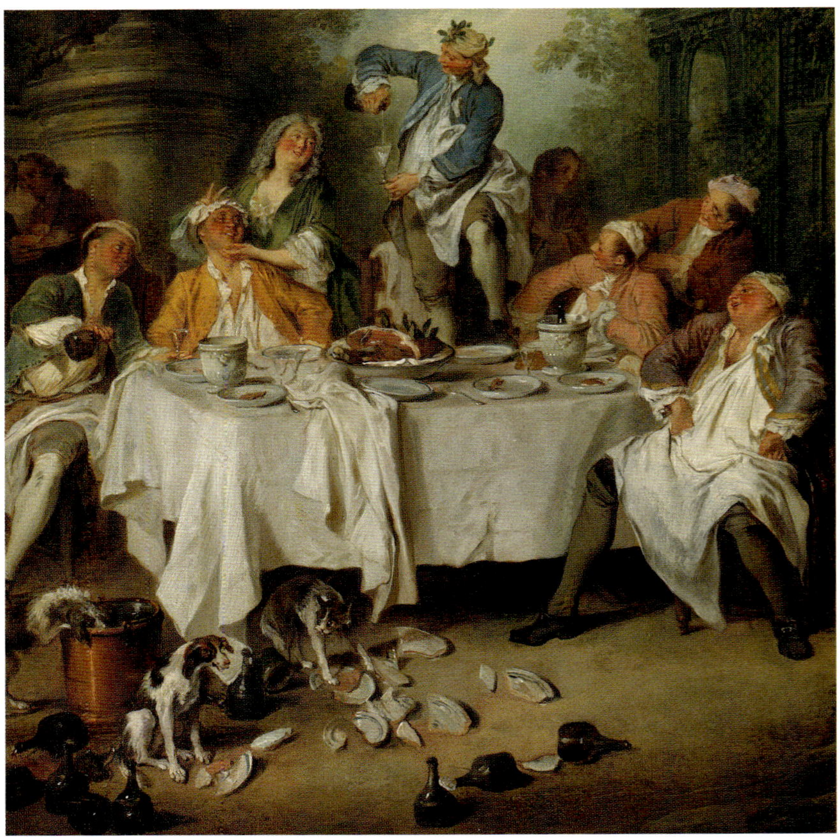

Figure 5 Nicolas Lancret, *Luncheon Party in a Park* (detail), France, ca. 1735. Oil on canvas. 21⁵⁄₁₆" x 18⅛". (Courtesy, Museum of Fine Arts Boston, Bequest of Forsyth Wickes.)

Figure 6 *A Macaroni Dressing Room* (detail), published by Matthew Darly, London, England, 1772. Black-and-white etching and line engraving with period hand color. (Colonial Williamsburg Foundation, Museum Purchase, 1941-13.)

Figure 7 *A Young Hussey Charging Old Toothless with an Imposibility, or the Cracked Pitcher* (detail), published by R. Sayer and J. Bennett. London, England, 1778. Black-and-white etching and line engraving with period hand color. (Colonial Williamsburg Foundation, Museum Purchase, 1990-172.)

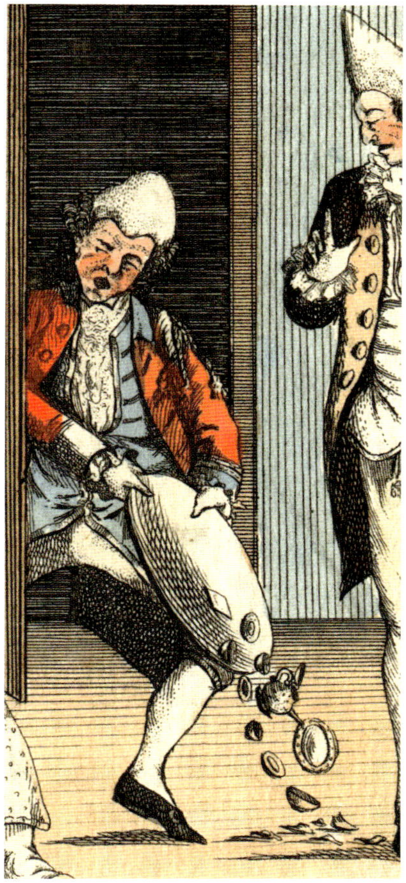

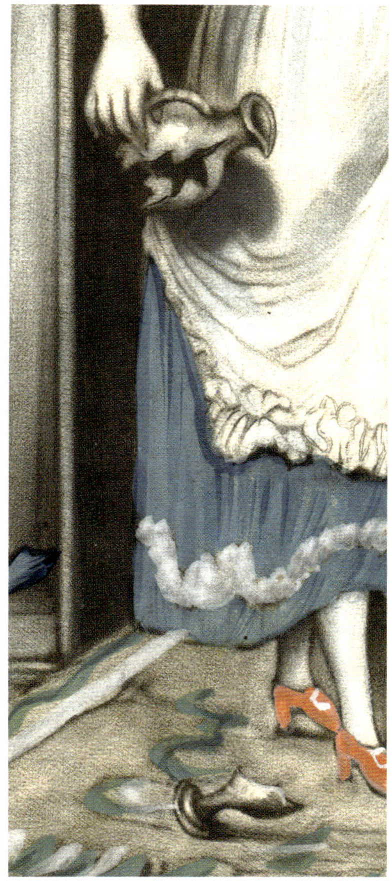

startled footmen who end up with a dishpan full of fragments (fig. 4). Those "below stairs" were not the only ones to break dishes, however. Rowdy behavior at the dinner table and clumsiness at the tea table often resulted in broken dishes, well illustrated in numerous comedic graphics of the period (figs. 5–7).

The purposeful breaking of dishes could convey a moral or political statement. The *Pennsylvania Gazette* reported one memorable incident in March 1742:

> [An eccentric Quaker named Benjamin Lay] bore a publick Testimony against the Vanity of Tea-drinking, by devoting to Destruction in the Market-place, a large Parcel of valuable China, &c. belonging to his deceased Wife. He mounted a Stall on which he had placed the Box of Ware; and when the People were gather'd round him, began to break it peacemeal with a Hammer; but was interrupted by the Populace, who overthrew him and his Box, to the Ground, and scrambling for the Sacrifice, carry'd off as much of it whole as they could get. Several would have purchas'd the China of him before he attempted to destroy it, but he refused to take any Price for it.[16]

The reaction stirred by Lay's extreme behavior is not surprising as Chinese porcelain was a very expensive and a highly desired commodity.

To Repair or Not to Repair

Whether broken by shipment, cataclysmic weather event, clumsy servant, rowdy guest, unsupervised children or animals, or impassioned religious or political zealots, the owner of the ill-fated ceramic object then had to decide whether to repair the dish or leave it broken. Even if it was left broken, however, either because the object was not valued enough to warrant repair or because the owner did not have the means to have it repaired, it could still find new life through repurposing—the eighteenth-century version of "reduce, reuse, recycle." After all, a chipped slipware dish that

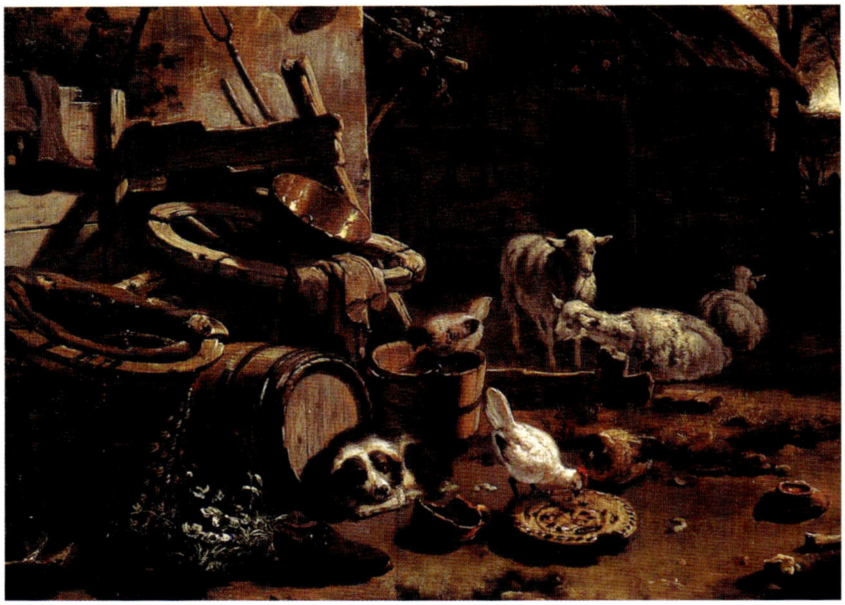

Figure 8 Egbert van der Poel, *Barnyard Scene* (detail), Netherlands, 1658. Oil on canvas. (Private collection; photo © Christie's, Amsterdam.)

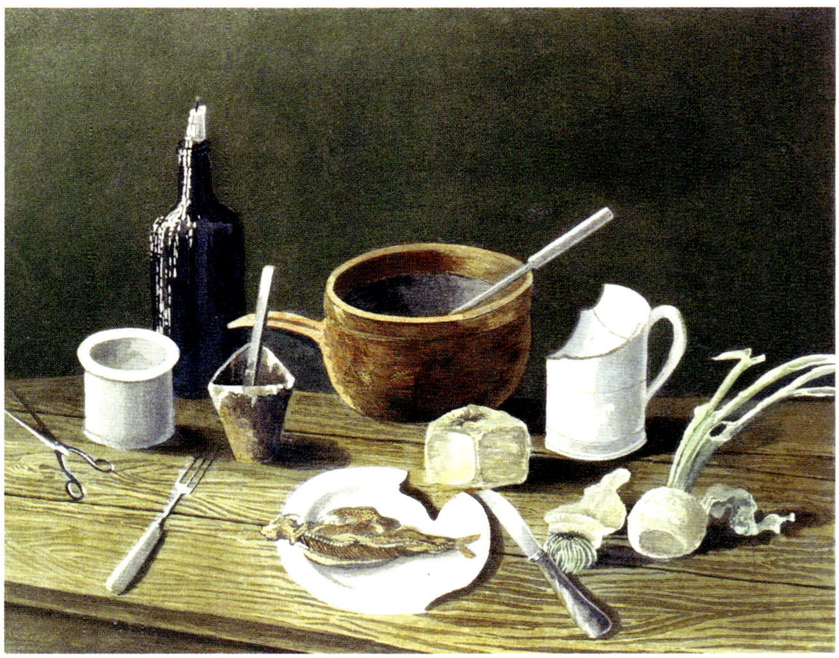

Figure 9 *Penury*, published by Peter del Pasquin, London, England, 1796. Black-and-white etching and line engraving with period hand color. (Colonial Williamsburg Foundation, Museum Purchase, 1964-64.)

once graced the corner cupboard or table could always find its way to the barnyard to hold chicken feed (fig. 8).

In his "Meditation on a quart mugg" Benjamin Franklin wrote that, instead of attempting to repair a shattered mug, its base could be used to serve other functions: "If thy Bottom-Part should chance to survive, it may be preserv'd to hold Bits of Candles, or Blacking for Shoes, or Salve for kibed Heels; but all thy other Members will be for ever buried in some miry Hole; or less carefully disposed of, so that little Children, who have not yet arrived to Acts of Cruelty, may gather them up to furnish out their Baby-Houses. . . ."[17]

If a broken object was not destined for repurposing and the owner could not afford to have it mended (fig. 9), it still did not necessarily follow that it was discarded. The penniless country schoolmaster was the subject of a poem that appeared in the *Virginia Gazette* in 1770. The poem describes the teacher's fireplace mantel where "broken tea cups, wisely kept for show, Rang'd o're the chimney, glist'ned in a row," confirming that broken dishes had value for display if they were no longer serviceable on the table.[18]

The Mender and the Method

If the eighteenth- or early-nineteenth-century owner did decide to repair the broken object in question, several options were available. One was that the owner could personally repair the piece. Numerous English receipt books and housewifery publications included recipes for china glues and East India pastes. Colonial American publications offered similar formulas. *Hutchin's Improved Almanac*, published in New York, issued a recipe for "an Exceeding fine Cement to mend broken China or Glasses." The recipe came with the guarantee to be "very useful to all Families" and able to be mixed and applied "with the greatest Ease."[19] Although one could wonder about the smell, the recipe calls for "Garlick stamped in a stone Mortar; the Juice whereof, when applied to the Pieces to be joined together, is the finest and strongest Cement for that Purpose."[20]

Another almanac offered a recipe for the "Chinese method of mending China" that was likely less pungent than the one offered by Hutchins but more labor intensive to mix: "Boil a piece of white flint-glass in river water 5 or 6 minutes; beat it to fine powder, and grind it well with the white of an egg, and it joins China . . . so that no art can break it again in the same place."[21]

In addition to making one's own concoction from a prescribed receipt, the owner of the broken ceramic object could purchase a glue or paste from merchants, craftsmen, and dry-goods sellers. The *South-Carolina Gazette* for June 18, 1763, advertised just such a product. Benjamin Hawes of Charleston offered for sale and newly imported from London, "a fine cement for joining of china or glass."[22] In similar fashion up the East Coast, Anthony Lamb, a mathematical instrument maker, advertised in the *Connecticut Journal* in 1779 that he had for sale "Putty for mending China."[23] A New York paper posted in the same year an ad for "INDIA GLUE" that possessed "strength superior

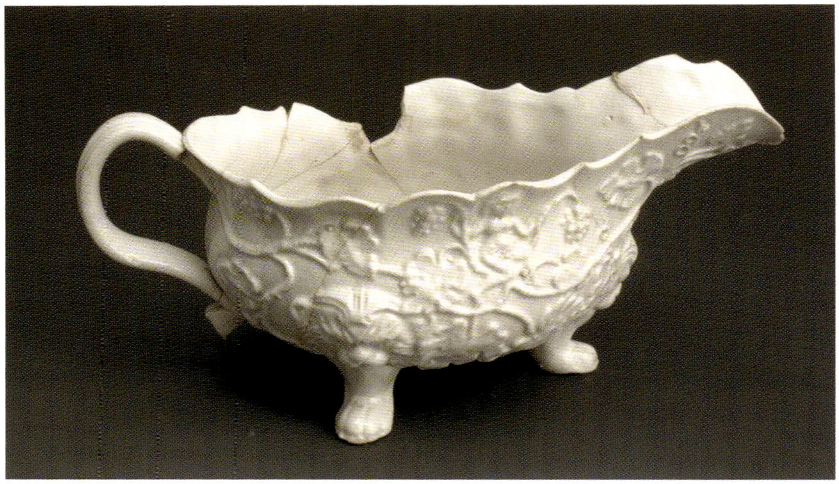

Figure 10 Sauceboat, Staffordshire, England, ca. 1750. Slip-cast, white salt-glazed stoneware. H. 2⅝". (Courtesy, Strawbery Banke, 1985.15; photo, Craig McDougal.) Excavated from the Hart-Shortridge House site in Portsmouth, New Hampshire.

to any thing of the kind for mending ornamented China, even basins for common use."[24]

Archaeologically recovered ceramics fragments sometimes bear evidence of the use of china glues and pastes. At least one fragment unearthed at the Anderson site in Williamsburg, Virginia, confirms a glue repair to a saucer or small dish.[25] Recent excavations at George Washington's Ferry Farm have found numerous damaged creamware tea- and punch wares associated with Mary Washington's tenure that were mended with glue repairs.[26]

The white salt-glazed stoneware sauceboat illustrated in figure 10 was mended using another eighteenth-century type of glue repair, this one recovered at a house site in Portsmouth, New Hampshire. The substance used to fix the break in the lip was made with what tested to be white lead house paint.[27] This appears to have been a common method for repairing ceramics in the eighteenth century. Probate inventories show other ceramics glued using the same technique. The inventory taken at the death of Richard Mitchell of Lancaster County, Virginia, for example, lists in the Dairy "2 Earthen [dishes] Stuck with White Lead."[28]

Newspapers and trade cards reveal that the owner of a broken dish could easily find a trained craftsman to assist in the repair. As early as 1725 American newspapers endorsed the skills of craftsmen like Robert Phillips, who set up shop in Boston, where he "cement[ed] China, Glass, or Delf."[29] He may have picked up his ceramic cementing techniques from potters in Liverpool, the city from which he proudly hailed.[30] England had a thriving business of China menders in the eighteenth century. Trade cards for people such as Edmund Morris of London advertised mending "all sorts of China Wares with a Peculiar Art . . . so as a Rivetted Piece of China will do as much Service as when New."[31] Morris also touted his methods for mending china in newspapers and boasted that he mended pieces for nobility in addition to selling repaired pieces at low cost. China burner and mender Edward Coombs of Bristol, England, not only promoted his skills on trade cards and in newspapers but also signed some of the pieces he repaired (figs. 11, 12).

On both sides of the Atlantic the types of repairs fell within three broad groups: the object was stuck back together through the use of some paste or glue, as evidenced by the white salt-glazed stoneware sauceboat and the multiple recipes for pastes previously mentioned; it was riveted together with various materials; or it was fastened together with some kind of binding. Additionally, many repairers advertised the ability to restore an object's usefulness when a piece of it was missing, for example a mug without a handle, a teapot without a spout, or a bowl without a piece of its rim. Almost every mender described his methods as unique or new and improved. But their techniques were often rooted in much older traditions derived from early cultures.

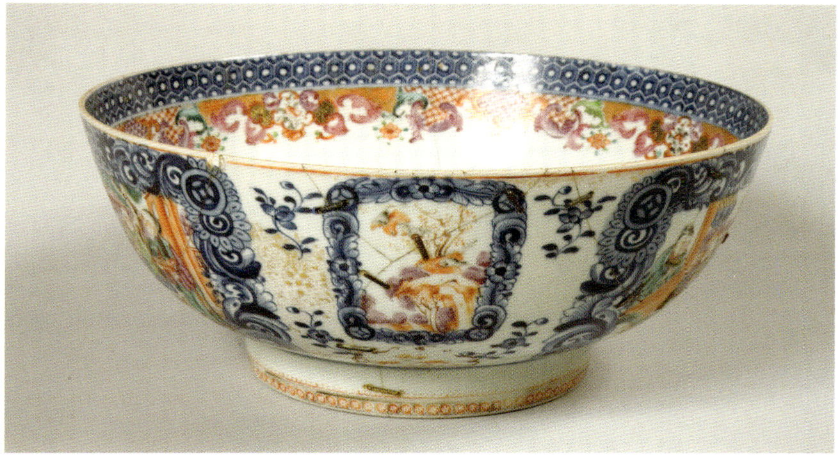

Figure 11 Punch bowl, Jingdezhen, China, ca. 1780. Hard-paste porcelain. D. 11¼". (© Bristol Museums, Galleries & Archives.)

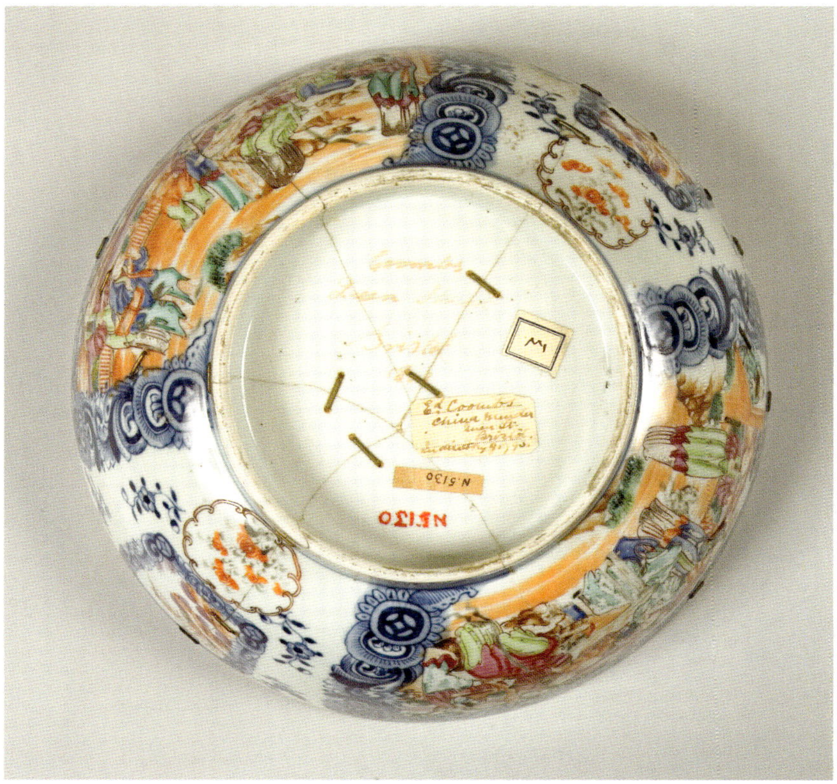

Figure 12 Base of the punch bowl illustrated in fig. 11. Inscribed by the china mender in low-fired iron gall ink under the foot: "Coombes / Queen Street / Bristol / 1802"

Figure 13 *A Porcelain Mender* (and detail), possibly by Puqua, Guangzhou, China, ca. 1790. Watercolor. 14¼" x 18⅛". (© Victoria and Albert Museum.)

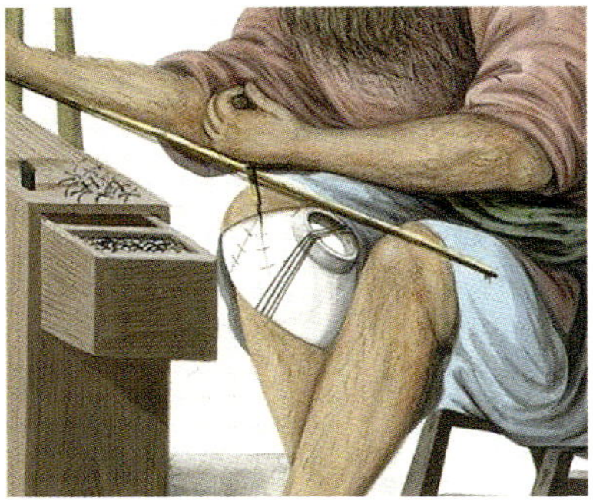

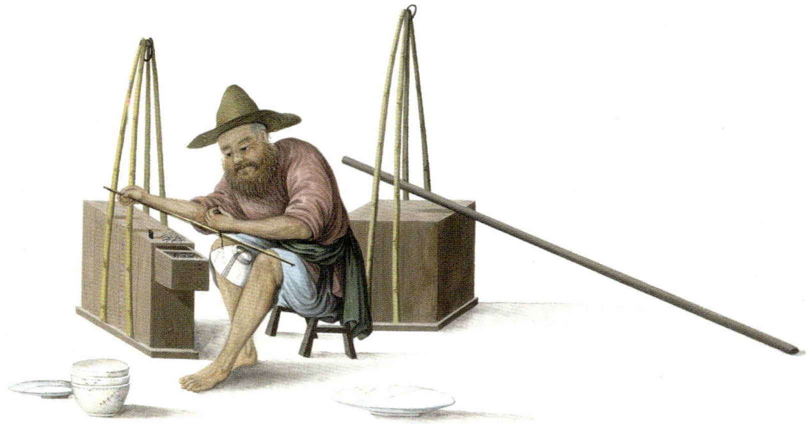

The Chinese wrote down recipes for porcelain glues at least as early as the seventeenth century and the recipes at that time were purported to be ancient.[32] The Chinese had also practiced the art of riveting for centuries. Riveting required great hand skill. A piece was typically tied together, as seen in figure 13, then a drill was used to pierce holes on either side of the crack. The holes had to be offset approximately fifteen to twenty degrees so that the pressure placed on the object from the rivet would actually hold the broken parts together and not cause further stress on the piece or the repair.[33] A bow drill (fig. 14) was often the tool of choice for creating such holes. The little dish illustrated in figure 14 was used to warm the metal rivets in order to make them flexible and able to be bent and inserted into the holes. The heated metal rivets or staples would then contract as they cooled, gripping the two pieces of the object together as one.

Repair techniques similar to those used in China were implemented throughout the colonies and England. James Walker from London advertised in *Parker's New York Gazette and Post Boy* that he repaired pieces "in the same Manner as done in the East-Indies." But he also added his own signature to certain repairs, stating that he performed "a new and neat Method of riming and Sewing China."[34] Exactly what this sewing entailed

Figure 14 Bow drill with accessories, China, nineteenth century. Wood, iron, bronze, steel. (Private collection; photo, courtesy private collection.)

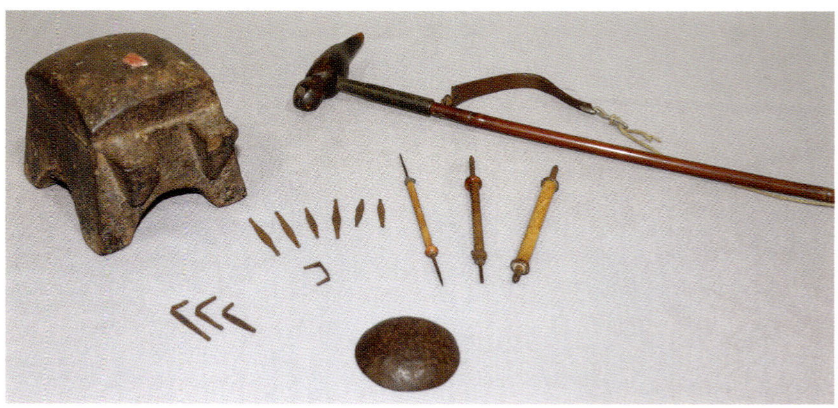

is a mystery, but the Bristol tin-glazed earthenware plate illustrated in figure 15 is an example of the so-called sewing china technique, which is perhaps the service Walker offered.[35] The dish is bound together with a cotton thread or twine woven between alternating holes that have been drilled through the object in the same manner that would have been carried out prior to the insertion of metal rivets. In the place of the usual metal staples, however, cotton twine was laced between the holes and knotted on the

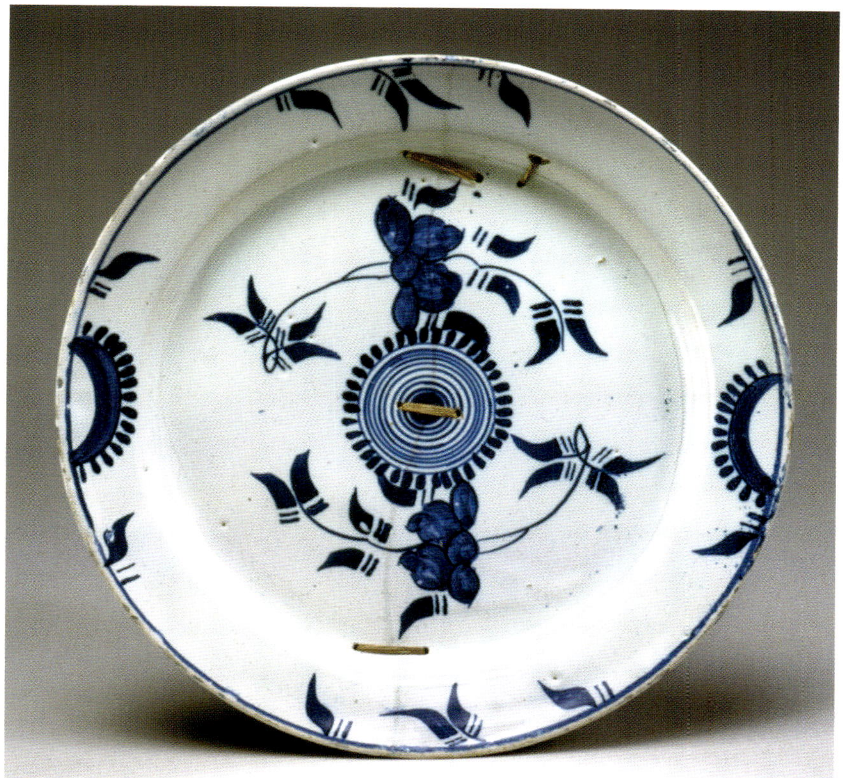

Figure 15 Plate, Bristol, England. Tin-glazed earthenware. D. 8⅝". (Courtesy, Historic Deerfield, Inc.)

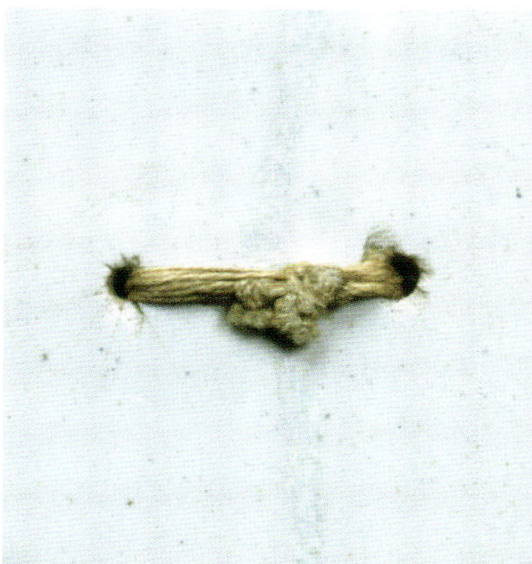

Figure 16 Detail of the thread repair on the back of the plate illustrated in fig. 15.

reverse of the dish with a strong double overhand knot (fig. 16).[36] James Walker's sewing technique might also have referred to sewing with wire. Fragments of Chinese porcelain found in an eighteenth-century archaeological context in Philadelphia exhibit threaded wire rivets made by lacing thin-gauge wire multiple times through alternating holes until the desired stability was reached (fig. 17).[37] This was an extremely common practice in the eighteenth century, although it does not always survive belowground. The repairs made using this method usually were further strengthened by filling the holes with glue (figs. 18–19).

There were many different methods for repairing pieces, and various china menders on both sides of the ocean advertised their skills and assured their work. But in addition to promising that their repairs would hold, many china menders guaranteed their methods to be reasonably priced. The cost of repairs varied depending on the type of repair and materials used. Nathaniel Lane, a china and glass mender from New York who had a thriving business in the 1760s, advertised the price of riveting according to the type of metal used and how it was supplied (fig. 20). If he provided the silver for a rivet, it cost two shillings; if the customer provided the silver, it cost one shilling. If the rivet was made of brass, it also cost one shilling. And if it was made of white metal—a period term for pewter—the rivet cost six pence.[38]

George Beattie, a Londoner who set up shop in Boston, promised his Massachusetts clients that he, too, would mend their broken china as reasonably as in London.[39] The prices of colonial China menders were comparable to the cost of services offered to London residents. London china

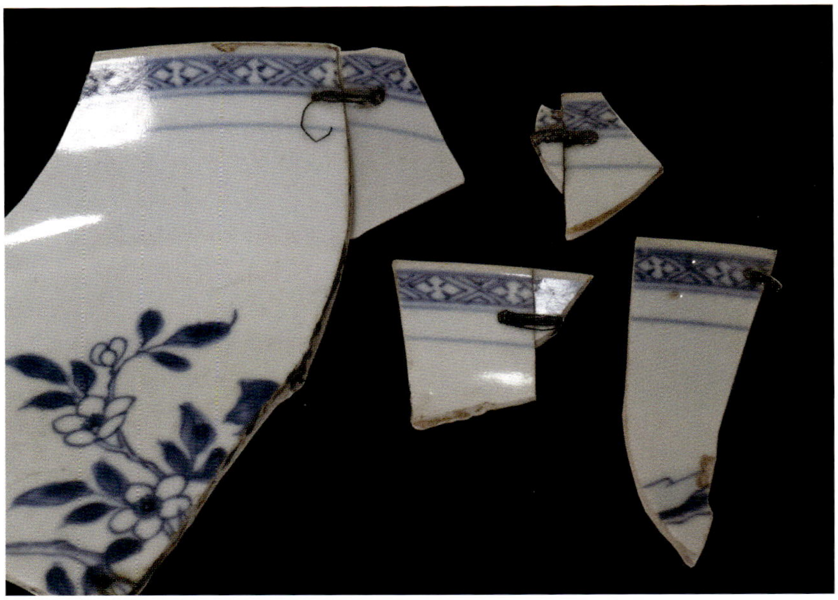

Figure 17 Threaded wire rivets on punch bowl fragments, Jingdezhen, China, ca. 1750. (Courtesy, Independence National Historic Park; photo, Robert Hunter.) Excavated in Philadelphia, Pennsylvania.

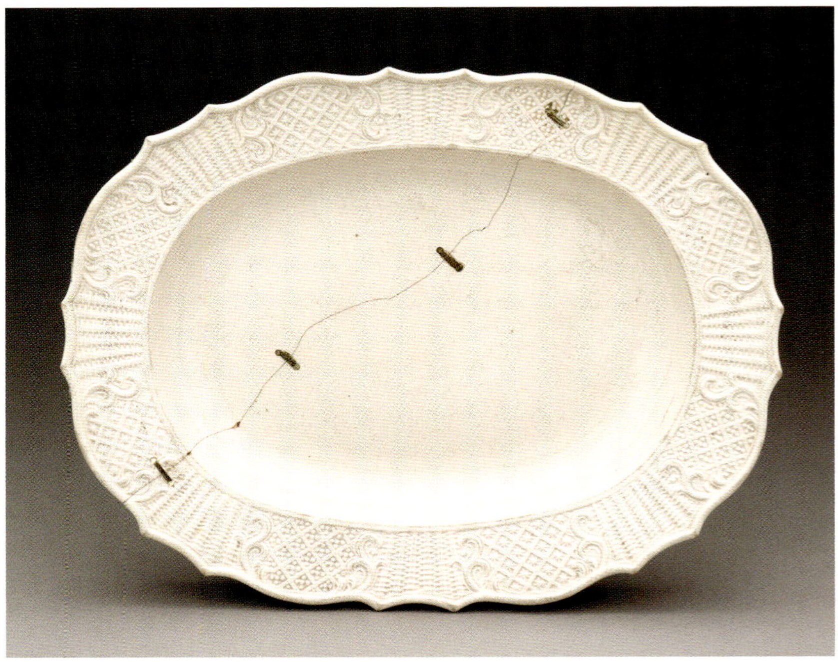

Figure 18 Dish, Staffordshire, England, ca. 1760. White salt-glazed stoneware. L. 11¼". (Courtesy, Colonial Williamsburg Foundation, Museum Purchase, 1976-115.)

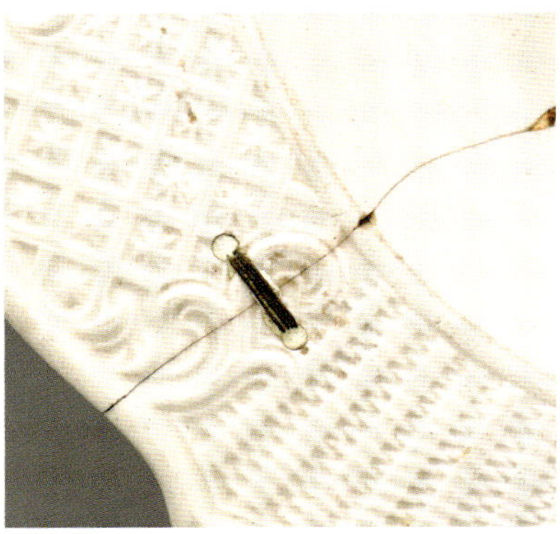

Figure 19 Detail of the dish illustrated in fig. 18. Note the threaded wire rivet with holes filled with glue.

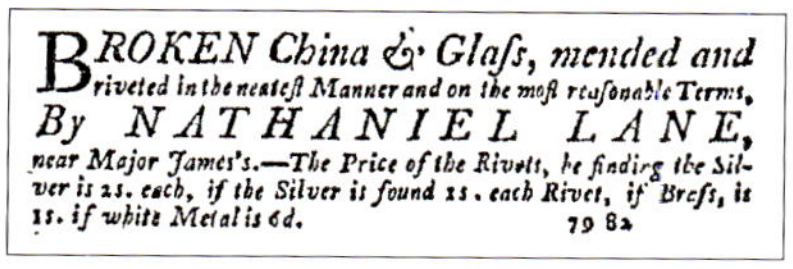
BROKEN China & Glaſs, mended and riveted in the neateſt Manner and on the moſt reaſonable Terms, By NATHANIEL LANE, near Major James's.—The Price of the Rivets, he finding the Silver is 2s. each, if the Silver is found 1s. each Rivet, if Braſs, is 1s. if white Metal is 6d. 79 82

Figure 20 Advertisement, *New York Journal*, July 23, 1767. (Courtesy, American Antiquarian Society.) China and glass mender Nathaniel Lane itemized his prices for rivet repairs.

riveter James Harris left a rare bill of sale on the reverse of his trade card.[40] For a repair made in 1770 he charged one shilling for two rivets. Unfortunately it is unclear whether the cost was for silver or brass, since he offered repairs in both.

For some individuals the business of mending ceramics was their sole occupation, or at least their primary advertised occupation. For others the business of china repair was a logical supplementary service offered to their customers. James Rutherford in Charleston, South Carolina, for example, advertised himself as a "regular-bred gold and silver smith . . . who . . . likewise works in jewelry, and clasps broken china in the neatest manner."[41] James Proby, also from Charleston, ran a scrap metal and tinsmith business in the 1760s and advertised "mending China, at the lowest Rates, and quickest Dispatch."[42]

Silversmiths in Philadelphia and Boston mended dishes as well. Surviving daybooks for Philadelphia silversmith Thomas Shields and Boston silversmith Paul Revere record numerous instances of repairing china dishes. With the exception of rimming a silver punch bowl, most of Revere's ceramic mends seemed to incorporate the use of silver rivets.[43] In addition to riveting numerous dishes, Shields occasionally mended porcelain coffeepot and teapot handles and covers. Sometimes he added a chain from the cover to the spout, perhaps in an attempt to prevent the cover from dropping off the object and incurring further damage (fig. 21).[44]

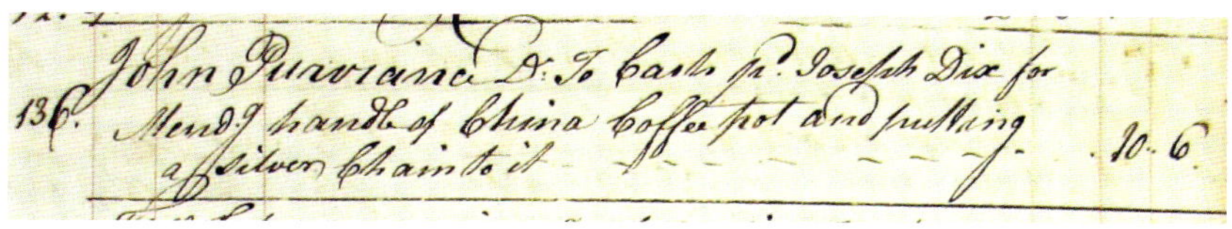
136. John Purviance Dr. To Cash pd. Joseph Dix for Mend'g handle of China Coffee pot and putting a Silver Chain to it 10.6

Figure 21 Entry from daybook kept by Philadelphia silversmith Thomas Shields, November 1, 1775. (Courtesy, The Joseph Downs Collection of Manuscripts and Printed Ephemera, Henry Francis du Pont Winterthur Museum.)

Tinsmiths also repaired ceramic objects. Virginia tinsmith John Beale, for example, advertised that, in addition to making and mending copper and tin objects, he repaired broken dishes.[45] Owned by Christina Spach when she was a child, the miniature teapot illustrated in figure 22 retains a history of having been repaired by tinsmith Philip Reich in Salem, North Carolina.[46] The teapot's missing ceramic handle was replaced with a handle composed of four pieces of tin. One piece wraps the foot, another encircles the rim, a third loops around the spout and connects to the piece at the foot, and a fourth crimped strip makes the handle, the crimping encasing and hiding wire for added stability.

The ceramic repair business was not the exclusive domain of men, however. Mrs. Barbary Opperman had a china repair business out of a Philadelphia tailor's shop.[47] Elizabeth Flagg, a widow in Boston, advertised in 1795 that she and her daughters "rivet[ed] and mend[ed] china and glass, and Needlework of all kinds, at their house. . . ."[48] Approximately ten months later Flagg placed another advertisement in the paper that revealed her progress and an unexpected problem: not all of her clients were collecting their mended wares. She stated that she had "a number of Pieces of repair'd

China in her Possession," which she requested be picked up by their owners.[49] Maria Warwell was another female china mender. She set up what she initially thought would be a temporary business in Charleston. She and her husband were intending to leave South Carolina, so while awaiting passage she took up work "mending in the neatest and most durable manner, all sorts of useful and ornamental china, viz. beakers, tureens, jars, vases, bust's." She did not use rivets but employed the use of a type of china putty so that when a piece of an object was missing she could "substitute a composition in its room and copy the pattern as neigh as possible."[50]

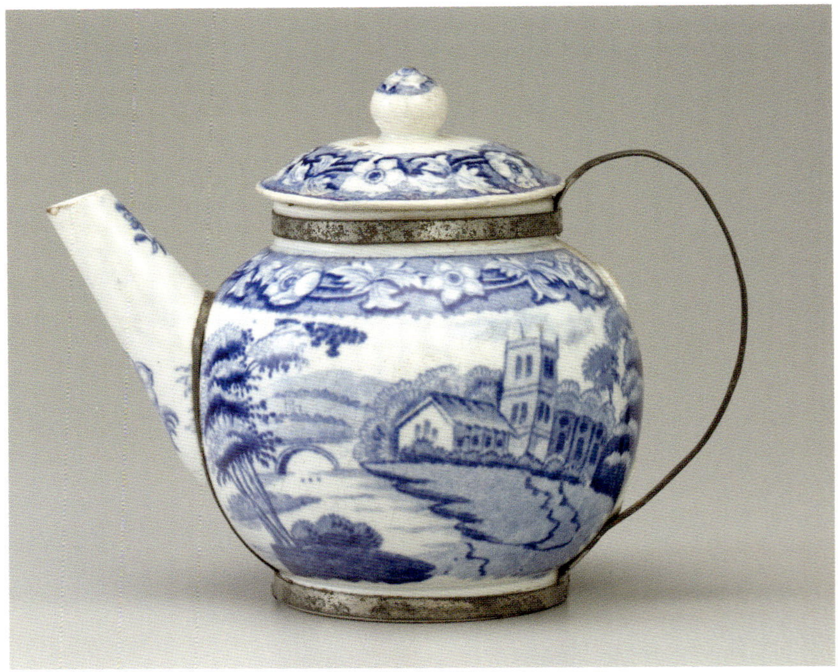

Figure 22 Teapot, England, ca. 1810. Transfer-printed pearlware with tin replacement handle. H. 4". (Courtesy, Collection of Old Salem Museums & Gardens, Gift of Emil Shaffner, 837.1.)

The Fate of Mended Ceramics

Throughout history the fate of the mended ceramic object has frequently been decided based on the object's usefulness, beauty, rarity, or a combination of all three. The way in which the owner or society qualified those characteristics defined the object's monetary value and intrinsic worth. Despite their condition, mended ceramics could still be used to convey status. For example, one husband bemoaned:

> I am never allowed to eat from anything better than a Delft plate, that the economy of the beaufait, which is embellished with a variety of China, may not be disarranged and indeed my wife prides herself particularly on her ingenious contrivance in the article, having ranged among the rest some old China not fit for use, but disposed in such a manner, as to conceal the streaks of white paint that cement the broken pieces together.[51]

Repaired and broken ceramics were often kept in the china closet or bowfat. The highly regarded Chinese porcelain carefully arranged in the bowfat enabled the inhabitants of the house to display their wealth, even if some of the objects were no longer functional. When intact dishes were removed

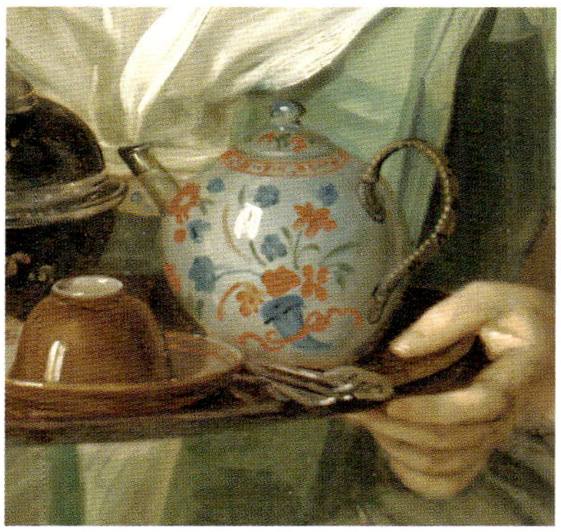

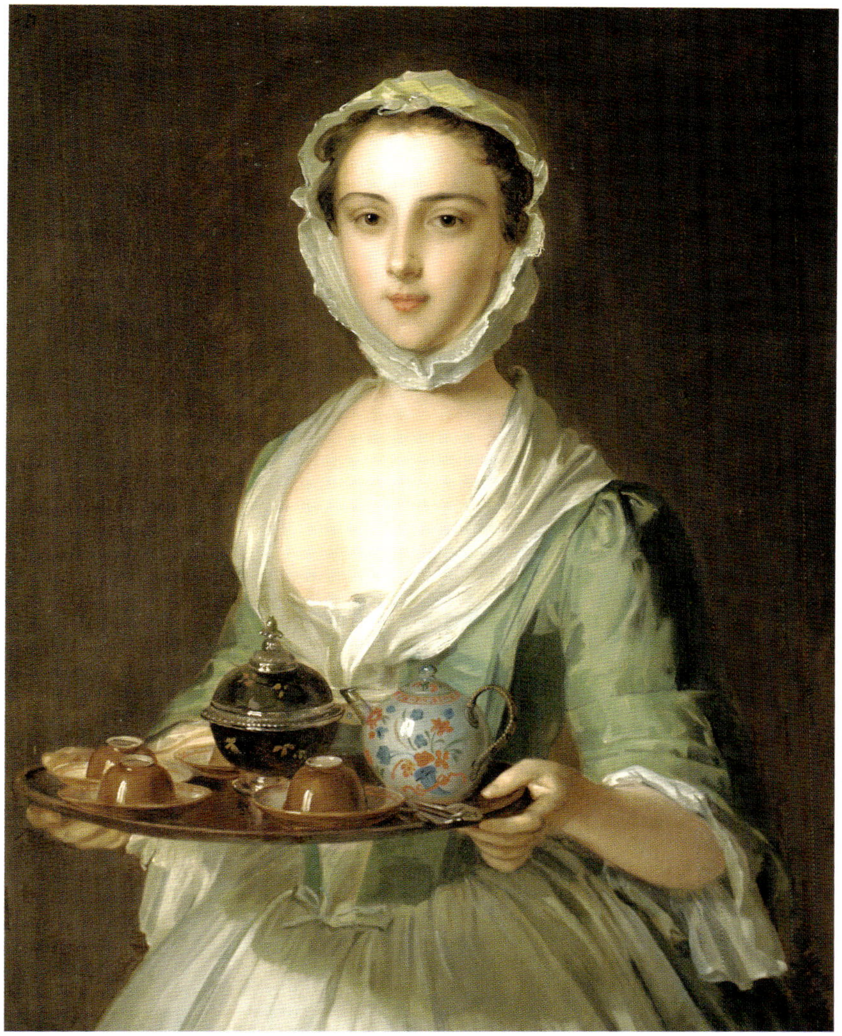

Figure 23 Philippe Mercier (1689–1760), *Portrait of a Young Woman Holding a Tea Tray*, France, eighteenth century. Oil on canvas. 35⅛" x 28". (Private collection; photo, © Christie's Images / Bridgeman Art Library.) Note the teapot in the detail with a silver-tipped spout and wicker-wrapped replacement handle.

for dining, the mended ones could remain, proudly suggesting to visitors that there was more wealth in the cupboard. This mind-set is further substantiated in the room-by-room inventories of prominent Virginians. The dining room of a Richmond County gentleman recorded, "in the Buffet [bowfat] / ½ Dozen Coffee Cups some crack'd and 2 of them without handles."[52] These broken ceramics were scattered amid the fine porcelain, glass, and silver listed in the same space. The probate inventory taken at the death of Thomas Nelson of Yorktown, Virginia, also listed broken ceramics in the dining room, including several punch bowls.[53]

The practice of spring cleaning was the subject of an account related in a Pennsylvania newspaper in 1785. Everything had to be removed from the house so that it could be scrubbed and whitewashed from the garret to the cellar. Off the kitchen "a closet has disgorged its bowels" and revealed the contents: "riveted plates and dishes, halves of China bowls, cracked tumblers, broken wine glasses . . . handfuls of old corks, tops of teapots, and stoppers from departed decanters."[54] Whether for storage or display, the broken and mended objects were crammed into the closet.

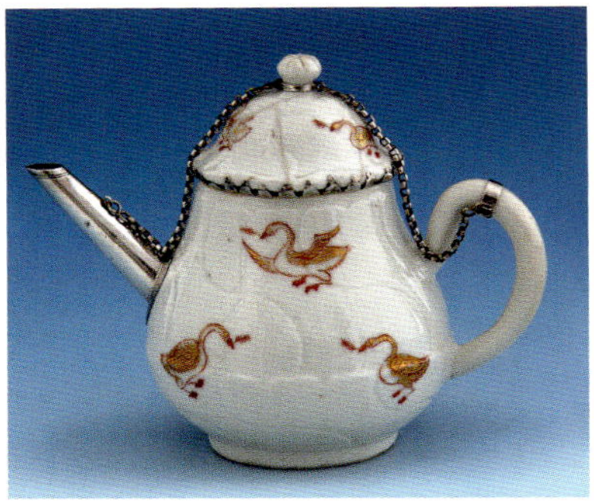

Figure 24 Teapot, Jingdezhen, China, ca. 1735. Hard-paste porcelain, silver replacement spout, chain, and mount for lid. H. 5⅛". (Colonial Williamsburg Foundation, Gift of Murray Nimmo, 2013-143.)

Figure 25 Saucer, Jingdezhen, China, ca. 1780. Hard-paste porcelain. D. 5¾". (Colonial Williamsburg Foundation, Gift of Catha Grace Rambusch, 2013-22.) An arc-shaped chip on the rim was ground down and fitted with a piece from a different Chinese porcelain dish decorated *en grisaille*. The replaced piece is attached to the dish with three metal rivets—only the holes are visible on the face of the dish.

Repairs to objects like teapots could enable them to return to service on the tea table, and repairs made of precious metals served the dual purpose of restoring the objects' usefulness while also enhancing their value and beauty. Some china menders offered to replace broken or faulty teapot spouts with silver replacements, often guaranteeing that the new spout would perform better than the original ceramic one, that is, not drip (fig. 23).[55] The fifty-plus-page probate for wealthy Virginian Robert "King" Carter of Corotoman, which was taken in 1733, listed among numerous objects "1 ditto [Chaney or china] Teapott with a Silver Spout."[56] The teapot, the whereabouts of which are unknown, could have been similar to the one illustrated in figure 24.

Conclusion

The tradition of mending ceramics is a long and interwoven story of value and worth, both sentimental and monetary. For example, the Japanese art of *kintsugi*—repairing ceramics with gold lacquer fill—has been practiced for centuries. This method of repair raises the humble teabowl from a utilitarian object to a revered work of art.

In 1857 Philadelphia hosted the Convention of Earthenware Dealers, where a gentleman from New York praised those in attendance and lauded the array of utilitarian and ornamental objects represented at the forum. He concluded his oration with the hopeful exclamation, "May we never break!"[57] His remarks referenced not only the camaraderie of the assembled audience but also the fragile stuff of their trade.

Indeed, today we often admire ceramic objects with old repairs. In recent years there has been a surge of interest among conservators and restorers to better understand repairs, recognizing their part in an object's history.[58] Mends and repairs can be beautiful, enhancing the otherwise ubiquitous cup or plate; they can also border on the bizarre, elevating it to the level of an "object of curiosity" (fig. 25). When such objects are viewed today they should remind us of the context in which those repairs were made for individuals who had particular reasons for retaining their fragile, imperfect wares.

ACKNOWLEDGMENTS I am grateful to many people for their assistance, support, and/or encouragement. Particular thanks go to Johanna M. Brown, Suzanne Findlen Hood, Robert Hunter, Amanda E. Lange, Nick Luccketti, Debbie Miller, Merry Outlaw, William R. Sargent, Janine E. Skerry, and my parents. "May we never break!" Also my appreciation to Andrew Baseman, for having attended my first lecture on mended ceramics. His blog *Past Imperfect* is an invaluable resource.

1. *Boston News-Letter*, August 12, 1773, p. 2. "Pencil China" probably refers to porcelain decorated *en grisaille*.

2. Nigel Williams, *Porcelain Repair and Restoration: A Handbook* (1983; new ed., London: British Museum, 2002), pp. 12–13.

3. Regina Krahl, *Chinese Ceramics in the Topkapı Saray Museum, Istanbul: A Complete Catalogue*, 3 vols. (London: Sotheby's Publications, in association with the Directorate of the

Topkapı Saray Museum, 1986), 1:297, no. 226a; 2:513, no. 603; 2:540, no. 652; 2:584, no. 790; 2:591, no. 817; and 2:592, no. 819.

4. Marcus Milwright, "Prologues and Epilogues in Islamic Ceramics: Clays, Repairs, and Secondary Use," *Medieval Pottery* 25 (2001): 72–83.

5. While many articles have discussed mended ceramics to some extent, the author is aware of only three published prior to this volume of *Ceramics in America* that specifically focus on the topic in American contexts. Although somewhat dated, these works are: Stanley A. South, "Archaeological Evidence of Pottery Repairing," in *The Conference on Historic Site Archaeology Papers 1967*, 2 vols. (Raleigh, N.C.: Conference on Historic Site Archaeology, 1968), 2:62–71; Dwight P. Lanmon, "Putting the Pieces Together: Ceramics and Glass Repairing in the Eighteenth and Nineteenth Centuries," *Delaware Antiques Show* (1969): 95–102; Bradford L. Rauschenberg, "Ceramic Menders and Decorators in Charleston, South Carolina, before 1820," *Journal of Early Southern Decorative Arts* 17, no. 2 (November 1991): 115–27.

6. For more on packing and shipping ceramics and the sometimes unsuccessful journeys of the wares, see Janine E. Skerry, "'A Tierce of Stone and Earthen Ware': Packing Ceramics for Shipment to America," in *This Blessed Plot, This Earth: English Pottery Studies in Honour of Jonathan Horne*, edited by Amanda Dunsmore (London: Paul Holberton Pub., 2011), pp. 172–80.

7. "From George Washington to Thomas Knox, January 1758," *Founders Online*, National Archives, last modified July 12, 2016, http://founders.archives.gov/documents/Washington/02-05-02-0060. [Original source: *The Papers of George Washington*, Colonial Series, vol. 5, *5 October 1757–3 September 1758*, ed. W. W. Abbot. (Charlottesville: University Press of Virginia, 1988), pp. 87–88.]

8. "From George Washington to Richard Washington, 18 March 1758," *Founders Online*, National Archives, last modified July 12, 2016, http://founders.archives.gov/documents/Washington/02-05-02-0078. [Original source: *The Papers of George Washington*, Colonial Series, vol. 5, *5 October 1757–3 September 1758*, ed. W. W. Abbot. (Charlottesville: University Press of Virginia, 1988), pp. 105–6.]

9. "Abigail Adams to James Lovell, 15 November 1781," *Founders Online*, National Archives, last modified July 12, 2016, http://founders.archives.gov/documents/Adams/04-04-02-0164. [Original source: *The Adams Papers*, Adams Family Correspondence, vol. 4, *October 1780–September 1782*, ed. L. H. Butterfield and Marc Friedlaender (Cambridge, Mass.: Harvard University Press, 1973), pp. 244–45.]

10. *Virginia Gazette*, June 22, 1739, p. 2.

11. *Pennsylvania Gazette*, August 25, 1763, p. 1.

12. *South-Carolina Gazette*, May 10, 1760, p. 1.

13. *Pennsylvania Gazette*, July 4, 1745, p. 2. It is amusing to note that the same article reported the deaths of seven men, but the glass and china were listed first.

14. Jonathan Swift, *Directions to Servants in General* (London: Printed for R. Dodsley and M. Cooper, 1745), pp. 61–62.

15. John Metz, *Archaeological Investigation of the Garden Terrace, Kitchen Dependency and Corner Terraces*, Monticello Archaeology Report, November 2000; and Suzanne Findlen Hood, "China of the Most Fashionable Sort: Chinese Export Porcelain in Colonial America," lecture, New York Ceramics Fair, January 2013.

16. *Pennsylvania Gazette*, March 25, 1742, p. 3. Some may recall similar activity during World War II, when some Americans broke their Japanese porcelain in the streets as a statement against the bombing of Pearl Harbor. See Otto Friedrich, *City of Nets: A Portrait of Hollywood in the 1940s* (Berkeley: University of California Press, 1986), p. 116.

17. Benjamin Franklin, "Meditation on a quart mugg," *Pennsylvania Gazette*, July 19, 1733.

18. *Virginia Gazette*, October 25, 1770, p. 4.

19. John Nathan Hutchins, *Hutchin's Improved: Being an Almanack and Ephemeris of the . . . Year of Our Lord, 1769* (New York: Printed and sold by Hugh Gaine, at the Bible and Crown, in Hanover Square, 1769), p. 9.

20. Ibid. For further discussion of period glues, see Mara Kaktins, Melanie Marquis, Ruth Ann Armitage, and Daniel Fraser, "Mary Washington's Mended Ceramics: A Study of Eighteenth-Century Glues," in this volume.

21. Benjamin West and Nathan Daboll, *The North-American Calendar: or, The Rhode-Island Almanack, for the Year of Our Lord Christ 1788* (Providence, R.I.: Printed by Bennett Wheeler, 1787), p. 17.

22. *South-Carolina Gazette*, June 18, 1763, p. 1.

23. *Connecticut Journal*, June 16, 1779, p. 1.

24. *Royal Gazette* (New York), October 16, 1779, p. 3.

25. Fragment excavated from the Anderson site, 1790–1810 context.

26. For more on this, see Kaktins et al., "Mary Washington's Mended Ceramics: A Study of Eighteenth-Century Glues," in this volume.

27. Janine E. Skerry and Suzanne Findlen Hood, *Salt-glazed Stoneware in Early America* (Williamsburg, Va.: Colonial Williamsburg, 2009), pp. 145–46.

28. Inventory for Richard Mitchell, December 20, 1781, Lancaster County Deeds and Wills, number 20, pp. 213a–219.

29. *New England Courant* (Boston), May 17–24, 1725, p. 3.

30. Ibid.

31. Edmund Morris Trade Card, 1740–1760, © Trustees of the British Museum.

32. Williams, *Porcelain Repair and Restoration*, pp. 13–14.

33. Ibid., pp. 16–19.

34. *Parker's New York Gazette and Post Boy*, November 20, 1760, p. 3.

35. Amanda Lange, *Delftware at Historic Deerfield, 1600–1800* (Deerfield, Mass.: Historic Deerfield, 2001), p. 74.

36. Ibid., and author's conversation with Colonial Williamsburg Foundation textile curator Linda Baumgarten, January 19, 2013, who identified the type of knot used.

37. Email correspondence between the author and Debbie Miller, January 20, 2013.

38. *New York Journal*, July 23, 1767.

39. *Boston Post-Boy*, August 18, 1766, p. 3.

40. James Harris Trade Card, ca. 1770, © Trustees of the British Museum.

41. *South-Carolina Gazette*, March 16, 1752, p. 4.

42. *South-Carolina Gazette*, July 1, 1766, p. 2.

43. Paul Revere Waste and Memorandum Books, entry for Captain William Torrey, "Riming a China Punch Bowl w/ Silver," September 12, 1762, Massachusetts Historical Society, Boston, Massachusetts.

44. Thomas Shields Daybook, 1775–1791, The Joseph Downs Collection of Manuscripts and Printed Ephemera, Henry Francis du Pont Winterthur Museum, Winterthur, Delaware.

45. Thomas Shields Daybook, 1775–1791, The Joseph Downs Collection of Manuscripts and Printed Ephemera, Henry Francis du Pont Winterthur Museum, Winterthur, Delaware.

46. According to the object file at Old Salem Museums & Gardens (acc. no. 837.1), Christina Spach married Salem, North Carolina, silversmith John Vogler in 1819. Since John Vogler was a skilled metalsmith, it is also possible that he crafted the replacement handle.

47. *Pennsylvania Gazette*, July 17, 1766, p. 5.

48. *Columbian Centinel*, Boston, April 15, 1795, p. 3.

49. *Boston Gazette and Weekly Republican Journal*, February 1, 1796, p. 3.

50. *South-Carolina Gazette and Country Journal*, July 28, 1767, p. 4.

51. *The Connoisseur*, vol. 3 (London, 1757).

52. Inventory for Daniel Hornby, April 11, 1750, Richmond County Wills No. 5, pp. 670–73.

53. Inventory for General Thomas Nelson, June 2, 1789, York County Wills and Inventories No. 23, pp. 181–83.

54. *Pennsylvania Packet and Daily Advertiser*, June 18, 1785, p. 2.

55. Edward Coombs China Burner and Mender Trade Card, Bristol Museums, Galleries & Archives. On that trade card, Edward Coombs advertises, "Tea Pots that pours indifferently, may be made to pour smoothly by tipping them neatly with Silver, otherwise by taking off their own Spouts and substituting new ones of Silver or any other Metal."

56. Transcription of the inventory of Robert Carter's Estate, 1733, available online at http://carter.lib.virginia.edu/html/C33inven.mod.html.

57. *The Sun* (Baltimore, Md.), June 27, 1857, p. 1. For the entire speech, see Robert Hunter and Oliver Mueller-Heubach, "The Captain George W. Russell Presentation Pitchers," in this volume.

58. Stephen Koob, "Obsolete Fill Materials Found on Ceramics," *Journal of the American Institute for Conservation* 37, no. 1 (1998): 49–67; and Isabelle Garachon, "From Mender to Restorer: Some Aspects of the History of Ceramic Repair," in *Glass & Ceramics Conservation* (International Council of Museums, 2010), pp. 22–31.

Mara Kaktins, Melanie Marquis, Ruth Ann Armitage, and Daniel Fraser

Mary Washington's Mended Ceramics: A Study of Eighteenth-Century Glues

▼ ARCHAEOLOGISTS interpret human lives using broken, cast-off, lost, or abandoned items. In today's world of mass production and disposable material culture, we sometimes forget the value objects had to people in past generations. It is generally assumed that broken items were discarded immediately. Those who study ceramics, however, know better. Even with this knowledge, archaeologists at George Washington's boyhood home, Ferry Farm, were surprised when a substance resembling glue was noted on fragments of an archaeologically recovered creamware punch bowl and other vessels from a context definitively associated with Mary Washington and her family.

To date, fragments of seven distinct vessels from the Washingtons' time at Ferry Farm have been found that show repair using adhesives. This reveals an attitude toward repairing damaged objects, a specific activity taking place in the Washington household that was previously unknown and not documented in the historical record. Given that every aspect of George Washington's life has been studied in great detail, with books written on the subject numbering in the thousands, this seems notable. The discovery of these residues has prompted several questions: Are the same glues present on all of the fragments? What is the composition of the glues? Was the mending done professionally or by a resident on the property? Were the mended vessels functional after their repair or used only for display?

Unfortunately, the use of historic glues on ceramics remains poorly documented and archaeological research on this sticky subject is scarce. Hence, we attempted to immerse ourselves in all things glue-related, which often took us into the unfamiliar waters of carpentry, art restoration, and book binding.[1] Chemistry played a large role in our analysis, as did experimental archaeology: intensive document research yielded eighteenth-century adhesive recipes, which we subsequently reproduced and used to mend modern ceramics in an attempt to better understand the properties and effectiveness of the various glues. Our research ultimately led to a better understanding of how and why the Washingtons and their contemporaries might have been repairing their ceramics in this fashion.

Background

Located across the Rappahannock River from Fredericksburg, Virginia, Ferry Farm was home to the Washington family from 1738 to 1772. The progenitor of the family, Augustine Washington, died in 1743, leaving a will that divided his property among his sons, with Ferry Farm going to

George.[2] Augustine's widow, Mary Ball Washington, never remarried. She and her children remained on the plantation, farming it with slaves until she moved to nearby Fredericksburg in 1772. The property, now owned by The George Washington Foundation, has been the focus of intensive archaeological study since 2001 under the supervision of chief archaeologist David Muraca.[3] Excavations concentrated on the site of the Washington house and its associated outbuildings have yielded an assemblage of more than 750,000 artifacts. This collection includes tens of thousands of fragments of ceramic tea- and tablewares from Mary Washington's household that include creamware, Chinese porcelain, and white salt-glazed vessels.

The stone-lined cellar of the Washington house contained three fill episodes. The earliest deposit dates to when the cellar was in use from 1728 to about 1770. The second fill layers were deposited at the time the Washington family was leaving the site (the property sold in 1774). After that, the cellar was abandoned and the primary deposits were capped after the last fill episode took place when the house was torn down, circa 1830.[4]

The discovery of glue residues began when a creamware punch bowl, decorated with an enameled floral motif accented with cherries, was unearthed in the main basement of the house (fig. 1). The punch bowl and the majority of the other subsequently identified glue-repaired vessels were found in the second fill episode, when the family presumably was discarding household items in anticipation of their leaving. Additional creamware examples with similar floral motifs and an identical color palette to that of the punch bowl were recovered from another sealed cellar, located under the hall portion of the house.

Figure 1 Punch bowl, probably Derby, England, 1765–1772. Creamware. D. 6". (Courtesy, The George Washington Foundation; photo, John Earl.) Decorated with distinctive overglaze enamels, this bowl was recovered from the cellar of the Washington house.

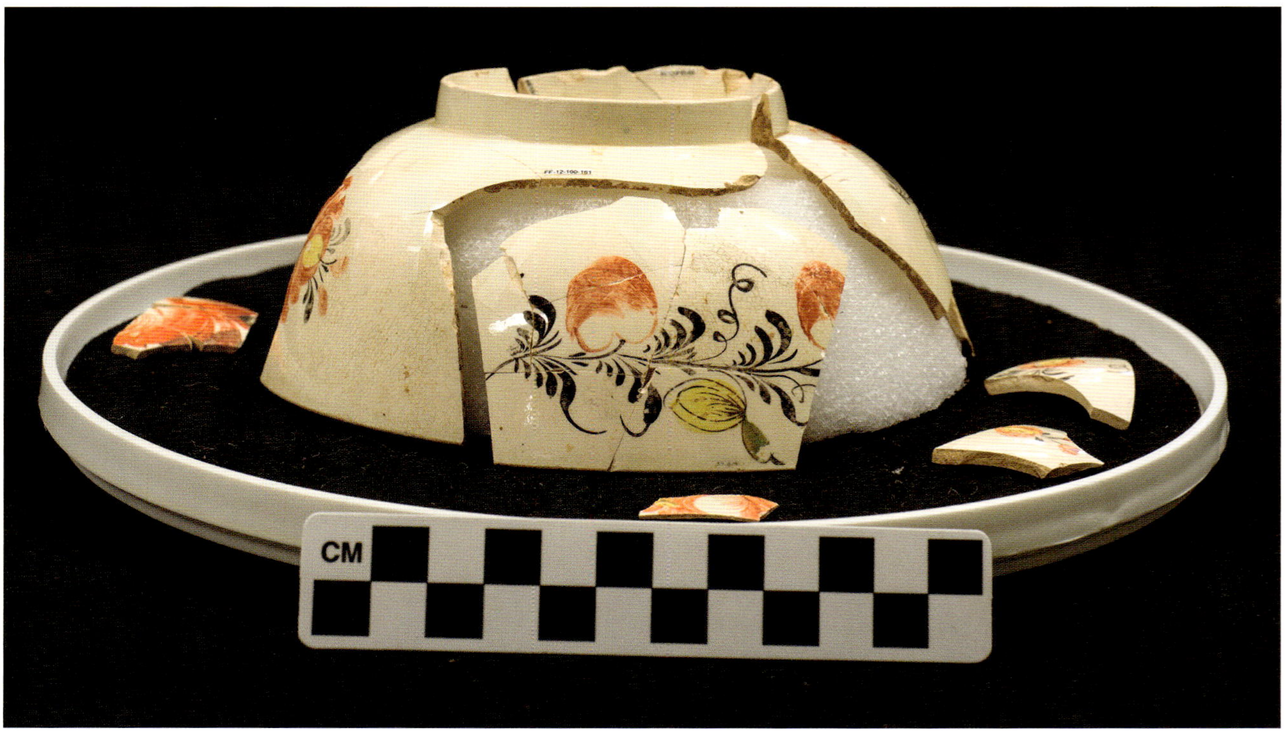

The English-made punch bowl is thought to have been manufactured between 1765 and 1772. Parallel antique creamware forms with cherry designs in the same palette have similar date ranges.[5] This small punch bowl, known as a "sneaker," would have been passed around from person to person at social gatherings with punch being consumed directly from it.[6] A privately owned intact creamware punch bowl (fig. 2) examined by archaeology staff

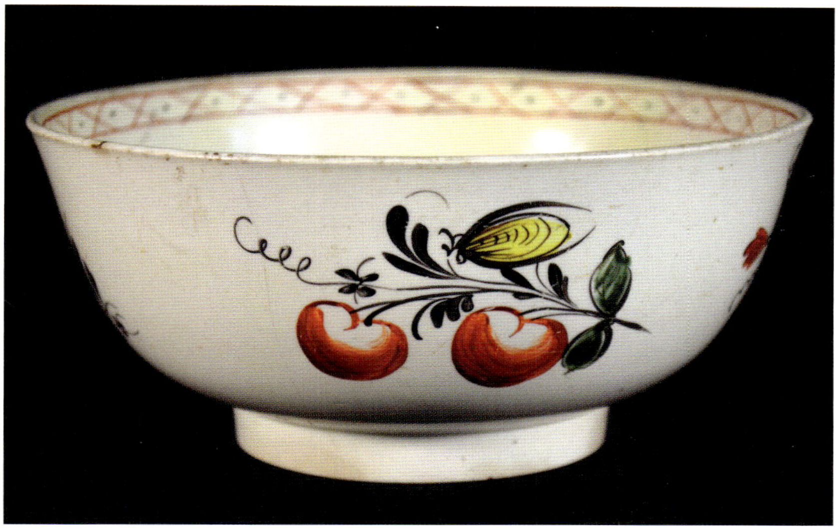

Figure 2 Punch bowl, probably Derby, England, 1765–1772. Creamware. D. 6¼". (Courtesy, Kimberley Overman; photo, Zac Cunningham.)

at Ferry Farm reveals marked similarities to the Washington punch bowl.[7] Color palette choice (red, black, green, and yellow), stylistic qualities, design placement, as well as corresponding individual motifs on the bowls (for example, roses, buds, flowers, stems, leaves, and cherries), appear on both examples. Analysis of these elements attributes the manufacture of the bowls to the Cockpit Hill Pottery in Derbyshire, Derby.[8]

Based on its interior wear patterns, the archaeologically recovered punch bowl appears to have been used extensively before being broken into at least four fragments. It was mended with glue before a presumed second breakage episode and its eventual discard. The glue residue can be seen as a light brown substance adhering to the broken edges of the vessel along what was clearly the original breakage pattern (fig. 3). There is a substantial amount of glue on the interior of the vessel as well. Microscopic examination confirms that the residues are not simply organic deposits, which are common in some depositional environments but have not been seen

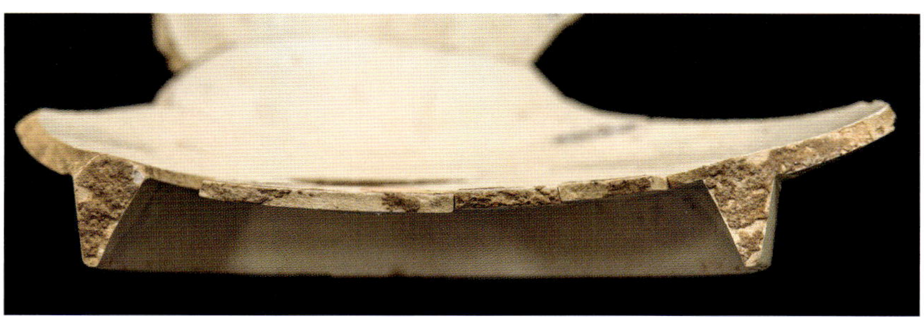

Figure 3 Cross section of the base of the bowl illustrated in fig. 1. (Courtesy, The George Washington Foundation; photo, John Earl.) Glue residue is clearly visible on the edges of the break.

archaeologically at Ferry Farm. Moreover, the interior glue on the punch bowl appears to exhibit brushstrokes from the application of the adhesive (fig. 4). These deep striations, which run parallel to each other and extend across adjacent sherds, are very apparent, but there is no indication of use wear on the surface of the glue, indicating that the vessel was not utilized post-mending.

Figure 4 Microscopic view of glue residue on the bowl illustrated in fig. 1. (Courtesy, Dovetail Cultural Resources Group.)

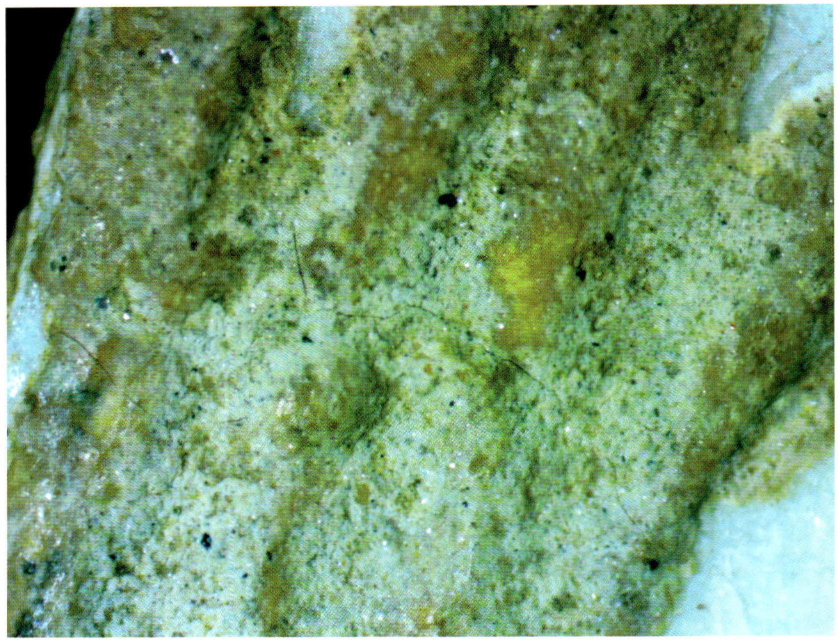

Following this initial discovery, we examined other ceramic fragments dating from Mary Washington's occupation of Ferry Farm. To date, at least seven additional vessels associated with the Washington family have been found to exhibit glue repairs. These include a creamware platter and plate, an enameled creamware teapot lid, white salt-glazed examples including

Figure 5 Close-up of a platter fragment, England, 1765–1772. Creamware. (Courtesy, The George Washington Foundation; photo, Zac Cunningham.)

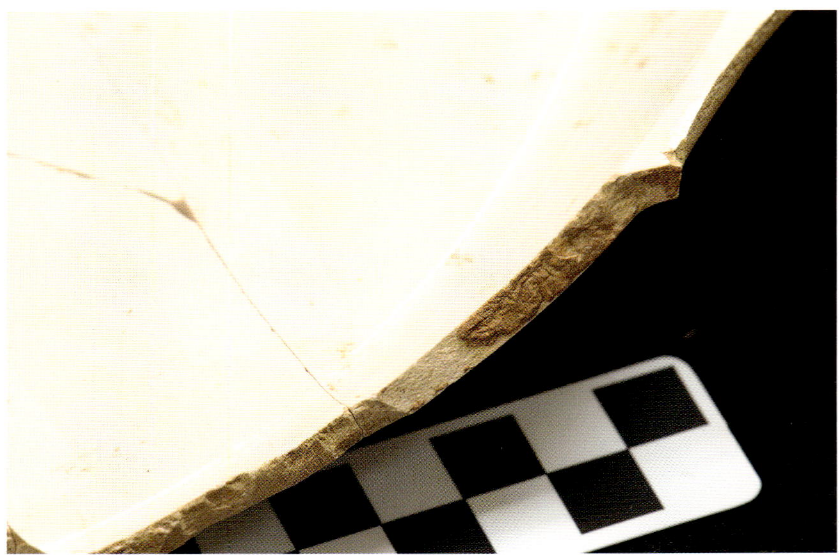

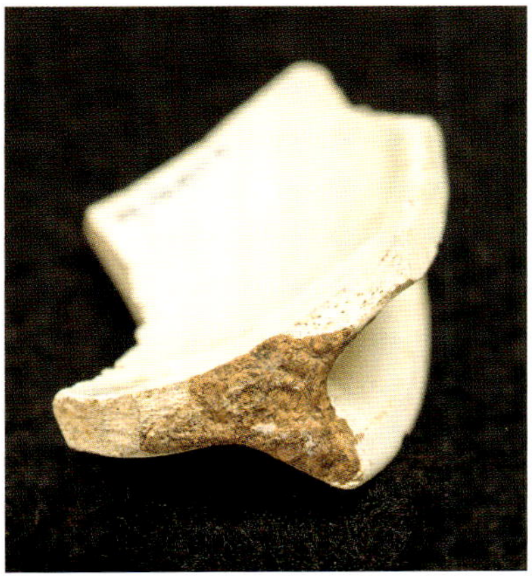

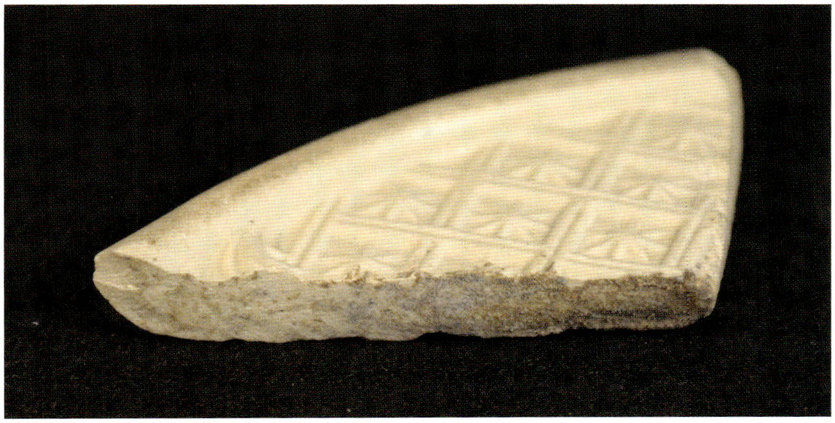

Figure 6 Close-up of a teapot lid, England, 1765–1772. Creamware. (Courtesy, The George Washington Foundation; photo, Zac Cunningham.)

Figure 7 Close-up of plate rim fragment, 1750–1760. White salt-glazed stoneware. (Courtesy, The George Washington Foundation; photo, Zac Cunningham.)

molded plate rims and a hollow-ware base, and a fragment of a Chinese porcelain bowl with an Imari palette (figs. 5–7). With the exception of the white salt-glazed plate fragment, which was recovered from a shovel test pit near the cellar in an early archaeological exploration, all examples exhibiting glue residues were recovered from the Washington house cellar.

Glues and Mending in the Written Record

A review of historical documents as well as dialogue with colleagues indicate that evidence of repaired vessels is not uncommon archaeologically. Yet literature within the field, especially with regard to the use of adhesives, is lacking, although it should be noted that very recently a number of scholars have begun researching this subject.[9]

It is understood that people have been producing mastics for various purposes—including mending ceramics—for thousands of years. Adhesives that contain bitumen and birch tars used to haft stone points have been found on archaeological materials dating as early as the Paleolithic period.[10] The earliest recorded pottery-mending techniques in prehistoric times involved drilling holes on either side of the break and lashing the pieces together with hide strips or plant fibers. Often this method was combined with a plant- or animal-based glue to further seal the break.[11] Ceramic menders have been performing this service for thousands of years. China menders often utilized metal rivets when making repairs, carefully drilling holes in the vessels and stabilizing or rejoining them with adhesive.[12]

A professional mender seems to have been readily available to Mary Washington, although there is no record that she used his services. A goldsmith operating in the 1760s and 1770s in Fredericksburg, just across the river from Ferry Farm, also engaged in china mending. While he did not mention this service in his advertisements, he is listed in an account book belonging to George Weedon as having mended "China Bowls."[13] To date, no artifacts excavated at Ferry Farm show any evidence of having been professionally mended with drill holes, metal hardware, or filling. Presumably Mary Washington chose the thriftier option of using homemade glues.

Eighteenth- and nineteenth-century all-purpose handbooks, cookbooks, and encyclopedias contain a multitude of glue and cement recipes, the examples of which are too numerous to list here. The "receipts," or recipes, range in difficulty from very simple and easily executable to extremely complicated. It is implied in most of this literature that these recipes can be made in a domestic setting and are effective enough to produce viable vessels from broken fragments. With the advent of synthetic glues and increasing access to inexpensive tablewares in the twentieth century, the need for professional china menders waned, and the concept of mixing one's own glue at home faded from memory—that is, until enterprising archaeologists and chemists combined their skills to resurrect the glue recipes of old.

Experimental Archaeology

To gain more insight into the practice of historic ceramic repair, Ferry Farm archaeologists attempted to replicate a number of eighteenth- and nineteenth-century home glue recipes and subsequently assessed their effectiveness. To date, three basic glue types found in the period literature have been replicated: cheese, hide, and resin. Samples were sent to chemist Ruth Ann Armitage of Eastern Michigan University and chemist Daniel Fraser of Lourdes University for chemical characterization. Such analysis would provide comparative data for similar studies conducted on the archaeological residues, which will be discussed later.

The initial difficulty of replicating the historic glues lay in obtaining all of the necessary ingredients. The materials for the glues included various mastics and gums, beeswax, brandy, cheese, duck eggs, garlic, isinglass (a gelatin derived from the air bladders of fish), ox gall, pickling lime, and our own tar created from pine resin, to name a few. Some recipes had to be rejected due to our inability to easily obtain items such as bull's blood and the "white slime of large snails."[14]

Cheese glue was the first to be reproduced. Based on a recipe published in *The Scots Magazine* in 1765, it involved only three ingredients: grated hard cheese, milk, and pickling lime.[15] The original type of cheese called for was "Suffolk Cheese," the cheese equivalent of hardtack and eaten primarily by sailors in the British Navy; those who consumed it at the time complained bitterly of its taste.[16] Suffolk Cheese, as it existed in the eighteenth century, is now thankfully extinct, so a shredded block of dried Romano cheese had to be substituted. Our initial attempt at replicating this glue resulted in a lumpy, foul-smelling concoction that set too quickly (within minutes) for efficient mending and was somewhat caustic, partially removing the outer layer of skin on our hands. With slight alterations to measurements and minor substitutions gleaned from similar recipes (such as substituting egg whites for milk), subsequent iterations of this glue produced a much-improved adhesive that effectively mends ceramics.

Two hide-based recipes were reproduced, both taken from *Handmaid to the Arts*, an all-purpose guide for housewives published in 1764.[17] The first one was fairly simple, containing dried chips of basic glue rendered from animal hide and combined with brandy and quicklime. This resulted in a functional adhesive. The second glue was much more complicated—using hide, brandy, garlic juice, ox gall, various resins, vinegar, quicklime, and isinglass—but was by far the most effective in terms of its ease of application, the strength of the repair, and its relatively inconspicuous nature. As a result, we chose to use it to mend a pearlware child's mug owned by a member of our staff (fig. 8) and, as of this writing, the glue has held for more than three years with no signs of deterioration.

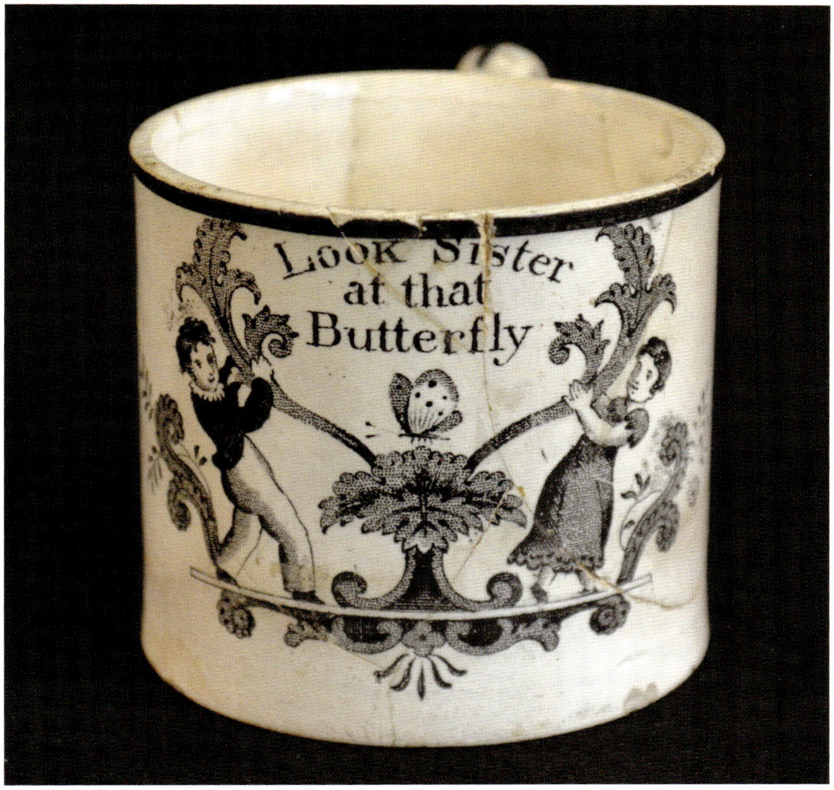

Figure 8 Mug, England, ca. 1820. Pearlware. H. 2⅜". (Courtesy, Laura Galke; photo, Zac Cunningham.) Antique mug mended using historic glue recipe.

The last glue reproduced was a resin glue recipe taken from *The Encyclopædia Britannica: A Dictionary of Arts, Sciences, and General Literature* (1888).[18] The recipe calls for only three ingredients: beeswax, resin (although the exact type was not specified), and quicklime. It became apparent as our research progressed that many of the recipes presumed a certain knowledge of ingredients and technique: we manufactured the tar from readily available pine resin, which proved a difficult and somewhat dangerously flammable endeavor. The resulting glue was almost impossible to work with and was too dark to be applied with any subtlety on white-bodied wares such as those Mary Washington was mending. However, it is entirely possible that lighter plant materials, such as sandarac and gum Arabic, were utilized in her day.

To ascertain the properties and functionality of our reproduced glues we obtained ceramics at a local thrift store to be broken and mended. This selection was purposeful in order to simulate the efficacy of our historic glue recipes on differing ware types with porosities similar to those of historic ceramics. Each vessel was dropped on a wide-planked wooden floor to replicate likely breakage conditions and mended with one of the experimental glues. Initially, our primary interest was simply in how effectively the glues mended a vessel. Although all of the glues functioned to some extent, the resin glue proved difficult to apply and did not produce many finished products worthy of subsequent use or even display. The hide and cheese glues, however, mended all ceramics well, with only a minimal amount of glue visible on the breaks after they had dried. Moreover, they were easy to apply with a paintbrush.

After drying, the mended vessels underwent experimentation to see how well they withstood the tests of burial, temperature, humidity, pressure, and liquid retention—all simulations designed to replicate the likely use and discard of ceramics during the colonial period. Some vessels were buried in the back dirt pile on our archaeological site for six months to determine how post-depositional processes affected the glue. The results were surprising. The resin glue had altered significantly in color and texture, turning from almost black to a medium brown, and was subject to bubbling and embrittlement or flaking from the broken edges of the vessel. The cheese glues remained cemented to the edges of the vessels after burial and changed from a white color to a mottled gray. Of all the glues, the cheese glue appeared to be the one most likely to survive archaeologically and resembled most closely the residues we observed on Mary Washington's wares. Both hide glues seemingly disappeared from the vessels when inspected with the naked eye; however, placing our buried hide-glue examples under short-wave ultraviolet light revealed that trace amounts of the glue remained and fluoresced bright blue.

In order to determine what effects fluctuating temperature and humidity had on the glues, representative mended vessels were placed in a non-climate-controlled building on the property for six months, from late spring to early winter. Two glues appeared to have been altered by their exposure in the shed. A hide glue was observed to be yellowed and peeling from the surface of the cup, although the vessel itself stayed intact. A saucer mended with cheese glue appeared to be unaltered, although it fell apart shortly after being brought inside the archaeology lab, ultimately succumbing to the infamous Virginia heat and humidity. To test the strength of all of the glues, jars of identical volume were filled with water and placed on suspended mended flatwares. All passed this test, demonstrating that every glue type re-created produced a suitably strong adhesive, as claimed. Ultimately, however, the water test proved that none of the glues could mend a vessel to the point that it was completely functional. Hollow wares repaired with each of the experimental glues were filled with cool water to test the mends. Of the vessels mended with the two versions of hide glue, only one did not leak when it was filled with cold water, although it did fail the subsequent hot water test. The pine-resin glue resisted cold water well but became pliable in hot water, significantly weakening the mends. With its black appearance and brittle texture, the resin glue makes an unlikely candidate for mending refined wares such as those used in Mary Washington's household.

Our experiments in adhesives indicate that the cheese and hide glues were the most effective in terms of ease of use and efficacy in joining fragments, and produced the most inconspicuous mends, although the resulting vessels did not hold water. The cheese glue, while the lightest in appearance and very good for mending white-bodied wares, was significantly weakened by moisture.

In general, our experimental attempts at understanding home ceramics mending yielded significant and promising results. Manufacturing adhesives for home use, while common, was a difficult, malodorous, and time-consuming task potentially exposing the mender to hazardous substances and, for these reasons, likely would have been accomplished by one of Mary Washington's enslaved people. Our reproduction and use of period adhesives revealed that the glued ceramics were not functional and as such would not have been utilized post-mending for practical purposes by the family or any other plantation residents. This is further supported by the lack of even microscopic use wear on the ample glue residue adhering to the interior of the creamware punch bowl (see fig. 1). Although cast off or damaged ceramics are thought to have been passed down to servants (and, in this case, the enslaved), the location of the mended items exclusively within and around the house cellar indicates that they remained in the possession of the Washingtons. Furthermore, no repaired ceramics have been recovered from the vicinity of slave dwellings or activity areas. Glued vessels were likely employed to decorate the mantel or fill out a china closet or bowfat, a possibility also suggested by Angelika Kuettner in this volume.[19]

Residue Analysis

Chemical analysis of the archaeologically recovered eighteenth-century glue residues by Ruth Ann Armitage and Daniel Fraser have shed further light on the composition of the glues used to mend Mary Washington's dishes. The samples were analyzed utilizing Direct Analysis in Real Time (DART) mass spectrometry and Fourier transform infrared (FTIR) spectroscopy. DART mass spectrometry is fast, requires only a small sample, and gives a molecular fingerprint of the material with little or no sample preparation. The mass spectrometer allows for determination of the composition of the samples and their comparison with other samples and materials. FTIR is a nondestructive method that has potential to inform us about the mineral composition of the residues, as well as any organic material that remains. As previously mentioned, replicated glues and raw ingredients were tested to provide a control for the glues sampled from Mary Washington's ceramics. The mass spectrometer could clearly differentiate between the glues and raw materials.

Preliminary results from both tests show that all of the archaeological glue residues contained high concentrations of calcium carbonate with the exception of the glue from the porcelain fragment, which was contaminated with indeterminate inorganic substances. Lime was a major ingredient in most of the glue recipes we researched and replicated, likely because it provided a stabilizing aplastic much like temper in clay used for ceramics. The FTIR analysis of historic glues indicated that there was little to no remaining organic material relative to the mineral content. The replicated cheese glues that were buried for six months showed a similar decomposition of the organic material. Before burial, the cheese glues showed signatures characteristic of fats from cheese as well as calcium carbonate. After burial, all of the organic signals were lost, with only the calcium carbonate remaining.

More sensitive analytical methods might be able to detect traces of remaining organic compounds that are characteristic of the other ingredients in the historic glues. Previous work by Armitage and Fraser has shown detectable, statistically significant quantities of hydroxyproline in historic glues. This amino acid is present in large quantities in the protein collagen and might be a marker for either hide- or bone-derived glue. Additionally, neither DART-MS nor FTIR indicated the presence of resin compounds. While it is possible that resin compounds had completely decomposed during burial, it is equally possible that no resins were present in the original glues at all. Moreover, there was variation between archaeological glues, indicating several types of glues were present and, therefore, multiple mending episodes. This would seem to indicate a pattern of behavior in the Washington household rather than a onetime activity.

Conclusion

Our goal going into this study was to better understand the seemingly idiosyncratic attitude of Mary Washington toward mending ceramics. Through documentary research and experimental archaeology, we determined that ceramic repair at home was commonplace until the early twentieth century. Mary Washington's decision to repair broken pieces may suggest a frugal woman who valued her possessions. It is telling that she chose to repair them at home, forgoing a professional mender. Through our experiences replicating eighteenth-century glue recipes available to her, and mending dozens of pots with the glues experimentally, it is clear to us that the vessels she had repaired at home would not have been functional. Given this, it is even possible that she had a sentimental attachment to one or more of these vessels, which resulted in her decision to salvage them. The punch bowl, as stated, did exhibit an extensive amount of use wear on the interior and may have been a favorite item of hers. Or perhaps these pots were valuable enough as showpieces that the demonstration of their ownership took precedence over utility.

In order to fully understand ceramics, we must have knowledge not just of ware and form types and dates of manufacture but also of how the pieces were utilized and the relationships people formed with them. Future research at Ferry Farm will be aimed at revealing more about the adhesives employed by eighteenth-century households. The hope is to gain more insight not only into the life of the Washingtons but also into how ceramics were used historically, their changing functions throughout their use-life, and how their utility could be both functional and communicative.

1. Patrick W. Edwards, "Why Not Period Glue?," *Journal of the Society of American Period Furniture Makers* (November 2001): 2–3; John S. Mills and Raymond White, *The Organic Chemistry of Museum Objects*, 2nd ed. (London: Routledge, Taylor & Francis Group, 2011).

2. Augustine Washington's will can be viewed online at www.kenmore.org/genealogy/washington/augustine_will.html (accessed May 5, 2016).

3. David Muraca, Paul Nasca, and Philip Levy, *Report on the Excavation of the Washington Farm: The 2004–2005 Field Seasons* (Fredericksburg, Va.: George Washington Foundation, 2011), p. 14.

4. David Muraca, Paul Nasca, and Philip Levy, *Report on the Excavation of the Washington Farm: The 2006–2007 Field Seasons* (Fredericksburg, Va.: George Washington Foundation, 2010), p. 23.

5. Donald C. Towner, *English Cream-Coloured Earthenware* (London: Faber and Faber, 1957), p. 14.

6. Susan Gray Detweiler, *George Washington's Chinaware* (New York: Harry N, Abrams, 1982), p. 209.

7. Personal communication with Kimberley Overman, February 3, 2016. Overman is the owner of the complete, repaired creamware punch bowl (H. 3", D. 6¼") that is enameled in black, red, green, and yellow paint with flower, cherries, and bud motifs. This non-archaeological whole example from the period represents exceedingly similar characteristics of the excavated bowl sherds from Washington's Ferry Farm in size, design, and color choice.

8. Towner, *English Cream-Coloured Earthenware*, p. 14.

9. Angelika R. Kuettner, "Simply Riveting: Broken and Mended Ceramics," in this volume.

10. Eric Boeda, Jacques Connan, Daniel Dessort, Sultan Muhesen, Norbert Mercier, Helene Valladas, and Nadine Tisnerat, "Bitumen as a Hafting Material on Middle Palaeolithic Artefacts," *Nature* (1996): 380, 336–38; Marin Cârciumaru, Rodica-Mariana Ion, Elena-Cristina Nitu, and Radu Stefanescu, "New Evidence of Adhesive as Hafting Material on Middle and Upper Palaeolithic Artefacts from Gura Cheii-Râsnov Cave (Romania)," *Journal of Archaeological Science* 39, no. 7 (2012): 1942–50; Paul Peter Anthony Mazza, Fabio Martini, Benedetto Sala, Maurizio Magi, Maria Perla Colombini, Giana Giachi, Francesco Landucci, Cristina Lemorini, Francesca Modugno, and Erika Ribechini, "A New Palaeolithic Discovery: Tar-hafted Stone Tools in a European Mid-Pleistocene Bone-bearing Bed," *Journal of Archaeological Science* 33 (2006): 1310–18.

11. S. Charters, R. P. Evershed, L. J. Goad, C. Heron, and P. Blinkhorn, "Identification of an Adhesive Used to Repair a Roman Jar," *Archaeometry* 35, no. 1 (February 1993): 91–101; Rosamund Cleal, "The Occurrence of Drilled Holes in Later Neolithic Pottery," *Oxford Journal of Archaeology* 7, no. 2 (July 1988): 139–45; Renske Dooijes and O. P. Nieuwenhuyse, "Ancient Repairs: Techniques and Social Meaning," in *Konservieren oder restaurieren: Die Restaurierung griechischer Vasen von der Antike bis Heute*, 3rd suppl. to the CVA Germany (Munich: Beck Publishers, 2007), pp. 15–20; R. Michael Stewart, *Ceramics and Delaware Valley Prehistory: Insights from the Abbott Farm*, ed. Charles A. Bello ([South Orange]: Archaeological Society of New Jersey, 1998).

12. George L. Miller and Emily Brown, "A Harry A. Eberhardt–Repaired Chinese Porcelain Saucer," in this volume; Maria Warwell, advertisement for mending china published in *South-Carolina Gazette; and Country Journal*, July 21, 1767.

13. George Weedon, Ledger of George Weedon, Fredericksburg City Archives, folio 13, 1773.

14. Arnold James Cooley, *Book of Useful Knowledge: A Cyclopaedia of Six Thousand Practical Receipts and Collateral Information in the Arts, Manufactures, and Trades, including Medicine, Pharmacy, and Domestic Economy, Designed as a Compendious Book of Reference for the Manufacturer, Tradesman, Amateur, and Heads of Families* (New York: D. Appleton and Co., 1851), p. 172.

15. Anonymous recipe, *The Scots Magazine*, vol. 27 (Edinburgh: Printed by W. Sands, A. Murray, and J. Cochran, 1765), p. 228.

16. Samuel Pepys, *The Diary of Samuel Pepys—Complete*, ed. Richard Griffin Braybrooke, Henry B. Wheatley (EBook-No. 4200, public domain in U.S.A., October 31, 2004), October 4, 1661.

17. John Nourse and Robert Dossie, *The Handmaid to the Arts...*, 2 vols., 2nd ed. (London: Printed for J. Nourse, Bookseller in ordinary to His Majesty, 1764).

18. Thomas Spencer Baynes, ed., *The Encyclopædia Britannica: A Dictionary of Arts, Sciences, and General Literature*, 9th ed. (New York: Henry G. Allen and Co., 1888).

19. Kuettner, "Simply Riveting: Broken and Mended Ceramics."

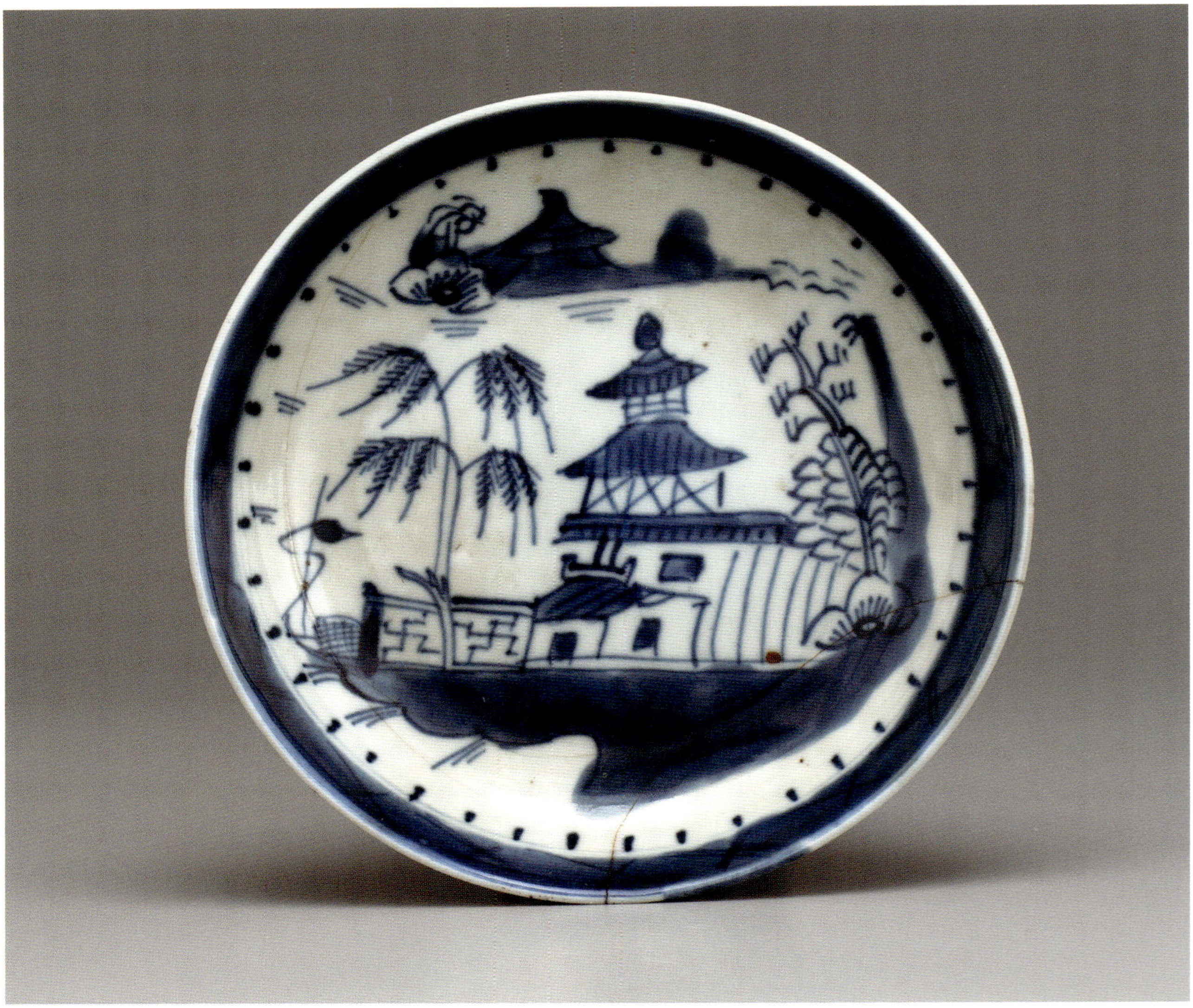

Figure 1 Saucer, China, ca. 1790–1820. Porcelain. D. 5¾". (George L. Miller collection; photo, Robert Hunter.) The break can be seen in the lower right section of the plate.

George L. Miller and Emily Brown

A Harry A. Eberhardt–Repaired Chinese Porcelain Saucer

▼ SEVERAL YEARS AGO George Miller purchased a group of painted Staffordshire saucers at the flea market in Columbus, New Jersey. The lot included a "Canton"-pattern Chinese porcelain saucer, which had a break that had been repaired with three rivets (fig. 1). The back of the saucer has a printed paper label (fig. 2) that reads "HARRY A. EBERHARDT / Art & China Repairing / 213 S. 11th St., Phila. Pa." Below the printed label is a faded handwritten note, which, when viewed under longwave ultraviolet light, is seen to read "17228 Cauffman / 3 Riv." The number "17228" is the Eberhardt record of the repair; "3 Riv" is probably shorthand instructions for three rivets to be placed along the break for the repair. The name "Cauffman" most likely refers to Joseph Cauffman, the proprietor of the Benjamin Franklin Antique Shop in Philadelphia from the 1920s.[1]

After acquiring the saucer George ran across an advertisement for Harry A. Eberhardt & Son in the July 1930 issue of *Antiques*.[2] The advertisement pictured several broken ceramic vessels with the caption "MISSING PARTS of Antiques china and pottery replaced; this work does not show. / Expert Repairing of Art Goods / Established 1888." The advertisement had the same address as the one that appears on the paper label. Given that the 1930 advertisement added "& Son," the repair of the saucer most likely was done before that date. The company was in business until December 2015, trading under the name H. A. Eberhardt & Son Inc. at 2010 Walnut Street in Philadelphia. Its website (eberhardts.com) bears the claim "The Antiques Shop and Restoration Studio came from Europe to America on, or before earliest records of 1865. The firm began trading as Eberhardt's in 1888, when Harry A., grandfather of William R. Eberhardt, the current owner, became a partner."

While researching the Eberhardt company, George Miller was joined by Emily Brown, a Graduate Fellow, Class of 2015, in the Winterthur/University of Delaware Program in Art Conservation. Prior to entering the Winterthur conservation program Emily had worked for the Eberhardt company for fourteen years, performing restoration work on ceramics, glass, and other types of decorative objects.

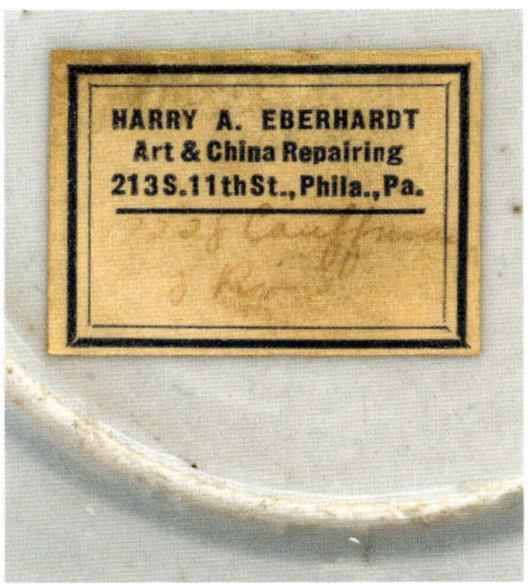

Figure 2 The label on the back of the Chinese porcelain saucer illustrated in fig. 1.

Rivet Repairs, Some History of the Practice

Father Martinus Martini, a Jesuit missionary in China, published one of the earliest descriptions of riveting ceramics in his 1655 publication, *Novus Atlas Sinensis*.[3] When riveting began to be practiced in Europe is not clear. One of the earliest advertisements for riveting ceramics in the American colonies occurred in the *Boston Gazette* on October 20, 1755. The advertiser offers to repair broken china "by riveting with Silver or Brass Rivets."[4] An advertisement in the *Leeds Intelligencer* on May 12, 1761, placed by Robinson and Rhodes for enameling of ceramics mentions mending porcelain by firing broken pieces with a glass or glaze composition between the breaks as an alternative to repairing by riveting the china.[5] By the 1760s there were china repairers advertising in New York, Philadelphia, and Charleston, South Carolina. David Hall, a Philadelphia jeweler, advertised in the *Philadelphia Gazette* of February 20, 1766, that, in addition to making all sorts of gold and silver work, he "drilled and clasped" china bowls and added wooden handles to teapots.[6] Nathaniel Lane advertised in *The New-York Journal or General Advertiser* on July 9, 1767, that "Broken China and Glass, Mended and riveted in the neatest manner and on reasonable Terms. . . . The Price of Rivets, if he finds the Silver is 2s [shillings] each if Silver is found 1s each Rivet, if Brass, is 1s if white metal 6d [pence]."[7] Joseph Dix, a Philadelphia watchmaker, advertised in the *Pennsylvania Journal* on December 14, 1769, "China Ware Rivetted. Very secure and in a neat manner, different to what is commonly practiced in riveting. . . ."[8]

Thus it is clear that, at least by the 1750s but probably earlier, ceramic repair was being done by riveting in the American colonies. That tradition continued into the 1950s, as indicated by the correspondence between Henry Francis du Pont and Harry A. Eberhardt.

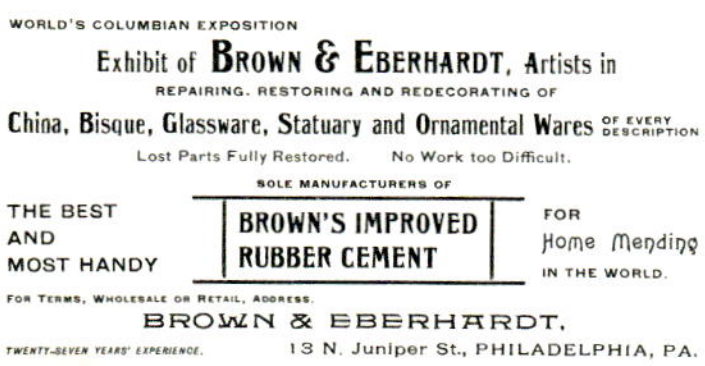

Figure 3 Trade card, Brown & Eberhardt, Philadelphia, ca. 1892. (Joseph Downs Collection of Manuscripts and Printed Ephemera, Collection 46, Winterthur Library.)

Researching the Eberhardt company at Winterthur proved productive. There is extensive correspondence between du Pont and Eberhardt concerning restoration of ceramics and glass for Mr. du Pont's collection. In addition, the Downs Collection has a trade card for Brown and Eberhardt from the Columbian Exhibition that was held in Chicago in 1892 and 1893 (fig. 3).[9] The trade card suggests that Brown and Eberhardt began in business in about 1865. Later printed invoice forms for Harry Eberhardt illustrate a medal that was awarded to various entrants in the Columbian Exposition (fig. 4).[10] The medal had a blank space for the engraving of the awardee's name. The illustrated medal has "H. A. EBERHARDT" engraved in that space.[11] A Google search of "Brown's Improved Rubber Cement" produced a reference to a patent for that product being taken out by John J. Brown on October 31, 1876.[12]

We are fortunate to have the extensive correspondence between Henry Francis du Pont and Harry A. Eberhardt that has been preserved as part of the Winterthur Archives in the Winterthur Library. This correspondence has carbon copies of letters sent by Mr. du Pont to Mr. Eberhardt and the response letters written on printed letterhead along with invoices for work completed. The earliest correspondence is a letter from du Pont to Eberhardt Brothers dated June 2, 1924, mentioning a "small Lowestoft stopper" to a tea caddy that had been badly restored. A great deal of the ceramics sent for restoration was referred to as Lowestoft, which was the common name for Chinese porcelain during that period.[13] The last correspondence in the Winterthur Archives is an invoice from William A. Eberhardt & Son to du Pont that is dated April 16, 1968. The bulk of their correspondence occurred

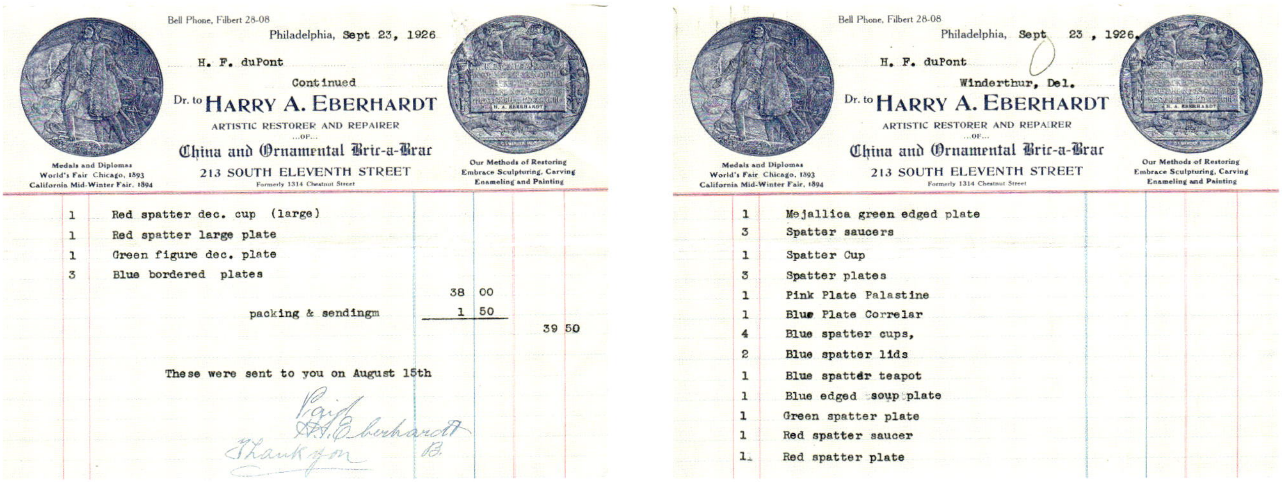

Bell Phone, Filbert 28-08

Philadelphia, Sept 23, 1926

H. F. duPont

Continued

Dr. to HARRY A. EBERHARDT

ARTISTIC RESTORER AND REPAIRER

...OF...

China and Ornamental Bric-a-Brac

213 SOUTH ELEVENTH STREET

Formerly 1314 Chestnut Street

Medals and Diplomas
World's Fair Chicago, 1893
California Mid-Winter Fair, 1894

Our Methods of Restoring
Embrace Sculpturing, Carving
Enameling and Painting

1	Red spatter dec. cup (large)		
1	Red spatter large plate		
1	Green figure dec. plate		
3	Blue bordered plates		
		38 00	
	packing & sendingm	1 50	
			39 50

These were sent to you on August 15th

Paid
H.A. Eberhardt
Thank you

Bell Phone, Filbert 28-08

Philadelphia, Sept 23, 1926

H. F. duPont

Winderthur, Del.

Dr. to HARRY A. EBERHARDT

ARTISTIC RESTORER AND REPAIRER

...OF...

China and Ornamental Bric-a-Brac

213 SOUTH ELEVENTH STREET

Formerly 1314 Chestnut Street

Medals and Diplomas
World's Fair Chicago, 1893
California Mid-Winter Fair, 1894

Our Methods of Restoring
Embrace Sculpturing, Carving
Enameling and Painting

1	Mejallica green edged plate
3	Spatter saucers
1	Spatter Cup
3	Spatter plates
1	Pink Plate Palastine
1	Blue Plate Correlar
4	Blue spatter cups,
2	Blue spatter lids
1	Blue spatter teapot
1	Blue edged soup plate
1	Green spatter plate
1	Red spatter saucer
1	Red spatter plate

Figure 4 Invoices, Harry A. Eberhardt, Philadelphia, September 23, 1926. (Winterthur Archives, Winterthur Library.)

between 1925 and 1940, a period during which du Pont was very active in collecting ceramics. While some of the correspondence concerns ceramics sold to him, most of the letters and invoices relate to orders for ceramics to be repaired and cleaned. In many cases, the detail is vague and does not mention what type of repair was being commissioned. In other cases the interchange of letters provides excellent details of what is to be done and the cost of the restoration.

Comments on Ceramics to Be Riveted

A letter written to du Pont on Harry A. Eberhardt letterhead dated February 13, 1926, reads as follows:

> Dear Sir
>
> Your chauffer brought to my place for repairs
> 7 Lowestoff [*sic*] plates all have been broken & painted over
> 2" Tureens no lids both have breaks painted over
> 3" After dinner coffee cups handles have been made of composition & painted over
> 1" Tea caddy no lid otherwise perfect
> 2 spatter cups & saucers
>
> I am writing to find out whether you want these pieces riveted together for table use ["yes" handwritten in pencil on letter] if so we will remove the paint take them apart clean the edges & join them & rivet on the bottom side the breaks will show slightly but there will not be any paint to scale off. If you Just want them for cabinet use we will clean them thoroughly & rejoin them by cementing only. awaiting further information & instruction on the above work.
>
> Respectfully,
> HAEberhardt

Du Pont responded in a letter dated February 16, 1926, saying that he wanted the vessels riveted together for table use. Parsons and Curl's 1963 book *China Mending and Restoration* states that riveting should be limited to tableware because it comes in contact with hot water, detergents, and steam. This type of repair did not use any adhesives, and vessels repaired with rivets would be watertight.[14] The cost for repairing du Pont's seventeen vessels was $49.30.

Unfortunately, the number of rivets involved is not listed. However, there is a cost on the riveting of cups in a line that reads: "3 Lowestoff [*sic*] cups only 1 riveted $1.20," which gives an indication of the price of riveting. Another invoice from Eberhardt to du Pont dated July 17, 1931, reads: "15 pieces of Lowestoft 159 rivets & restoring $115.00." That comes to $0.72 per rivet, not including the restoration of the wares or an average of $7.67 per vessel that was restored.

A handwritten note sent to du Pont on a postcard dated March 7, 1933, reads: "Now is the time to have those pieces of china repaired as we have reduced our prices to just meet the cost Riveting of china & glass [illegible] 35 a rivet & other work accordingly." It was signed "Eberhardt Bros."

The large volume of restoration work done for du Pont is recorded in the Eberhardt–du Pont correspondence from 1925 through the 1930s. A letter on Harry A. Eberhardt & Son letterhead and dated September 16, 1931, gives an idea of the quantities that might be sent for restoration: "The lot of china we were repairing for you has been completed, and I think it would be advisable to send the station car as it consists of 2 barrels & 3 cartons." Clearly, du Pont did not limit his collecting to perfect objects. The last invoice that mentions rivet repairs is dated March 13, 1950, and the last invoice from Harry Eberhardt & Son to H. F. du Pont in this collection is dated April 16, 1968.

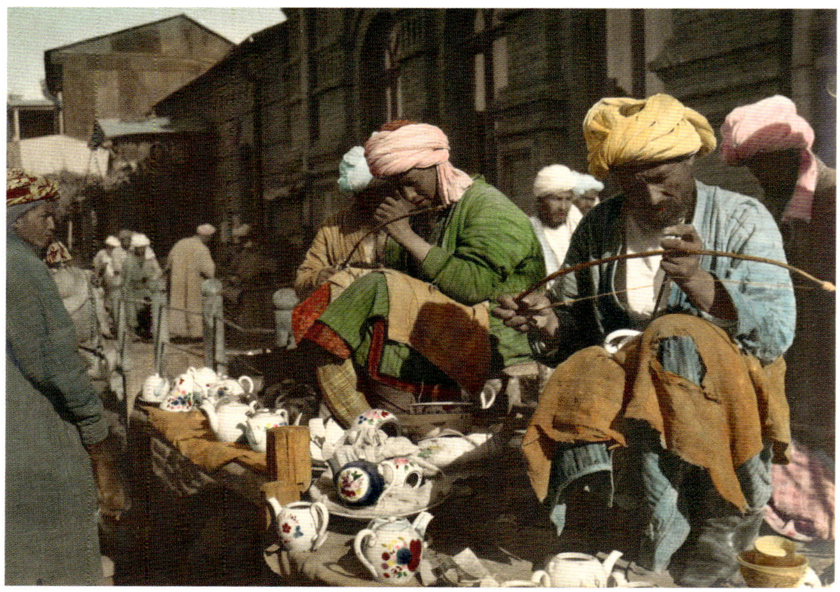

Figure 5 Workers repair broken china dishes with copper rivets, Samarkand, Uzbekistan, twentieth century. (photo, Maynard Owen Williams; National Geographic Stock: Vintage Collection / Granger, NYC—All rights reserved.)

The Winterthur Museum Objects catalog lists at least five vessels that still have Eberhardt repair labels. Only one of them has rivet repairs. It is a blue shell-edged basin by Andrew Stevenson that has ten rivets.[15] Unfortunately, the Winterthur catalog records do not indicate when Eberhardt repaired the pieces.

The End Use of Riveting for Ceramic Repair

According to William R. Eberhardt, the son of Harry A. Eberhardt Jr. and the grandson of the founder of H. A. Eberhardt & Son Inc., repairing ceramics and glass with rivets ended about sixty years ago.[16] Eberhardt said that the repair business had declined to a great extent. They used to have four people employed in restoring ceramics and other objects and in 2012 had only one. Eberhardt grew up in the business and was trained in the restoration of ceramics and artworks by his family. In his younger years he had done rivet repairing of ceramics. One of Eberhardt's employees who retired after forty years had done rivet repairs of ceramics, a task now done with better-quality cements such as epoxy. In *China Mending and Restoration*, Parsons and Curl state that "recent scientific developments in adhesives have led many people to believe that riveting is now obsolete."[17] The transition to epoxy for ceramic repair is echoed in Thomas Pond's book *Mending and Restoring China* that was published in 1970. Pond entered the restoration business in the 1960s, and he states his feeling about rivets in the following passage.

> In 1963 a 5 guinea book was published on china mending [clearly the Parsons and Curl book], and of its 420 pages over half were devoted to riveting and doweling. I have found riveting difficult and unsatisfactory and have always avoided it. In fact my greatest pleasure is to extract all rivets and re-mend and restore with modern glues, which were not at the disposal of restorers until ten years ago.[18]

Although the riveting of ceramics seems to have come to an end in America and probably England, it is still practiced in other countries. A recent color photograph by Maynard Owen Williams for *National Geographic* shows two men with bow drills repairing vessels in Uzbekistan with a collection of teapots and other vessels in front of them in an outdoor market (fig. 5). The photograph is labeled "Workers repairing broken china dishes with copper rivets."[19]

The Process of Riveting Ceramics

Fortunately, the process of riveting ceramics has been very well described in *China Mending and Restoration*.[20] A rivet looks like a staple with a long, center section and two short legs, one at each end. The legs are at a slight angle toward the break line so that the staple exerts pressure on the fragments to be joined together. Rivets were commonly made from half-round brass or German silver wire that ranged in size from 18 to 12 gauge. In this scale, the 18 gauge is the smallest size used and was commonly used to repair cups. Thicker gauge wire was used for larger vessels, such as platters.[21]

The process of riveting pottery or porcelain together involved cleaning the broken surfaces, putting the vessel together with tape, and marking places where holes would be drilled for the placement of the rivet—on the back of the vessel, especially if it was to be used after repair rather that just being on display. Rivets were placed a couple of inches apart, at right angles to the break line so that the clamping force would not distort the movement along the break. Once the locations were marked, small holes were drilled with a diamond-tipped tool that was rotated by a bow or pump drill.[22] The drilled holes were at a 15-degree angle, facing in toward the fracture, and they were drilled only partway into the body.

Once the holes were drilled, the half-round brass or German silver wire was cut off and one end bent at the 15-degree angle, which was placed into the hole to make sure it was a good fit. Then the half-round wire was laid across the break with one leg in the drilled hole, which established the position to bend the second leg of the rivet, which was also at a 15-degree angle to the break. Once the rivet was made and one end placed, the other end was pushed into place by a pair of needle-nose pliers. If the rivet was well made, the pressure it exerted on the fragments formed a solid bond. Once all of the rivets were placed, any open areas around the drill holes were filled with plaster of Paris or other material.[23]

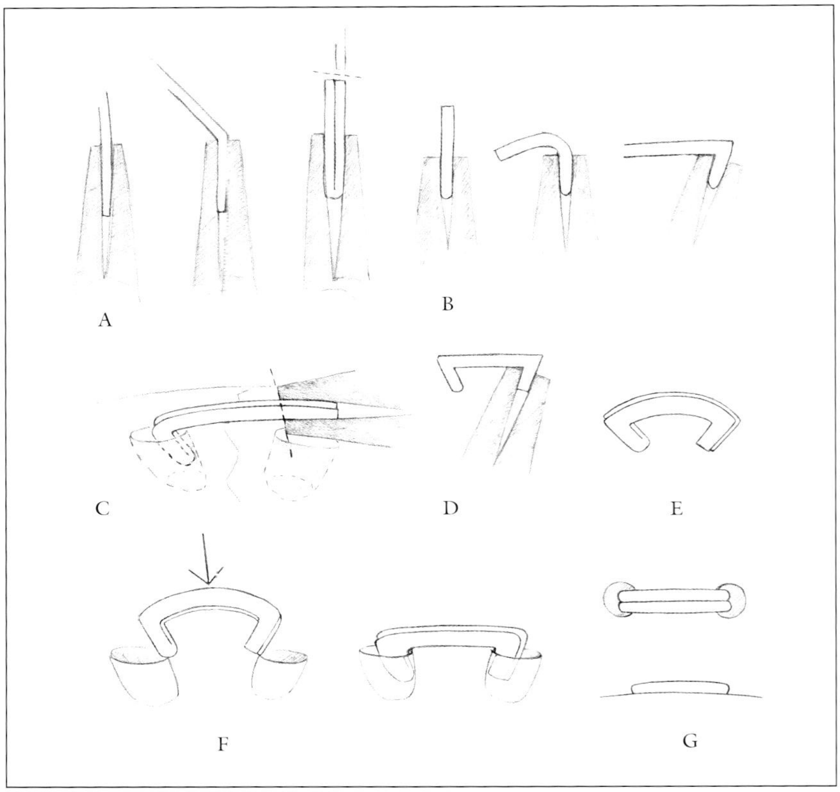

Figure 6 Illustration of riveting process described by William R. Eberhardt and based on images found in Claudia S. M. Parsons and Frederick H. Curl, *China Mending and Restoration* (London: Faber and Faber, 1963), pp. 49, 52, 55. The illustration shows how: (A) the wire is bent double; (B) the first rivet leg is bent and hammered into position; (C) the second leg is measured; and (D) the second leg is formed. The remaining images show: (E) the completed rivet with a bowed shape; (F) the rivet in position at the holes; and (G) the final appearance of the rivet when viewed from the top and side.

William R. Eberhardt said that he was trained in riveting as a young man. His description of the process is parallel to that described above with a couple of differences (fig. 6): They used German silver wire because he said that it "retained its springiness." The wire was bent double (a) and the doubled end was bent to form the first leg to be fitted into the hole (b and c). A slight bow was given to the rivet (e) so that when ready to be placed, the second leg could be maneuvered into place by pressing on the bowed section of rivet toward the vessel being repaired (f). This would extend the second leg of the rivet into the hole drilled for it. From this description, it appears that different gauges of wire were used.[24]

A drill used for riveting by Eberhardt & Son was recently rediscovered by Emily Brown after a visit to their offices (figs. 7, 8). She was given a sample of what is believed to be the wire used to make rivets, as described above. It is a round wire of a small gauge, which most certainly could have been used to make a rivet as described by Eberhardt.

The middle rivet on the saucer illustrated in figure 1 (fig. 9) was analyzed at the Scientific Research and Analysis Laboratory (SRAL) at Winterthur Museum using X-ray fluorescence (XRF) analysis.[25] XRF is a process whereby X-ray beams excite electrons in a material. These excited electrons emit X-rays of their own, which are characteristic of the atomic element present in a material. In this way scientists can determine the elemental composition of a work of art. XRF analysis on the saucer rivet detected the presence of copper, nickel, and zinc, positively identifying it as German silver.

From the correspondence between du Pont and Eberhardt it can be seen that there are different reasons for having ceramics repaired. One of the earliest reasons for such repair would be to restore vessels to usefulness. This is suggested by the eighteenth-century advertisements described above. In such cases the ceramic vessel might not be an antique. When one vessel of a set was broken, the repair might have been to maintain a complete set. The Canton-pattern porcelain saucer discussed here was not an item of great value, and one wonders if the repair was undertaken to maintain an intact set of teaware. The correspondence shows that riveting was not a very expensive process. Another factor in paying to have a vessel repaired might be because it has emotional value, in terms of who gave the vessel or from whom it was inherited. Antique vessels of value would make a natural candidate for repair. Clearly this was part of the motivation for repairing many of the vessels that du Pont sent to Eberhardt for repair.

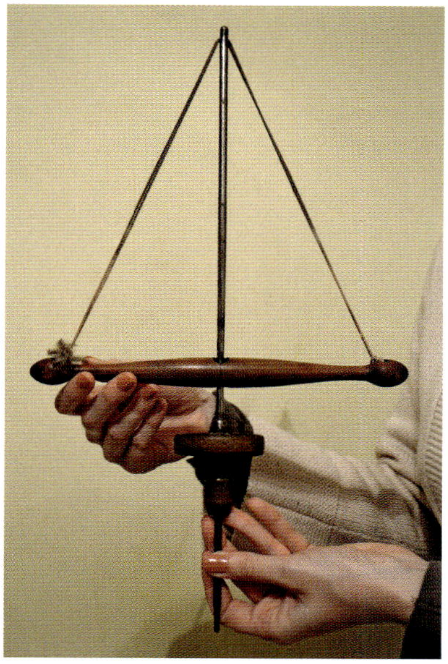

Figure 7 Drill stock with solid metal flywheel. (Photo, William David Brown.)

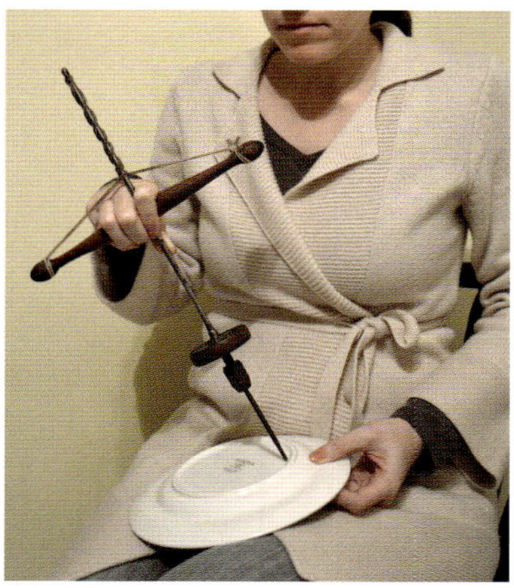

Figure 8 Emily Brown holding the wooden crossbar wound in position around spindle to begin drilling movement. (Photo, William David Brown.)

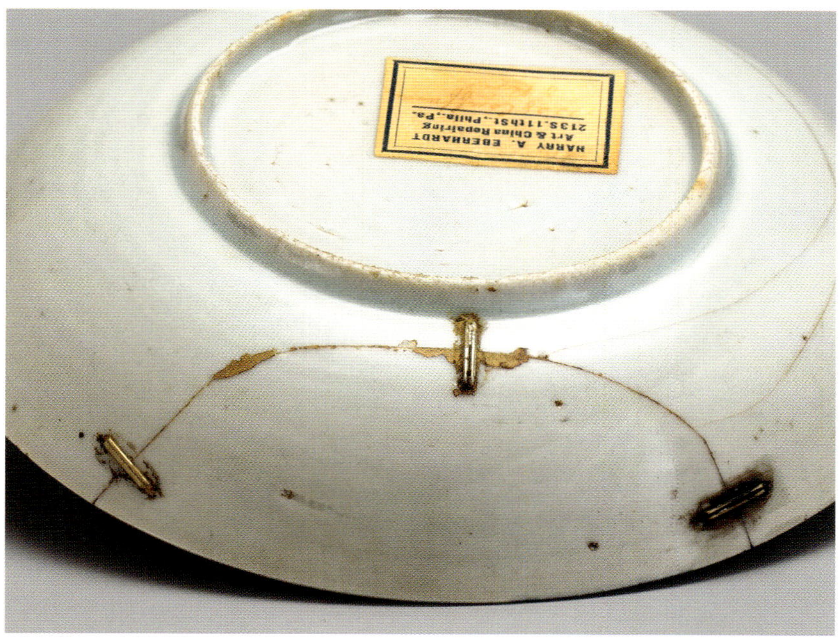

Figure 9 Detail showing the rivets in the saucer illustrated in fig. 1. (Photo, Robert Hunter.)

From the correspondence between du Pont and Eberhardt, it does not seem possible to determine the ratio of repairs that were done with rivets versus cements or which objects were in for cleaning or covering over chipped areas. In addition, some of the invoices from Eberhardt were for ceramics sold to du Pont. Beyond that there is the question of what proportion of the business was for antique ceramics versus repairing household ceramics for continued usage once they had been restored. Clearly, the repairs, cleaning, and restorations done for Henry Francis du Pont are not typical of the range of materials repaired by H. A. Eberhardt & Son.

An insight into the repair business can be gotten from Emily Browns's daily work log at Eberhardt's, which sheds some light on the nature of the business. From August 4 to August 27, 2010, she worked a total of 100 hours, during which period she handled 101 objects for various types of treatment, ranging from cleaning up old mends, restoring vessels back together, touchup painting of patterns, and other activities. These objects belonged to 77 clients, and 59 were completed during that work period. Major portions of these items were ceramic, but there were also some glass and wood items, and one resin object was repaired. Many of the items were not antiques; figurines were the most common item being repaired. These appear to be modern collectible and/or art objects, which include 22 Llado figures out of the 33 figurines that were restored. This record of what was typically being restored helps bring a balance to the record of the restoration work done by Harry A. Eberhardt & Son for Henry Francis du Pont.

While riveted ceramics can be found in antique shops and flea markets, they rarely bear the labels of who did the repairs. Thus, vessels with repair labels provide an insight into the vessel's history and are to be treasured. Those labels will be valued even more in the future as the firm of Harry A. Eberhardt & Son, after being in business continually since 1888, closed its doors in 2015, leaving a legacy of repaired china dishes throughout the Philadelphia region.

ACKNOWLEDGMENTS We are grateful for the information and comments on our paper provided by William R. Eberhardt, third-generation owner of H. A. Eberhardt & Son Inc., who grew up in the family business and repaired ceramics with rivets when he was younger. His description of the process was most helpful. We extend our appreciation to Winterthur librarians Heather Clewell, Jeanne Solensky, and Laura Parrish for their assistance in locating Eberhardt materials and securing copies of them. I would also like to thank my wife, Amy C. Earls, for proofreading earlier versions of this paper.

1. Some letters to Henry Francis du Pont in the Winterthur Archives are from Joseph Cauffman, the proprietor of the Benjamin Franklin Antique Shop in Philadelphia. He might well be the person who had the saucer repaired. The letters date from March 1927 to July 1934.

2. Advertisement, *Antiques* 18, no. 1 (July 1930): 83.

3. Bevis Hillier, *Pottery and Porcelain, 1700–1914: England, Europe, and North America* (New York: Meredith Press, 1968), pp. 301–2.

4. George Francis Dow, *The Arts & Crafts in New England, 1704–1775: Gleanings from Boston Newspapers Relating to Painting, Engraving, Silversmiths, Pewterers, Clockmakers, Furniture, Pottery, Old Houses, Costume, Trades and Occupations &c...* (Topsfield, Mass.: Wayside Press, 1927), p. 88.

5. Donald C. Towner, *The Leeds Pottery*, 1st American ed. (New York: Taplinger Publishing Co., 1965), p. 164.

6. Alfred Cox Prime, comp., *The Arts and Crafts in Philadelphia, Maryland, and South Carolina: Gleanings from Newspapers* (Topsfield, Mass.: Walpole Society, 1929), p. 67.

7. Rita S. Gottesman, comp., *The Arts and Crafts in New York, 1726–1776: Advertisements and News Items from the New York City Newspapers*, Da Capo Press Series in Architecture and Decorative Art 35 (1938; reprint, New York: Da Capo Press, 1970), pp. 86–87.

8. Prime, comp., *Arts and Crafts*, p. 237.

9. Brown and Eberhardt trade card, Joseph Downs Collection of Manuscripts and Printed Ephemera, collection 46, 68 x 155.92, Winterthur Library.

10. Brown and Eberhardt trade card, Winterthur Archives, HF 153, Winterthur Library.

11. Printed invoice from Harry A. Eberhardt to H. F. du Pont, dated June 7, 1927, in the collection of invoices to H. F. du Pont in the Winterthur Archives, Winterthur Library.

12. Patent No. 859: *Annual Report of the Commissioner of Patents for the Year 1876* (Washington, D.C.: Government Printing Office, 1877), p. 235 (*Official Gazette*, vol. 10, p. 705).

13. Lowestoft was an English porcelain factory. William Chaffers writing about this factory claimed that Chinese porcelains were being decorated there and that situation gave rise to collectors and antiques dealers talking about "Oriental Lowestoft." His *Marks & Monographs on European and Oriental Pottery and Porcelain*, 15th ed. (London: William Reeves, 1965), states that the British section was edited by Geoffrey A. Godden. Under "Lowestoft" in volume 2, p. 211, it states: "Previous editions have closely followed William Chaffers' original writings, in which he attributed a class of hard paste Chinese porcelain (decorated for the European market, with armorial bearings, etc.) to the Lowestoft factory. Former editors have added notes in an endeavor to correct Chaffers's serious error but confusion still arises. The time has not arrived to rewrite the Lowestoft section and in so doing to mention the confusing hard paste myth has been deleted." Chaffers's claims were rejected by L[ouis] M[arc] Solon's book *A Brief History of Old English Porcelain and Its Manufactories, with an Artistic, Industrial, and Critical Appreciation of Their Productions* (London: Bemrose & Sons, 1903), pp. 210–11. Because collectors and dealers nevertheless continued to use the term "Oriental Lowestoft" into the 1950s, mentions of Lowestoft in the Eberhardt–du Pont correspondence are most likely references to Chinese porcelain.

14. Claudia S. M. Parsons and Frederick H. Curl, *China Mending and Restoration* (London: Faber and Faber, 1963), pp. 27, 63.

15. Winterthur Museum Object Catalog no. 1969.0357, Eberhardt repair no. 10746.

16. William R. Eberhardt and George Miller, telephone conversation, August 14, 2012.

17. Parsons and Curl, *China Mending*, p. 20.

18. Thomas Pond, *Mending and Restoring China* (New York: Avenel Books, 1970), introduction.

19. Available online at www.corbisimages.com/Search#q=42-32917977&p=1 (accessed August 21, 2013).

20. Parsons and Curl, *China Mending*, n. 12. The authors dedicated 37 pages (28–64) to the process, including excellent line drawings that take the reader through drilling the holes, creating the rivet, and the process of pushing it into the prepared holes.

21. Ibid., p. 28.

22. For a long description of setting an industrial diamond chip into the end of a iron shaft such as a nail and finishing it into a bit for drilling ceramics, see ibid., pp. 144–51.

23. This description was taken from ibid., pp. 58–64.

24. William R. Eberhardt and George Miller, telephone conversation, August 14, 2012.

25. Analysis was performed using a Bruker ARTAX XRF spectrometer equipped with a rhodium tube at 500 A current, 40 kV voltage, and 100s live time, and beam spot size of 70–100 m. Spectra were collected and analyzed using Bruker ARTAX computer software (version 7.2.5.0).

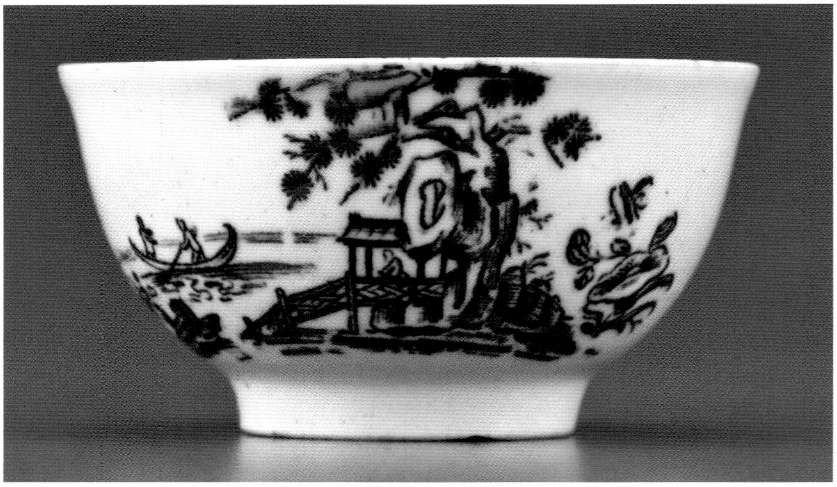

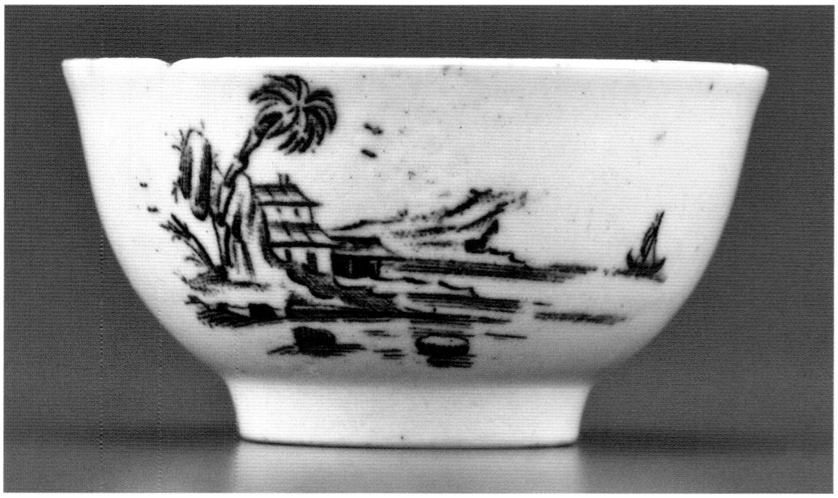

Figure 1 Teabowl, attributed to John Bartlam, Cain Hoy, South Carolina, 1765–1772. Soft-paste porcelain with transfer-printed decoration. H. 1⅝", D. 2⅞". (Private collection; photos, Gavin Ashworth.)

J. Victor Owen, John D. Greenough, and Nick Panes

Statistical Evaluation of Analytical Data for Eighteenth-Century American and British Sulphurous Phosphatic Porcelains

▼ THE PAST FEW decades have seen a proliferation of analytical (compositional) data for historical ceramics, most notably eighteenth-century American and British porcelains. These data provide an objective means of characterizing these diverse wares and therefore are useful in provenance and authentication studies. Moreover, when coupled with archival information, analytical data can help trace the dispersion of trade secrets to early porcelain manufacturers not only across Great Britain but also to its North American colonies.[1] Information about ceramic paste and glaze ingredients, their proportions, and other technical aspects of the manufacture of porcelain was transferred from factory to factory by workers changing employment or setting up their own potworks. Some of these workers are identified by historical documents.[2] The influence they had on the wares produced by established or new enterprises is confirmed by comparing the type and composition of porcelain made by the new and previous factories where these migrant artisans worked.

Perhaps the most significant recent discovery concerning the dissemination of porcelain technology and the production of new wares centers on the realization that Bonnin and Morris in Philadelphia (ca. 1770–1773) were likely not the first to make porcelain in what is now the United States. Rather, it seems that a Staffordshire immigrant named John Bartlam produced both fine earthenware and phosphatic (bone ash) porcelain in South Carolina starting in the mid-1760s. His enterprise there ended in about 1770. To date, four intact examples of this porcelain—all teabowls with a printed pattern (fig. 1) matching that from the several teabowl fragments (fig. 2) excavated at the site of his potworks—have been discovered.[3] Although obvious porcelain wasters (biscuit examples or fragments showing firing flaws such as body distortion) do not appear to have been discovered at the site, the general consensus presently holds that John Bartlam

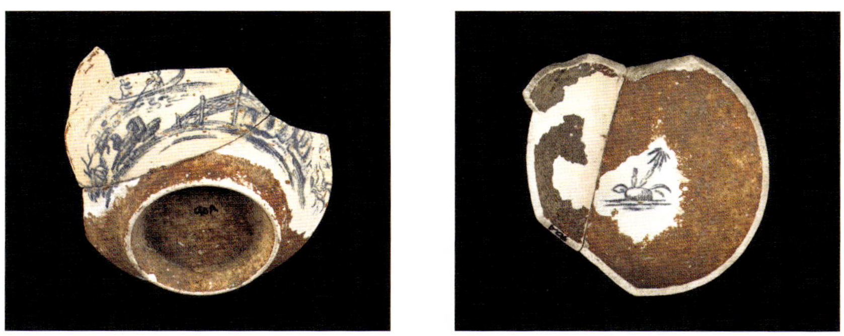

Figure 2 Teabowl fragment, recovered from the John Bartlam site, Cain Hoy, South Carolina, 1765–1769. Soft-paste porcelain. (South Carolina Institute of Archaeology and Anthropology; photos, Gavin Ashworth.)

produced porcelain in what is now the United States several years prior to Bonnin and Morris.

Analyzed porcelain sherds from the Bartlam and the Bonnin and Morris sites have sulphurous phosphatic compositions and lead-rich glazes.[4] The body of some of the Bonnin and Morris fragments contains small amounts of lead, a component not yet detected in analyzed Bartlam pastes.[5] A dated (1773) and marked ("Philadelfia") openwork porcelain basket attributed to Bonnin and Morris, however, has a silicious–aluminous–calcic (SAC) composition. If it was indeed made by Bonnin and Morris, as seems likely since no other maker of porcelain is known in Philadelphia at that time, this artifact demonstrates that at least two types of porcelain pastes were produced at this factory.[6]

Diverse types of artificial (soft-paste) porcelains were produced in Britain starting in the 1740s (or perhaps earlier), including those with phosphatic and SAC compositions. Of the many porcelain factories that manufactured phosphatic porcelains during the mid-eighteenth century, only two—Isleworth and Bow (both of London)—are known to have included sulphur and/or lead in their ceramic pastes. The presence of either or both of these components in early American porcelains therefore points to the influence that London factories had on their counterparts across the Atlantic. In the case of Bonnin and Morris, it appears that a William Ball (a former employee at the Bow factory but later manager of the former William Reid works in Liverpool), together with a number of ex-Bow workmen, provided knowledge to Bonnin and Morris at their Philadelphia factory about how to make Bow wares.[7] Just how John Bartlam became familiar with the use of sulphur (probably in the form of gypsum, a common calcium sulphate mineral), however, is not known.

This paper focuses on the composition of eighteenth-century American and British porcelains with variably lead-bearing, sulphurous phosphatic compositions: those made by Bow, Isleworth, Bartlam, and Bonnin and Morris. Published analytical data for sherds of this material, supplemented by the compositions of authenticated and documentary pieces, are compared with the compositions (1) of a dozen intact, blue-and-white decorated objects that have been assigned, on aesthetic and compositional grounds, to either the Bow or Isleworth factory, and (2) two intact dishes and two excavated but reconstructed sauceboats suspected to have originated at the Bonnin and Morris factory.[8] Unlike the first-cited study, which uses binary (X-Y) plots of selected components to distinguish between products of the two London factories in question, in the present contribution we use an exploratory statistical method that collectively appraises all components simultaneously. New discrimination diagrams have then been constructed using these analytical data.

Sample Description

The information compiled here comprises published data for both paste and glaze compositions. The former outnumber the latter because at some sites many of the excavated porcelain sherds are biscuit and, in the case of glazed objects, the glaze may in rare instances flake away from its ceramic

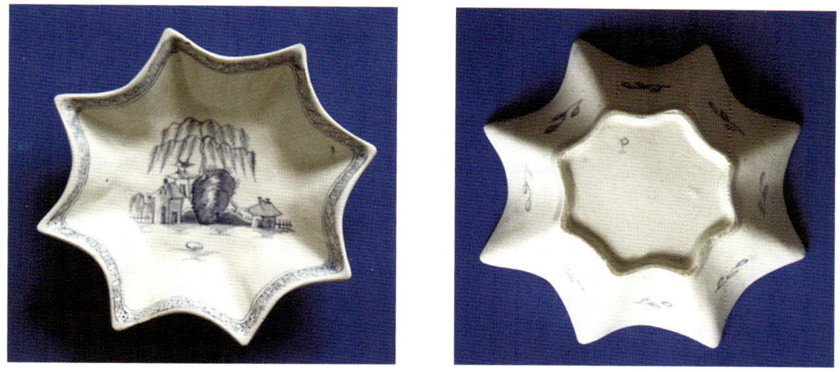

Figure 3 Dish, possibly the American China Manufactory, Philadelphia, Pennsylvania, 1770–1772. Soft-paste porcelain. D. 5⁷⁄₁₆". (Private collection; photo, Nicholas Panes.) This dish is designated "Sample STR."

substrate during sample preparation (polishing of glass-mounted ceramic fragments).[9] To avoid clutter on the statistical plots, data reported for glazes only and not for the associated pastes (e.g., the averaged Bow glaze compositions reported by Ross Ramsay and coworkers) were avoided but are included in discrimination diagrams described below.[10]

The two intact porcelain objects suspected to be American in origin are (1) a star-shaped dish (fig. 3) marked on its base with a P, the symbol commonly used by Bonnin and Morris, and (2) an unmarked saucer (fig. 4)

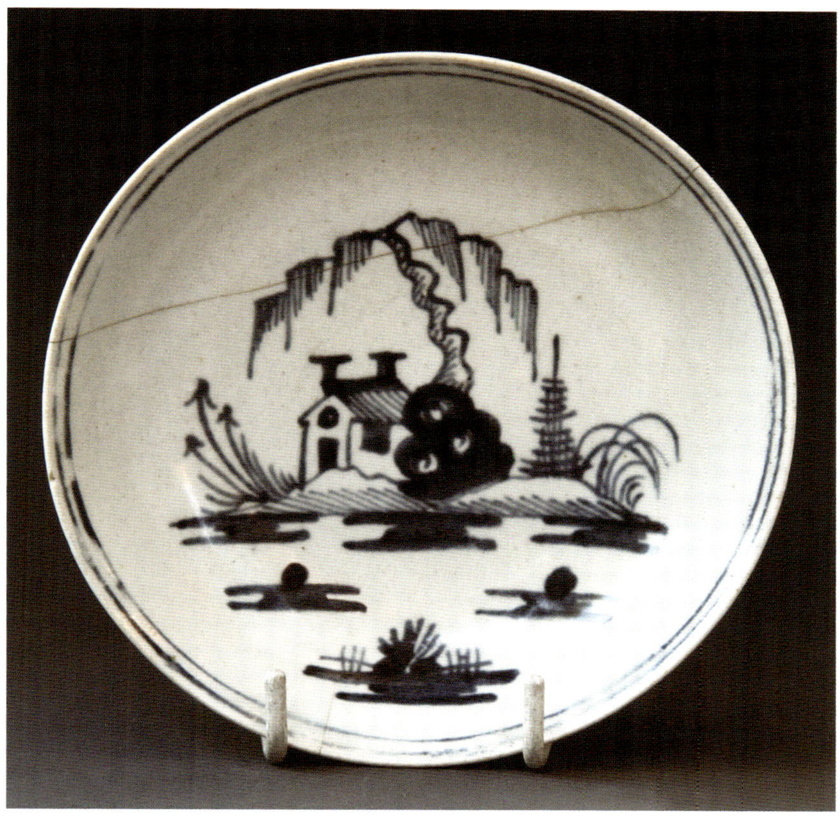

Figure 4 Saucer, possibly the American China Manufactory, Philadelphia, Pennsylvania, 1770–1772. Soft-paste porcelain. D. 4¹⁵⁄₁₆". (Private collection; photo, Nicholas Panes.) This saucer is designated "Sample DB1."

painted in underglaze blue with a house, a willow tree growing from a rock, and curling reeds. The compositions for both are reported in Table 1.

Two sauceboats (fig. 5) excavated from Independence National Historic Park in Philadelphia have been assigned to the Bonnin and Morris factory on the basis of their (1) similarity to a Bonnin and Morris sauceboat in the Brooklyn Museum, and (2) composition and the presence in their bodies of heavy minerals that match those in clay deposits at White Clay Creek, a tributary of the Delaware River, known to have been exploited by Bonnin and Morris.[11]

Table 1. Compositions of two intact porcelain objects (see figs. 3 and 4) suspected to originate at the Bonnin and Morris factory.

	STR paste	STR glaze*	DB1 paste	DB1 glaze*
SiO_2	49.8	45.04	54.9	52.35
TiO_2	0.8	0.23	0.4	0.34
Al_2O_3	7.7	1.19	10.6	8.06
FeO	0.5	0.41	0.5	0.80
MgO	0.6	0.08	0.3	0.00
CaO	22.7	2.51	18.3	2.76
Na_2O	0.6	0.68	0.4	0.92
K_2O	2.2	2.39	1.7	3.76
P_2O_5	13.7	0.16	10.5	0.00
PbO	0.0	45.28	1.9	31.02
SO_3	1.3	0.14	0.2	0.00
SnO_2	0.0	1.27	0	0
Total	99.9	99.38	99.7	100.01

*unzoned

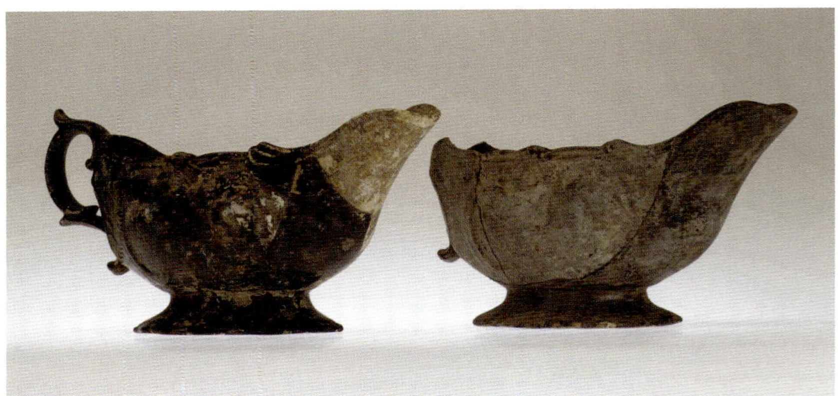

Figure 5 Sauceboats, probably the American China Manufactory, Philadelphia, Pennsylvania, 1770–1772. Soft-paste porcelain. H. 3⅜", W. 3", L. 6½". (Independence National Historic Park, Philadelphia; photo, Jed Levin.) These two reconstructed sauceboats (samples 2028A, 2028B) were recovered from Independence National Historic Park in Philadelphia. Based on their composition and mineralogy, they have been attributed to the Bonnin and Morris factory.

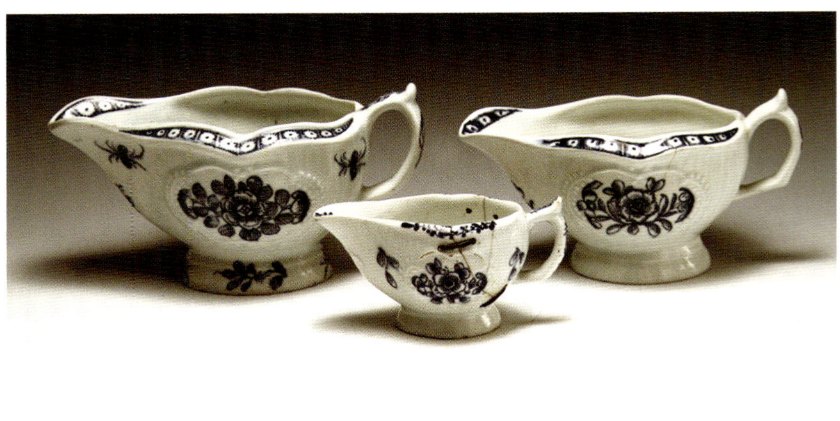

Figure 6 Three sauceboats of a class sometimes attributed to Isleworth (London) factory, England, 1770–1775. *Left to right*: (U3) L. 6⅛"; (U4) L. 3¹⁵⁄₁₆"; (U5) L. 5½". Mark: crescent on U5; others unmarked. (Private collection; photo, Nicholas Panes.)

Twelve intact, sulphurous phosphatic objects variously attributed to Bow and Isleworth were included in the sample. These include a saucer and two baskets that have been illustrated elsewhere.[12] Illustrated in this article are three sauceboats (fig. 6), three plates (fig. 7), and three dishes (fig. 8).

Figure 7 (*top row*) Plates and dish of a class sometimes attributed to the Isleworth (London) porcelain factory, England, 1765–1775. Soft-paste porcelain. *Left to right*: (U6, plate) D. 7¼"; (U7, dessert dish) L. 8 15/16"; (U8, plate) D. 8". Mark (on all): mock Chinese characters. (Private collections; photos, Ray Howard [U6] and Nicholas Panes [U7, U8].)

Statistical Methods

Paste and glaze analyses were subjected to exploratory statistical analysis (multidimensional scaling [MDS]) to identify compositional relationships between the different samples.[13] MDS resembles principal component analysis except that the results are usually easier to interpret because vari-

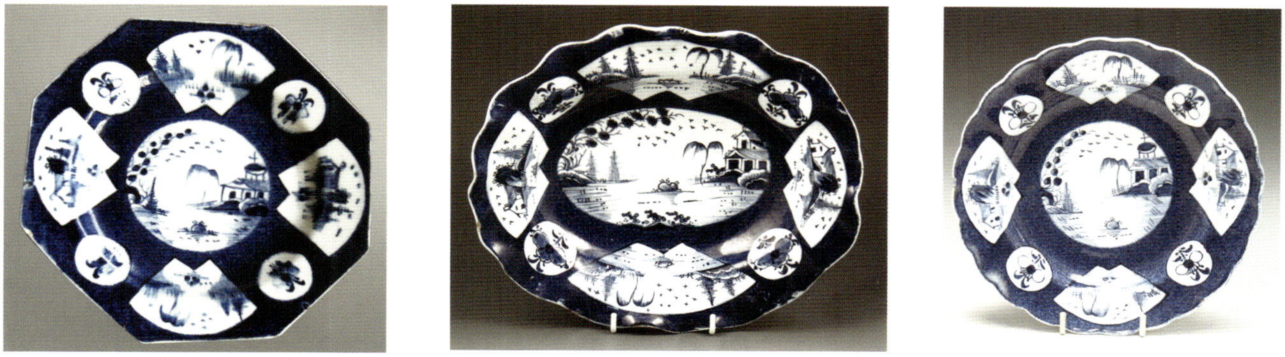

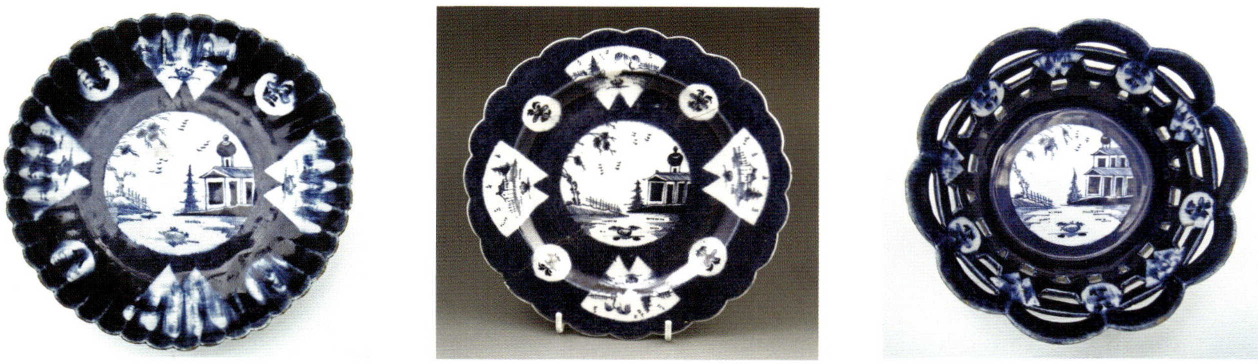

Figure 8 (*bottom row*) Plates with a different style of decoration from those illustrated in fig. 7; tentatively attributed to the Isleworth (London) porcelain factory, England, 1765–1775. Soft-paste porcelain. *Left to right*: (U9, fluted plate) D. 7⅛"; (U10, lobed plate) D. 8⅛"; (U11, basket) D. 8¾". Marks: none. (Private collections; photos, Ray Howard [U9, U11] and Nicholas Panes [U10].)

ance is summarized in fewer dimensions. The MDS calculations were performed using SYSTAT software. This software does not distinguish one sample (or chemical component) from another and does not seek correlations supporting preconceived models for relationships between samples or chemical components.

The clustering of data points on MDS diagrams identifies which elements are geochemically linked in the analyzed samples and, likewise, which samples are related. In the present study, only sample compositions are examined, since the behavior of particular components is largely irrelevant to distinguishing the products of different porcelain factories.

In essence, MDS diagrams can be thought of as "maps" showing "distances" between samples or chemical constituents. In three dimensions, MDS data tend to plot as a spheroidal cloud of points that can be viewed in three, mutually perpendicular planes (dimensions 1 versus 2, 3 versus 2, and 3 versus 1). Generally, only the first of these two-dimensional plots is required to interpret the analytical data, but if a relatively imperfect circle is defined by the plotted data (i.e., some points plot in its interior), then additional planes should be consulted.

Samples (or chemical constituents) that are similar, or that behave similarly, plot close together on MDS diagrams. Points on opposing sides represent samples (or chemical constituents) that are dissimilar or behave differently. These diagrams can be used to interpret compositional similarities and dissimilarities in a variety of media, including earthenware, glass, rocks, and fluids such as wine and even maple syrup.[14] In the present application of the method, the analytical data are restricted to major and minor elements (Si, Ti, Al, Fe, Mn, Mg, Ca, Na, K, S, and Pb, expressed as oxides). Trace element data (and isotopes) for these porcelain wares have not yet been determined (or, at least, reported in the literature).

The same paste and glaze data used to generate MDS plots provided a basis for generating new discrimination diagrams using F-statistics as a guide for identifying diagnostic components that best distinguish the products of the different porcelain factories considered here.[15]

Results

PASTE COMPOSITIONS An MDS plot for sulphurous phosphatic porcelain paste compositions is shown in figure 9. The plotted compositions of samples excavated from the Bonnin and Morris site cluster in a tight field in the lower right of this diagram. In contrast, four of the five Bartlam samples

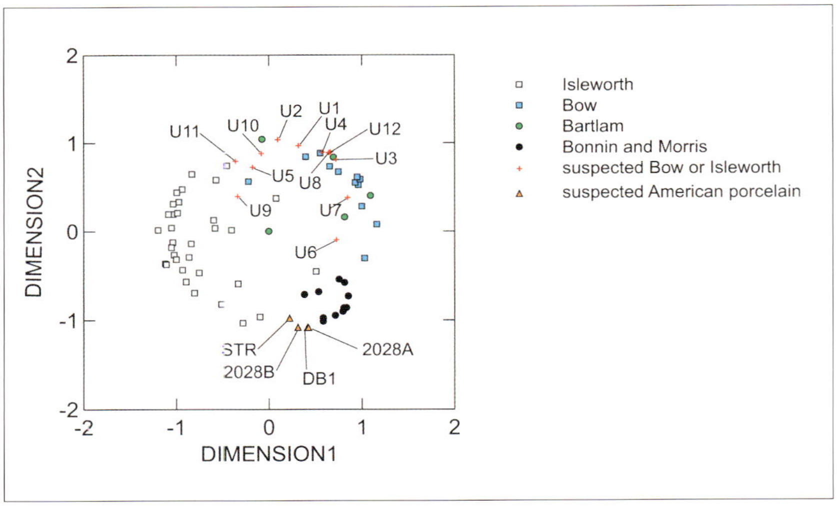

Figure 9 MDS plot for paste compositions of Bartlam, Bonnin and Morris, Bow, and Isleworth porcelain, plotted against data for four suspected American porcelain objects (two intact and two excavated), and twelve intact porcelain objects thought to be either Bow or Isleworth.

plot in a broader field in the upper right, with one sample (B10) plotting in the interior of the diagram. The main cluster of Bow samples overlaps the field defined by most of the analyzed Bartlam sherds. Significantly, all four suspected American porcelain samples cluster at the bottom, between the Bonnin and Morris and the Isleworth fields; two (DB1 and 2028A) plot atop one another on this diagram.

The plotted compositions of most of the analyzed fragments from the Isleworth factory define an arc on the left side of figure 9, although five samples plot in the interior of the diagram, and one plots closer to the lower right, near the field for the Bonnin and Morris samples. The compositions of the Bow fragments plot as an arc on the right to upper-right part of figure 9. The average composition of Bow pastes circa 1747–1754, with a relatively aluminous composition, however, plots in the upper left of this diagram.[16]

The compositions of the twelve analyzed intact objects suspected to be either Bow or Isleworth porcelain plot as an arc along the top of figure 9, extending toward the right from one end of the Isleworth field, and overlapping the fields for Bow and Bartlam porcelain.[17] Their distribution is generally consistent with inferences made about their origin based on aesthetic criteria and the three ceramic-paste discrimination diagrams shown in Table 2 (Test 1, Test 2, and Test 3) proposed by Panes and coworkers.[18] Six samples (U1, U3, U4, U7, U8, U12) cluster with the Bow samples on figure 9, and another (U6) plots close to the Bow field. Three other specimens (U5, U9, U11) plot close to the Isleworth field. A blue-and-white basket with the Pinecone pattern (U2) and thought to be Bow, and a powder-blue plate (U10) considered to be Isleworth, plot approximately midway between the Bow and Isleworth fields.

TABLE 2. Description and appraisal on compositional grounds of 12 intact phosphatic porcelain objects suspected to be either from the Bow ("B") or Isleworth ("I") factory.

	Description	Test 1	Test 2	Test 3	Glaze Na_2O vs Al_2O_3	MDS (Fig. 5) Paste	MDS (Fig. 6A) Glaze: D1 vs D2	MDS (Fig. 6B) Glaze: D2 vs D3	MDS (Fig. 6C) Glaze: D1 vs D3
U1	saucer	B	B	B	B	B	(I)	i	(I)
U2	basket	B	B	B or I	B	(B) or (I)	(I)	i	?
U3	sauceboat	B	B	B	B	B	?	?	(I)
U4	sauceboat	B	B	B	B	B	?	i	B
U5	sauceboat	I	I	I	*	I	*	*	*
U6	Octagonal powder-blue plate	B	I	I	?	(b)	(I)	i	(I)
U7	Oval powder-blue dessert dish	B	B	B	B	B	B	B	?
U8	24 lobed powder-blue plate	B	B	B	?	B	B	?	(I)
U9	40 flute powder-blue dish	I	I	I	I	I	I	(B) or (I)	I
U10	23 lobed powder-blue plate	I	B or I	I	I	(B) or (I)	I	?	i
U11	Powder-blue basket	I	I	I	I	I	I	I	i
U12	Octagonal platter/meat plate	B	B	B	B	B	(I)	I	(I)

NOTES

Test 1, lead content of paste

Test 2, P2O5 versus Al2O3 content of paste

Test 3, SiO2 versus sum of MgO, Al2O3, Na2O, and K2O content of paste

These three tests and the soda vs alumina diagram used to distinguish Bow from Isleworth glazes are from Nicholas Panes and J. Victor Owen, "Attribution Enhanced: Possible American 18th-Century Porcelain Finds in England," *English Ceramic Circle Transactions* 23 (2012): 11–30.

* no glaze data are available

MDS PLOTS

(B), (I) Sample plots close to an isolated Bow or Isleworth sample (respectively)

? Sample plots close to neither the Bow nor Isleworth field (or individual samples)

B, i Sample plots close to but not in the Bow or Isleworth field (respectively)

GLAZE COMPOSITIONS Nick Panes and his colleagues noted that published analyses of Isleworth glazes tended to be enriched in both Na_2O (soda) and Al_2O_3 (alumina) compared with their Bow counterparts.[19] Three of the intact powder-blue objects described in that paper (U9, U10, and U11 in Table 2) plot in or near the Isleworth field, corroborating their attribution to that factory based on paste compositions. In contrast, neither of the suspected Bow powder-blue plates (U6, U8) plots in the Bow field on Panes et al.'s glaze diagram, since they are relatively enriched in alumina.[20] The glazes on the other intact objects plot in or near the Bow field on their soda versus alumina diagram (see Table 2), supporting their attribution to the Bow factory based on paste compositions. It therefore seems that these two glaze components have some potential in distinguishing the phosphatic wares produced by these two factories.

What if all analyzed glaze components for sulphurous phosphatic wares were appraised simultaneously? An MDS plot for glaze compositions is shown in figure 10a. Apart from two outlying samples (U4, U8), the compositions of the glazes on the analyzed intact objects plot in two fields. As in Panes et al.'s soda versus alumina plot, the glazes on powder-blue samples U9, U10, and U11 cluster in or near the field for Isleworth glazes on the left side of figure 10a. Six glazes plot on the opposing side of this diagram, but not with Bow glazes, most of which form an elongate field across the middle of figure 10a. Instead, they are associated with Bartlam glazes, some of the Bonnin and Morris glazes, and one Isleworth glaze. The glaze on a lobed, powder-blue plate (U8) plots near the Bow and the Bonnin and Morris fields, whereas the glaze on a blue-and-white sauceboat (U4) plots near a Bartlam glaze in the interior of figure 10b.

Most Bonnin and Morris glazes plot on the bottom of figure 10a, overlapping the Isleworth field, whereas others plot on the upper right. The two excavated sauceboats have glazes that plot close to one another on the top right of the diagram, close to several Bartlam glazes. The glaze on the suspected American saucer (DB1) plots at the left end of the lower Bonnin and Morris glaze field. The glaze on the P-marked star-shaped dish (STR) plots in the gap between the two Bonnin and Morris fields, coincident with the position of the glaze on a powder-blue dessert dish (U7).

The clustering of most Bow glazes in the interior of figure 10a suggests that their relation to other glazes should be considered in other dimensions. Consequently, all glaze data are plotted on two other MDS diagrams: dimensions 2 versus 3 (fig. 10b) and 1 versus 3 (fig. 10c). In contrast to figure 10a, glaze data on the dimensions 1 versus 3 plot (fig. 10c) define a clear circle, with relatively few points in its interior. Apart from one outlying glaze on the upper right, Bow glazes form a well-defined arc on the lower part of the diagram that overlaps with the main Isleworth field to the left. Several of the intact objects suspected to be Bow (see Table 2) plot as extensions of this field (U1, U3, U7, U12); only one (U4) plots within it. The three powder-blue specimens thought to be Isleworth (U9, U10, U11 in Table 2) are associated with the Isleworth field. Bartlam and Bonnin and Morris glazes overlap on the right, and their collective field encloses three of the four suspected

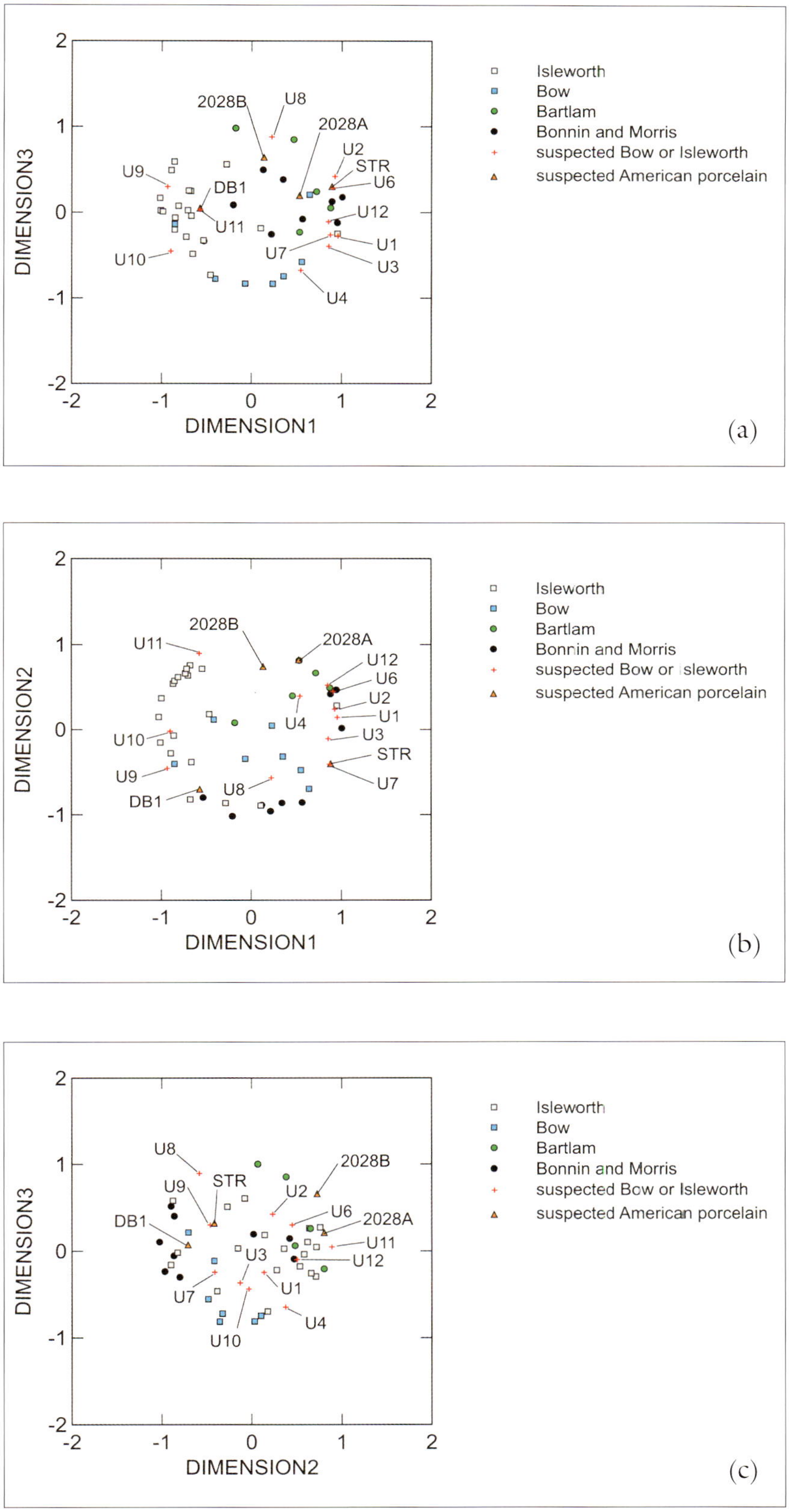

Figure 10 MDS plots for glaze compositions. (a) dimensions 1 versus 2; (b) dimensions 2 versus 3; (c) dimensions 1 versus 3.

American dishes (2028A, 2028B, STR). The fourth (DB1) is coincident with a suspected Isleworth sample (U11).

The glaze data are distributed as a poorly defined circle on the dimensions 2 versus 3 plot (see fig. 10b), suggesting that it is of relatively little use in distinguishing most samples. The Bow field is on the lower left and contains only one of the suspected Bow objects (U7). Most Isleworth glazes plot on the upper-right and middle-right part of the diagram. Of the three powder-blue dishes with strong Isleworth paste attributes (see fig. 7), only the glaze on U11 is associated with the Isleworth field on this diagram.

Some New Discrimination Diagrams for Sulphurous Phosphatic Porcelains

Statistical evaluation of published sulphurous phosphatic paste compositions shows that several elements can be used to distinguish these wares. This can be accomplished in two ways. Components can be sought which help distinguish the products of (1) all four factories (Bonnin and Morris, Bartlam, Bow, and Isleworth) from each other, and (2) one factory from the others. In the first instance, there invariably is some overlap of points representing the compositions of one type of porcelain (paste or glaze) with those of another. Consequently, it is more effective to use a series of diagrams designed to distinguish the products of two or three of the four factories

Figure 11 Discrimination diagrams for sulphurous phosphatic porcelain pastes. (a) K_2O versus Al_2O_3; (b) PbO versus P_2O_5; (c) SO_3 versus CaO; (d) Na_2O versus CaO.

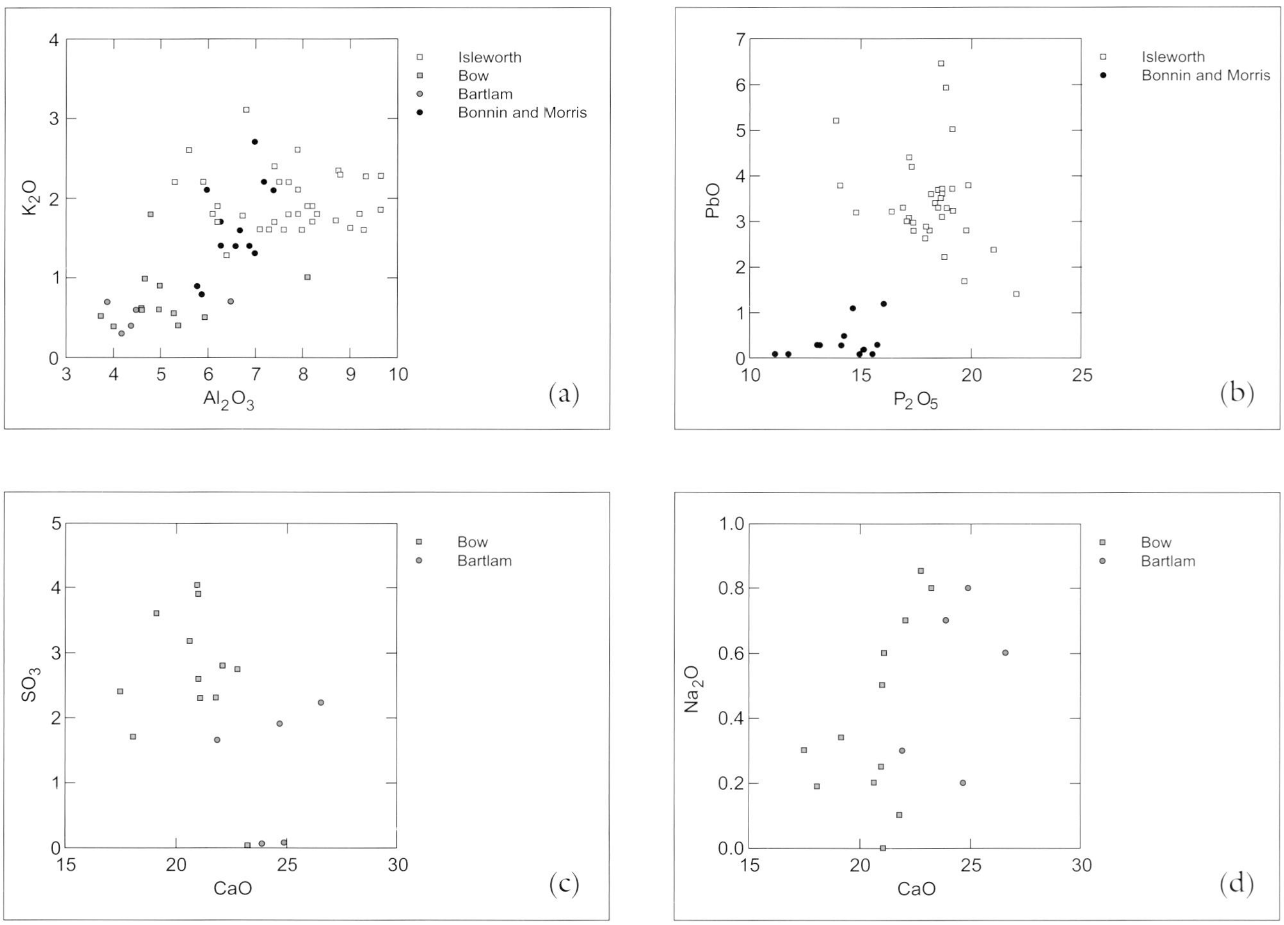

considered here, rather than all at once. Bartlam, Bonnin and Morris, Bow, and Isleworth pastes are best distinguished by Si Al, P, and the alkalis (or subsets thereof). Their glazes can be distinguished by Ti, Na, K, and Pb.

Examples of discrimination diagrams based on these various components are shown in figures 11a–d (pastes) and figures 12a and b (glazes). Figure 11a compares the potash and alumina contents of Bartlam, Bonnin and Morris, Bow, and Isleworth porcelain pastes. Most of the analyzed Bow samples have lower concentrations of these components than Isleworth, so this plot clearly distinguishes these two factories. It also distinguishes between products of the two American factories, but there is nearly complete overlap of Bonnin and Morris compositions with the Isleworth field (even though there is no known association between these two factories) and of Bartlam with the Bow field (even though Bow is linked to Bonnin and Morris). So, another plot must be used to distinguish between Bartlam and Bow, and between Bonnin and Morris and Isleworth. Isleworth pastes have higher lead contents than those of Bonnin and Morris, so a lead oxide versus phosphate plot (fig. 11b) diagram separates these two wares. Similarly, higher lime but lower sulphate contents tend to distinguish Bartlam from Bow porcelain, so the CaO versus SO_3 diagram (fig. 11c) helps separate these two wares. A lime versus soda plot (fig. 11d) can also be used for this purpose.

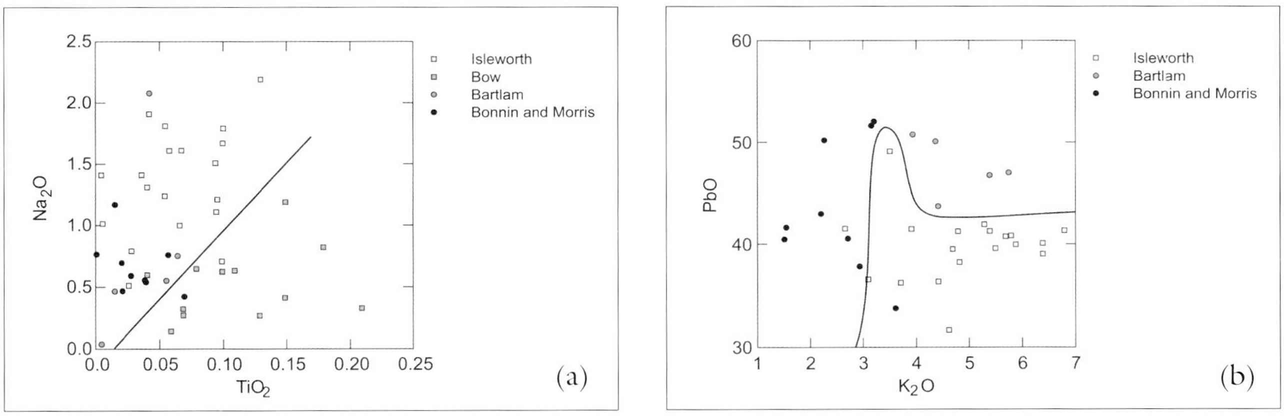

Figure 12 Discrimination diagrams for glazes on sulphurous phosphatic porcelains. (a) TiO_2 versus Na_2O; (b) PbO versus K_2O (the line encloses the Isleworth field).

As with the discrimination diagrams based on paste compositions, those constructed for glazes tend to distinguish two or three factories from one another, but not all of the four considered here. For example, a titania versus soda plot distinguishes most of the analyzed Bow from Isleworth glazes, but both American factories overlap the latter's field (fig. 12a). However, a lead oxide versus potash plot (fig. 12b) can be used to distinguish most Bartlam, Bonnin and Morris, and Isleworth glazes.

Discussion

The use of compositional data to identify or authenticate early American and British porcelain objects is limited by the small number of samples and the limited number of components that have been determined, and reported, for these wares. Despite the limited amount of data for these wares, attempts to determine the origin of both intact and archaeological specimens on compositional grounds have met with some notable successes. Perhaps best-known among these was the recognition of the first intact specimens of Bovey Tracey porcelain—first a fuddling cup[21] and later a sauceboat.[22] Composition-based attributions are best attempted in conjunction with aesthetic criteria but in and of themselves are inherently objective and avoid the subjectivity of connoisseurship. There nonetheless remains the strong possibility that variations in the pastes and glazes produced by a given porcelain factory are as yet unrepresented in the analytical database currently available for early American and British wares. For this reason, simply plotting the composition of the paste and/or glaze of an unidentified specimen on a discrimination diagram, or diagrams, might mislead the researcher. The result of this approach, to be reliable, should be consistent with what is known about the character of the wares of the factory suspected to have produced the sample. In other words, there should be no "loose strings."

An advantage of the statistical approach (MDS) advocated here is that the software evaluates all data simultaneously, rather than piecemeal. This reveals relationships between samples, or the constituents (chemical components) making up these samples, that would elude the researcher casually perusing the dataset.

The MDS plots reported here highlight both similarities and dissimilarities between the sulphurous phosphatic porcelains produced at different American and British factories. They supplement more traditional, binary (X-Y) plots comparing, for example, the Al_2O_3 and P_2O_5 concentrations in phosphatic porcelain pastes. Nick Panes and coworkers proposed additional binary plots to help distinguish among twelve intact objects suspected to be either Isleworth or Bow.[23] These diagrams, supplemented by aesthetic criteria, pointed to a Bow origin for most of these samples, and MDS of the paste data tend to support this interpretation (see fig. 9). It is encouraging that, with one exception, the paste compositions of analyzed sherds from the Bow and Isleworth factory sites plot on distinct parts (opposing sides) of figure 9. This largely reflects the relatively high lead content of Isleworth porcelain.

The paste compositions of the four suspected American porcelain objects plot in a tight cluster between the Bonnin and Morris and Isleworth fields (see fig. 9). Although it is known that former Bow employees found work at the Bonnin and Morris factory, to our knowledge there is no known connection between Isleworth and either American factory. Moreover, the excavated sauceboats (2028A, 2028B) contain some of the same heavy minerals that occur in a known clay source (White Clay Creek, a tributary of the Delaware River) used by Bonnin and Morris. The clustering of all four of these objects near Bonnin and Morris supports but does not prove a Philadelphia origin for these artifacts. The compositions of their glazes are broadly consistent with this interpretation. Bonnin and Morris glazes plot as a divided arc on the upper right/bottom of one of the MDS glaze diagrams (see fig. 10a), and as an irregular cluster on another (see fig. 10b). In the case of the former, glaze compositions of the suspected American artifacts plot as extensions of the

Bonnin and Morris fields; in the case of the latter, three of the four plot in the Bonnin and Morris field, with the intact saucer (DB1) being associated instead with the Isleworth field.

Apart from three of the powder-blue dishes thought on the basis of their paste compositions to be Isleworth, the glazes of the intact Bow/Isleworth artifacts show less clear affiliation with the fields of glazes for these two types of porcelain. As noted elsewhere,[24] the use of glaze compositions in provenance/authentication studies is potentially problematic. Glazes can be compositionally zoned, can contain irregular concentrations of refractory ingredients (notably tin oxide, a whitener and opacifier), and, in some instances (such as at unusually high glost kiln temperatures) can behave as a chemically open system, exchanging or taking up components (e.g., alumina) from the ceramic substrate,[25] and losing lead at high temperatures to the kiln atmosphere.[26] At extreme temperatures (approaching those at which true porcelains are fired), all the lead from the glaze can eventually be driven off.

Although peak temperatures in the glost kiln were intended to be much lower than those used to fire artificial porcelain biscuit (prior to glazing), the common observation that glazed wares (found in wasters at factory sites) suffered body distortion demonstrates that glost kiln conditions could be difficult to control. Indeed, it may well be that variations in the compositions of glazes (e.g., their lead content) noted on wares made during particular periods of a factory's production history may largely be a result of this effect rather than substantial differences in their recipes, although this inference could be disputed given the discovery of detailed factory records to the contrary.

It is not always the case that the degree, if any, of zoning in porcelain glazes is noted or even sought by analysts, and those using portable XRF to test glaze compositions are, by necessity, confined to analyzing glaze surfaces. The same is true for variations in the distribution of refractory particles, although this is of limited concern in the present instance because of the lack of tin oxide particles in most of the analyzed glazes (except, for example, the glaze on the star-shaped dish illustrated in fig. 3; see Table 1). The potential open-system behavior of glazes during kiln firing is of special concern, because a given starting composition of glaze could systematically change its composition with temperature and/or soaking interval. With these caveats in mind, it is nonetheless noteworthy that the glaze data presented by Ross Ramsay and coworkers[27] suggest minor but apparently systematic variations in the compositions of Bow glazes over time (such as an increase in K_2O), and the soda versus alumina plot comparing Bow and Isleworth glazes presented by Panes and coworkers shows good field separation.[28] These studies therefore point to the potential use of glazes in provenance studies of artificial porcelains, and indeed, the lead content of glazes are diagnostic of some phosphatic porcelains (e.g., analyzed Nantgarw glazes contain only ~15% PbO).[29] The data presented here suggest that, in the case of Bartlam, Bow, Bonnin and Morris, and Isleworth, titania and the alkalis can be diagnostic of the glazes used at these four factories (see figs. 12a, b).

Only major and minor elements numbering about a dozen are presently available in the literature for historical Anglo-American wares. In contrast, trace element packages offered by commercial labs equipped with a high-precision, mass spectrometer–based apparatus (e.g., laser ablation micro-probe—ICP-MS) typically include an additional forty or more components that can be determined at parts-per-million concentrations. Consequently, the analysis of American and British porcelain pastes and glazes for trace elements (and isotopes) could revolutionize not only our understanding of how these wares were made (especially with respect to the sources of their raw materials) but also our ability to distinguish them. Such work has been undertaken on ancient Chinese high-fired ceramics for more than a decade and could serve as a template for these types of investigations.[30]

Conclusions

Statistical analysis of compositional data for sulphurous phosphatic, eighteenth-century American and British porcelains supplements attempts to distinguish the products of different factories using individual pairs of components (or combinations thereof) identified by simply perusing the analytical data. Multidimensional scaling of published data for twelve intact porcelain objects suspected to have been made at either the Bow or the Isleworth factory corroborates their attribution using simple binary plots, but tend to support a Bonnin and Morris origin for two intact dishes for which this attribution initially seems premature.

Despite potential problems related to zoning, and the progressive lead loss and uptake of some components from the ceramic substrate during kiln firing, glaze compositional data also tend serve to distinguish between the sulphurous phosphatic procelains described here. In contrast to ongoing trace element (and isotopic) analysis of ancient Chinese high-fired wares, the analytical investigation of eighteenth-century American and British porcelains has, to date, been confined to major and minor elements. We therefore anticipate that the successful distinction between the products of some of these porcelains and a better understanding of the character and source of their raw materials can be augmented only when a comprehensive database including these additional components becomes available.

ACKNOWLEDGMENTS We thank Independence National Historic Park in Philadelphia for providing the image of the sauceboats shown in figure 5. Analytical work was supported by a Saint Mary's University Faculty of Graduate Studies and Research Grant to J. Victor Owen.

1. J. Victor Owen and Maurice Hillis, "From London to Liverpool: Evidence for a Limehouse–Reid Porcelain Connection Based on the Analysis of Sherds from the Brownlow Hill (ca. 1755–1767) Factory Site," *Geoarchaeology* 18, no. 8 (2003): 851–82 [DOI 10.1002/gea.10094]. Nicholas Panes, "Mr Ball the English Potter and the American China Manufactory," *English Ceramic Circle Transactions* 20, no. 3 (2009): 633–45.

2. Dr. Richard Pococke's widely cited mid-eighteenth-century correspondence describes some of the wares produced at the Lund's Bristol porcelain factory, whose assets were acquired by the proprietors of the Worcester factory. The names of those associated with the latter enterprise who had previous experience in this industry elsewhere are mentioned in R. Burt, "Profits, Politics and Philanthropy: The Origins of Worcester Porcelain, 1740–1751," *The Inter-*

national Ceramics Fair & Seminar: 2006 ([London]: [International Ceramics Fair], 2006), pp. 22–27.

3. Roland Arkell, "The First American Porcelain—in Britain," *Antiques Trade Gazette* (February 4, 2013), available online at http://www.antiquestradegazette.com/news/2013/feb/04/the-first-american-porcelain-in-britain/

4. As defined by J. Victor Owen, "A New Classification Scheme for Eighteenth-Century American and British Soft-Paste Porcelains," in *Ceramics in America*, edited by Robert Hunter (Hanover, N.H.: University Press of New England for the Chipstone Foundation, 2007), pp. 121–40.

5. J. Victor Owen, "Geochemical and Mineralogical Distinctions between Bonnin and Morris (Philadelphia, 1770–1772) Porcelain and Some Contemporary British Phosphatic Wares," *Geoarchaeology* 16 (2001): 785–802 [DOI 10.002/gea.1020]; J. Victor Owen, "Geochemistry of High-fired Bartlam Ceramics," in *Ceramics in America*, edited by Robert Hunter (Hanover, N.H.: University Press of New England for the Chipstone Foundation, 2007), pp. 209–18.

6. J. Victor Owen and Robert Hunter, "Too Little, Too Late: The Geochemistry of a 1773 Philadelphia Openwork Porcelain Basket," *Journal of Archaeological Sciences* 36 (2009): 333–42.

7. See Owen and Hillis, "London to Liverpool: Evidence for a Limehouse"; Panes, "Mr Ball the English Potter."

8. For those assigned to either the Bow or Isleworth factory, see Nicholas Panes, Ray Howard, and J. Victor Owen, "Attribution Enhanced—'Isleworth Porcelain' Re-examined," *English Ceramic Circle Transactions* 24 (2013): 89–116. For the intact dishes, see Nicholas Panes and J. Victor Owen, "Attribution Enhanced: Possible American 18th-Century Porcelain Finds in England," *English Ceramic Circle Transactions* 23 (2012): 11–30. For the reconstructed sauceboats, see J. V. Owen, A. Meek, and W. Hoffman, "Geochemistry of Sauceboats Excavated from Independence National Historical Park (Philadelphia): Evidence for a Bonnin and Morris (c. 1770–73) Provenance and Implications for the Development of Nascent American Porcelain Wares," *Journal of Archaeological Science* 38 (2011): 2340–51.

9. Sources of data are Owen, "Geochemical and Mineralogical Distinctions"; Owen, "Geochemistry of High-fired Bartlam Ceramics"; Panes, Howard, and Owen, "Attribution Enhanced—'Isleworth Porcelain'"; W.R.H. Ramsay and E. G. Ramsay, "A Classification of Bow Porcelain from First Patent to Closure: c. 1743–1774," in *Proceedings of the Royal Society of Victoria* 119, no. 1 (2007): 1–68; J. Victor Owen and T. E. Day, "18th-Century Phosphatic Porcelain: Bow & Lowestoft. Further Confirmation of Their Compositional Distinction," *English Ceramic Circle Transactions* 16, no. 3 (1998): 342–44; Owen, Meek, and Hoffman, "Geochemistry of Sauceboats Excavated from Independence National Historical Park"; I. Freestone, L. Joyner, and R. Howard, "The Composition of Porcelain from the Isleworth Manufactory," *English Ceramic Circle Transactions* 18, no. 2 (2003): 284–93; and J. Victor Owen, unpublished data.

10. W.R.H. Ramsay, K. Sutton, and E. G. Ramsay, "Bow Porcelain: Glaze Compositions Associated with the Phosphatic Wares ~1742–1774," *Proceedings of the Royal Society of Victoria* 123, no. 2 (2011): 161–71.

11. Owen, Meek, and Hoffman, "Geochemistry of Sauceboats."

12. Panes, Howard, and Owen, "Attribution Enhanced—'Isleworth Porcelain.'"

13. Ingwer Borg and Patrick J. F. Groenen, *Modern Multidimensional Scaling: Theory and Applications* (New York: Springer, 1997); and Leland Wilkinson, M. Hill, J. P. Welna, and G. K. Birkenbeuel, *SYSTAT for Windows, Version 5: Statistics* (Evanston, Ill.: SYSTAT, 1992).

14. See J. Victor Owen and John D. Greenough, "Mineralogical and Geochemical Characterization of Eighteenth-Century Moravian Pottery from North Carolina," in *Ceramics in America*, edited by Robert Hunter (Hanover, N.H.: University Press of New England for the Chipstone Foundation, 2010), pp. 168–87; and J. Victor Owen, Dorota Forfa, and John D. Greenough, "Geochemical and Mineralogical Constraints on the Provenance of L'sitkuk Bear River Pottery from the Annapolis Basin, Nova Scotia," *Canadian Journal of Archaeology* 38, no. 1 (2014): 76–105. J. Victor Owen and John D. Greenough, "Influence of Potsdam Sandstone on the Trace Element Signatures of Some 19th-Century American and Canadian Glass: Redwood, Redford, Mallorytown, and Como-Hudson," *Geoarchaeology* 23, no. 5 (2008): 587–607 [DOI 10.1002/gea.20232]. See John D. Greenough and J. Victor Owen, "A Petrochemical Study of Basaltic Layering at Henley Harbour, Labrador, Using Multidimensional Scaling," *Atlantic Geology* 38, nos. 2–3 (2002): 161–75. See John D. Greenough, Leanne M. Mallory-Greenough, and Brian J. Fryer, "Geology and Wine 9: Regional Trace Element Fingerprinting of Canadian Wines," *Geoscience Canada* 32, no. 3 (2005): 129–37; and John D. Greenough,

Brian J. Fryer, and Leanne Mallory-Greenough, "Trace Element Geochemistry of Nova Scotia (Canada) Maple Syrup," *Canadian Journal of Earth Sciences* 47, no. 8 (2010): 1093–110 [DOI 10.1139/E10-055].

15. An F statistic is a value obtained when a regression analysis or analysis of variance test is run to determine whether the mean between two populations of data is significantly different.

16. Panes, Howard, and Owen, "Attribution Enhanced—'Isleworth Porcelain,'" Table 1.

17. See ibid.

18. Ibid.

19. Ibid.

20. See the Na_2O versus Al_2O_3 plot in ibid., fig. 17.

21. M. White, "A Bovey Tracey Fuddling Cup," *English Ceramic Circle Transactions* 19, no. 3 (2007): 501–7.

22. Nicholas Panes and J. Victor Owen, "Attribution Enhanced—A Bovey Tracey Porcelain Sauceboat," *English Ceramic Circle Transactions* 21 (2010): 125–34.

23. Panes, Howard, and Owen, "Attribution Enhanced—'Isleworth Porcelain.'"

24. J. Victor Owen, "What Analytical Data Can (and Cannot) Tell Us about 18th-Century Fine Ceramics," *English Ceramic Circle Transactions* 22 (2011): 215–30.

25. This effect is commonly seen in natural petrological systems, for example, at the interface between layers of metamorphic rocks of different composition, across which components can migrate on a millimeter-to-decimeter scale. See J. Victor Owen, "Syn-metamorphic Element Transfer across Lithological Boundaries in the Port-aux-Basques Gneiss Complex, Newfoundland," *Lithos* 29, nos. 3–4 (1993): 217–33.

26. See Robert L. Hallse and Ralph L. Cook, "Volatility Studies of Lead Silicate Melts," *Journal of the American Ceramic Society* 41, no. 9 (1958): 331–36 [DOI 10.1111/j.1151-2916.1958.tb12927.x]; and Daniel Rhodes, *Clay and Glazes for the Potter*, 3rd ed. (Iola, Wisc.: Krause Publications, 2000).

27. Ramsay, Sutton, and Ramsay, "Bow Porcelain."

28. Panes, Howard, and Owen, "Attribution Enhanced—'Isleworth Porcelain,'" fig. 17.

29. See J. Victor Owen and Michelle L. Morrison, "Sagged Nantgarw Porcelain (ca. 1813–1820): Casualty of Overfiring or a Fertile Paste?," *Geoarchaeology* 14, no. 4 (1999): 313–32 [DOI: 10.1002/(SICI)1520-6548(199904)14:4<313::AID-GEA1>3.0.CO;2-H], and J. Victor Owen, John O. Wilstead, Rheinallt W. Williams, and Terence E. Day, "A Tale of Two Cities: Compositional Characteristics of Some Nantgarw and Swansea Porcelains and Their Implications for Kiln Wastage," *Journal of Archaeological Science* 25, no. 4 (1998): 359–75.

30. See Baoping Li, Jianxin Zhao, Kenneth D. Collerson, and Alan Greig, "Application of ICP-MS Trace Element Analysis in Study of Ancient Chinese Ceramics," *Chinese Science Bulletin* 48, no. 12 (2003): 1219–24 [DOI 10.1360/02wb0121]; and Baoping Li, Jianxin Zhao, Alan Greig, Kenneth D. Collerson, Zhen-Xi Zhuo, and Yue-Xin Feng, "Potential of Sr Isotopic Analysis in Ceramic Provenance Studies: Characterisation of Chinese Stonewares," *Nuclear Instruments and Methods in Physics Research B: Beam Interactions with Materials & Atoms* 240 (2005): 726–32 [DOI 10.1016/j.nimb.2005.06.003].

Robert Hunter and Juliette Gerhardt

An Eighteenth-Century American True-Porcelain Punch Bowl

▼ ONE OF THE MOST intriguing stories in world ceramic history is the search for the secrets of making hard-paste porcelain. First developed by Chinese potters around the seventh century A.D., these so-called true porcelains were later exported in huge quantities to a market of hungry consumers in the West, from the merchant class to the nobility.[1] The competition by Western potters to uncover the secret for making porcelain is usually told as a Eurocentric tale, centered on the romanticized exploits of Prussian alchemist Johann Friedrich Böttger under the patronage of August the Strong, elector of Saxony and king of Poland.[2] Böttger's research led to the opening of the Meissen porcelain factory near Dresden, which was producing hard-paste porcelain by 1710. Other European and British individuals and factories joined the race in the early part of the eighteenth century, but, lacking the knowledge of the materials and techniques discovered by Böttger, they had to make do with a variety of so-called soft-paste porcelain formulas.[3]

The primary distinctions between hard- and soft-paste porcelains are the ingredients used and the firing temperature necessary to vitrify the body, creating some degree of translucency. The formula for hard paste consists of kaolin clay mixed with feldspar and quartz, which is then fired to a temperature of about 1400°C (2552°F). Hard-paste porcelain is usually "once-fired," meaning the non-lead glaze is applied to the raw clay body and then fired in a single episode. Soft-paste porcelains, which were made with compounds of white firing clays with added bone ash, glassy frits, or ground soapstone, required lower firing temperatures (in the 1100–1200°C range). Usually they are fired first to bisque temperatures and then subjected to a second glaze firing. Soft-paste porcelain glazes are typically lead-based.

Dozens of soft-paste formulas were devised during the eighteenth century in England at the factories of Chelsea, Bow, Liverpool, Worcester, Vauxhall, and Limehouse, among many others.[4] The products and histories of these factories have been the subject of countless collector and scientific studies.[5]

In America, the archaeological evidence for John Bartlam's use of soft-paste porcelain at his manufactories in Cain Hoy and Charleston, South Carolina (ca. 1765–1772), has been recently published.[6] The soft-paste porcelain of the American China Manufactory of Philadelphia, ca. 1770–1772, continues to be the subject of scholarly research, articles, and exhibitions.[7] In his 1972 seminal study of the Philadelphia enterprise, Graham Hood suggested that the owners—Gousse Bonnin and George Anthony Morris, who were already successfully producing a soft-paste phosphatic porcelain—

were also experimenting with a hard-paste porcelain body in 1772.[8] As evidence, Hood cited a Philadelphia news blurb that appeared in the *Pennsylvania Packet* on August 3, 1772: "that the Proprietors of the China Manufactory in this city, have lately made experiments with some clay presented to them by a Gentleman of Charlestown, South-Carolina, which produces China superior to any brought from the East Indies, and will stand the heat beyond any kind of crucibles ever yet made."

Claims for an American discovery of the methods and materials required to make hard-paste or "true" porcelain have been associated as early as 1737 with potter and merchant Andrew Duché (ca. 1710–1778) and his ceramics exploits in Savannah, Georgia.[9] From the late 1730s onward, continued declarations emanated from Duché and other interested parties that the kaolin clay found in western North Carolina, known as "Cherokee clay," was the crucial ingredient necessary to produce true porcelain. At the time of Graham's writing, no domestically made specimens had surfaced to substantiate the claims of Bonnin and Morris or Duché, which led Graham to dismiss, cautiously, the claims of the former as hyperbole used in part to bolster public opinion for their failing soft-paste porcelain works. Exciting physical evidence has surfaced that reflects these "lately made experiments" in Philadelphia.

Figure 1 A white punch bowl is being examined and discussed in the archaeological laboratory of Commonwealth Heritage Group, Inc., West Chester, Pennsylvania, February 2016. The ware type appeared to be an anomaly for Philadelphia archaeological contexts. (Photo, Robert Hunter.)

A small, white bowl (fig. 1) was recovered in 2014 from an eighteenth-century archaeological context at the site of the proposed Museum of the American Revolution (MoAR) in the Old City section of Philadelphia. It was among nearly 85,000 artifacts found in excavations conducted by archaeologists from the Commonwealth Heritage Group, Inc.[10] Initially thought to have a stoneware body, subsequent material analysis by J. Victor

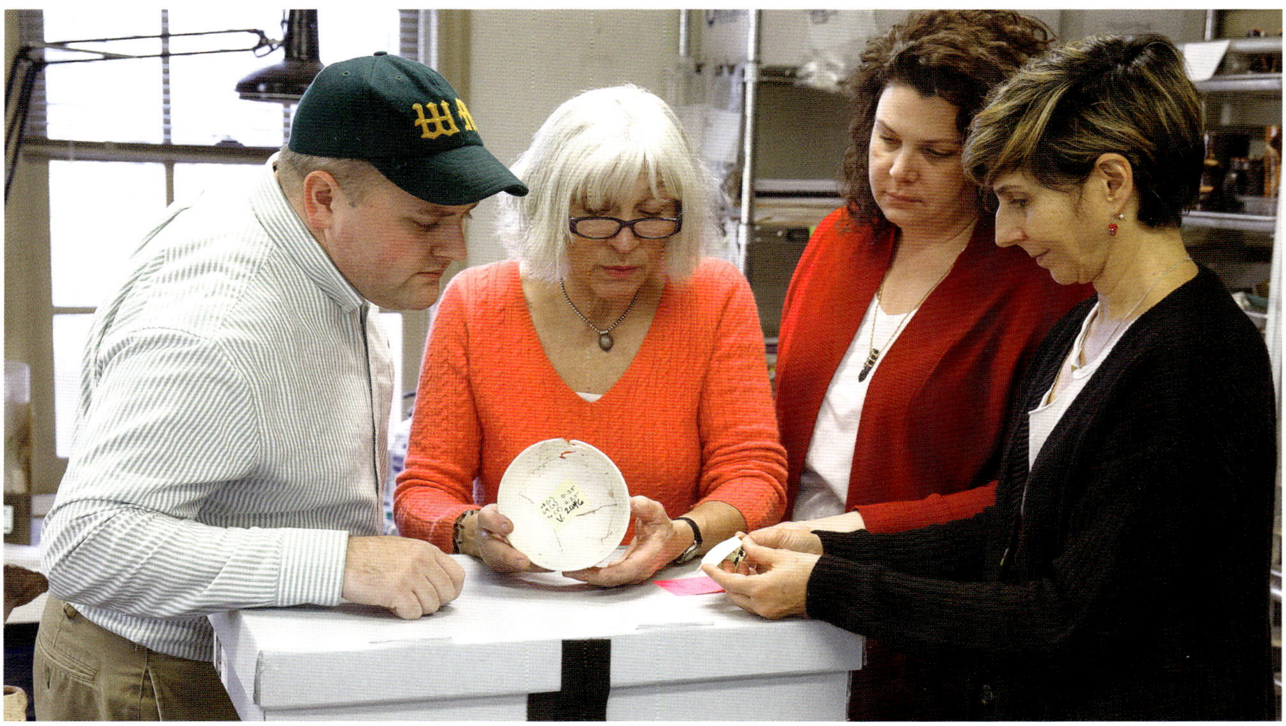

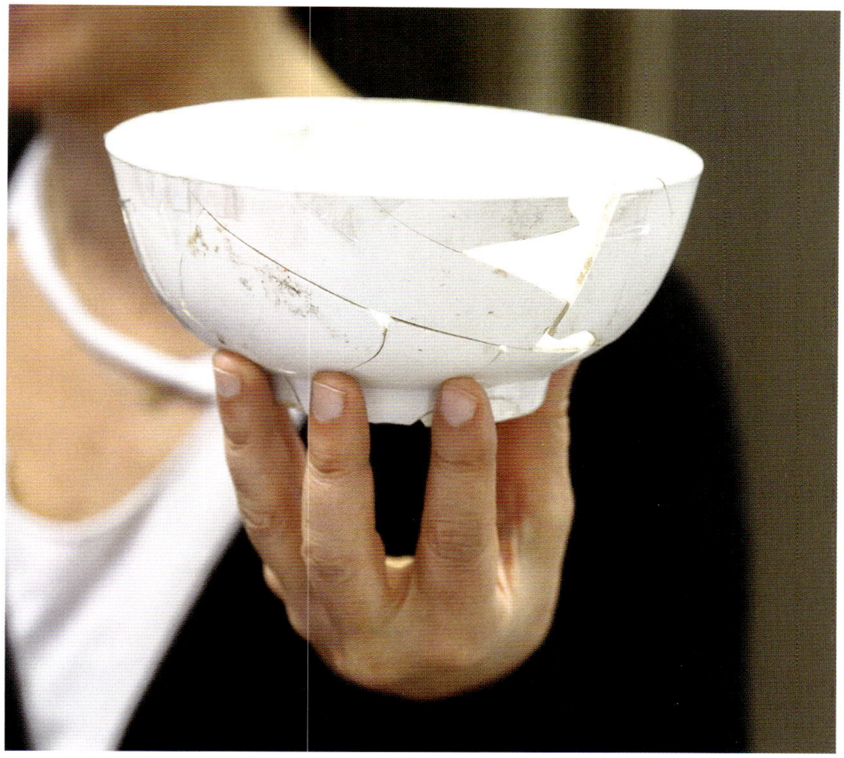

Figure 2 A close-up of the bowl illustrated in fig. 1. Originally catalogued as a white, salt-glazed stoneware slop bowl with an unusual matte finish, subsequent physical spectrographic analysis revealed its composition to be aluminous-silicic ware—that is, a true or hard-paste porcelain. (Photo, Robert Hunter.)

Owen and his colleagues has revealed that the bowl is true, hard-paste porcelain, and most likely of Philadelphia manufacture (fig. 2). Their findings are presented as a companion article, appearing next in this volume.[11] The present discussion chronicles the American quest for making true porcelain and frames the archaeological and historical context of the discovery of this incredibly significant artifact.

The American Pursuit of the Secrets of Hard-Paste Porcelain

The American porcelain story begins with the Duché family of Philadelphia. The family's progenitor, Anthony Duché, a Huguenot émigré, and his sons began making lead-glazed earthenware and salt-glazed stoneware in Philadelphia in the 1720s. His stoneware and earthenware products are known only through archaeological examples, but he was an extremely prolific and proficient potter.[12]

One of Anthony's sons, Andrew, took an interest in expanding the family pottery business southward and circa 1734 moved to the growing metropolis of Charleston, South Carolina. The history of Andrew Duché has been recounted by several ceramic historians throughout the twentieth century but is perhaps best summarized in the 1991 comprehensive article by Bradford Rauschenberg, former director of research at the Museum of Early Southern Decorative Arts.[13] His compilation of information known about Duché's life forms the basis of the summary presented below.

Once in Charleston, Duché wasted no time in building a kiln and advertising to the public in 1735 the availability of "Butter pots, milk pans, and all other sorts of Earthenware."[14] Unfortunately, no physical evidence of

Duché's kiln or pottery from those years has come to light; from records, however, we do get the sense that he was no mere potter but an astute businessman who clearly had an interest in creating more than pots and pans for the local market.

It was around this time that the Chinese secrets of porcelain manufacture were made known to the Western public. A series of letters written by Jesuit missionary Père d'Entrecolles, who had surreptitiously observed porcelain works in China from 1710 to 1722, was published by Jesuit historian Jean Baptiste Du Halde, first in a French edition in 1735, and a year later in English. Of crucial importance, d'Entrecolles identified the two key ingredients of the Chinese true porcelain: "Petuntse" and "Kaulin."[15] Petuntse, also known as china stone, was a widely available feldspar mineral. However, sources for the more essential kaolin were relatively unknown to English and American potters at this time.

That was to change in 1737, when Duché moved to New Windsor, Georgia, a settlement along the Savannah River located, notably, in the center of the lucrative Cherokee Indian trade. At this time, he probably became aware of the fine white clay from their territory in western North Carolina, a substance the Cherokee called *unaker*.[16] For the Cherokee, *unaker* had a ritual signficance and was used for making their white clay pipes and even used as a whitewash for their wattle and daub houses. Duché recognized the Cherokee clay as kaolin.

In its first two decades, 1732–1752, the Georgia Colony was governed by a board of trustees created by the British Parliament. The colony's commander, General James Oglethorpe, had enticed Andrew to bring his craft to the new colony of Georgia, giving him land and arranging for the trustees to provide him with funding. In 1738 Duché moved to the growing town of Savannah, Georgia, and set up a pottery there.

Much of what we know about Duché's activities in Savannah comes from a series of letters written by William Stephens, secretary to the trustees. One such letter, written in 1738 in support of the potter's efforts, confirmed that Duché had produced "2 Kilns of handsome Ware, of various kinds, Pots, Pans, Bowls, Cups, & Jugs."[17] Duché was successful at producing stoneware and earthenware, and archaeological evidence has confirmed that his salt-glazed stoneware was on the same high level as his family had produced in Philadelphia.[18] But he had greater aspirations than simply producing crockery. In the same 1738 letter, Stephens also wrote that Duché had made a "small Teacup of Which he shewed me when held against the Light, was very near transparent."[19] Of course this was big news for the trustees, since it suggested that Duché had made use of the Cherokee clay and had discovered the methods of porcelain making. While it is unclear whether Oglethorpe had actually seen these porcelain specimens or was just relying on the reports of Stephens, he nevertheless passed on to his superiors in London the information that "clay had been found here that a Potter has bak'd into China Ware."[20]

Those back in England were very excited, especially the president of Georgia's board of trustees, the Earl of Egmont, who resided in London.

Egmont was instrumental in encouraging Duché, enthusiastically but vastly overstating that "He is the first Man in Europe, Africa, or America that found the true material and manner of making porcelain."[21] Egmont sent samples of Chinese porcelain to Duché for use as models. In response, Duché asked for a patent for his product; he did not receive one, but did acquire additional funding. Duché continued to experiment for another two years but seems not to have produced a final porcelain product—much to the consternation of his supporters.

The story of Andrew Duché in Savannah is eventually one of frustration and self-destruction. Although the trustees gave him money and encouragement, he complained bitterly and made endless excuses for his lack of success with the porcelain body. He became a political malcontent who criticized the way the trustees managed the colony. Secretary William Stephens, initially a supporter, eventually lost patience with Duché, pronouncing that the potter had become "a little too much addicted to Politicks" and that, as a result, his porcelain pursuits had become inconsequential.[22]

In spite of all the rhetoric, hype, materials, and funding provided by the trustees, it has remained unproven whether Duché was ever actually successful in producing a true porcelain body while in Savannah. It is also uncertain whether Duché himself had visited the Cherokee kaolin mines or whether he had developed agents through his business as an Indian trader—without question, though, he had quantities of the Cherokee kaolin at his disposal. By the 1740s other interested parties in England and in the colonies had become aware of the valuable white clay deposits in the Cherokee region of the North Carolina mountains.

By 1742 Duché was still irritating the trustees, owing them money and porcelain, and making several unsuccessful attempts to leave Georgia. On one occasion he was caught sneaking out of Savannah on a boat loaded with "divers chests, casks, & other effects."[23] Duché finally left Georgia in February 1743 and arrived in England in May to meet with his primary supporter, the Earl of Egmont. On September 23, 1743, Duché delivered a long letter to the trustees, rehashing his complaints and political entanglements.[24] Little else is recorded about Duché's extended time in England, although two very important events suggest that he was sharing his secrets of porcelain making with all who would listen.

Most notably, on December 6, 1744, a British patent was filed by "Edward Heylin of Bow in the county of Middlesex, merchant, and Thomas Frye of the parish of West Ham in the county of Essex, painter . . . for the production of porcelain from an earthy mixture."[25] The material cited "is an earth, the produce of the Cherokee nation in America, called by the natives unaker."[26] Scholars have argued that this patent is directly connected both to Duché's porcelain experiments in Savannah and to his subsequent journey to England.[27] Evidence for the success of this patent has been ascribed to twenty-seven surviving examples referred to by scholars as the "A"-marked group.[28]

In May 1745 British potter William Cookworthy wrote that he had been visited by "the person who hath discovered the china-earth." Cookworthy added that his visitor "had several examples of the china-ware of their making with him, which were, I think, equal to the Asiatic." According to Cookworthy's letter, using the language of d'Entrecolles, this person had discovered "both the petunse and kaulin" while searching for mines "in the back of Virginia" and that this kaolin is "the essential thing towards the success of the manufacture."[29] Duché is generally identified as this individual who must have sought out Cookworthy during his 1744 trip.[30] Cookworthy would later discover his own source of china clay in Cornwall, and then, in 1768, his own method for making England's first successful hard-paste porcelain.[31]

Duché returned to Charleston by January 1745 and continued to develop his trade in the Cherokee territory ostensibly to secure a regular supply of the kaolin. He went so far as to petition the South Carolina Assembly for consent to build a 300-mile road between Charleston and the Cherokee territory, possibly to ease the overland shipping of clay.[32] The effort to obtain funding for the road was unsuccessful and Duché eventually left the Carolinas about 1748 and moved to Norfolk, Virginia, continuing as a merchant but apparently forsaking interest in the porcelain-making business. For the next fifteen years, however, other individuals picked up where Duché left off and made shipments of the Cherokee clay to England. A summary of those shipments is presented below.[33]

> 1747—Robert Pringle, a Charleston merchant, shipped "1 Bag White Clay" to his brother in London.[34]
> 1755—Dr. Alexander Garden, a Charleston physician and naturalist, "penetrated into the indian country and formed a treaty with the cherokees in their own mountains." He shipped samples to Worcester.[35]
> 1757—Henry Laurens, a wealthy and influential Charleston merchant and rice planter, wrote to Thomas Mears of Liverpool: "We will send by the first opportunity that shall present for 2 or 300 lb weight of the Cherokee Clay."[36]
> 1759—Dr. Alexander Garden procured some more of the "different layers of Savannah West Bluff" and sent a sample to London.[37]
> 1764—Henry Laurens wrote to Andrew Williamson of Ninety-Six, South Carolina: "I am applied to by a particular friend to procure a Specimen of our Cherokee Clay for making potters fine Ware. Can you help me at any reasonable expence & soon to the quantity of a Flower Barrel full."[38]
> 1765—Caleb Lloyd of Charleston sent his brother-in-law Richard Champion, a Bristol merchant and porcelain developer, "a box of Porcelain Earth which I have sent you by this vessel to be forwarded to Worcester to the Proprietors of the China Manufactory there, to have a few pieces of china made of it for me, agreably to a List I enclose. I have desired them to spare no expense about it, and to send it to you to be forwarded to me. It was at considerable pains and expence this Earth was procured. It comes from the internal part of the Cherokee Nations, 400 miles from hence, on mountains scarcely accessible."[39]

Clearly, the clays mined from the Carolinas were an important catalyst in the development of successful porcelain formulas for the earliest English porcelains. In the midst of various parcels of American clay being shipped to England, master Staffordshire potter John Bartlam arrived in Cain Hoy, South Carolina, circa 1765. Bartlam's arrival got the attention of English potter Josiah Wedgwood, who was worried that the defection might affect his American business. Although not known for his interest in porcelain, Wedgwood was fanatically concerned about any new ceramic materials that he did not have access to or control over. Not wanting to miss an opportunity, in 1767 he commissioned an English agent named Thomas Griffiths to sail to America and obtain a quantity of Cherokee clay.[40] Wedgwood did not find an immediate use for the kaolin, but eventually he incorporated it into some of his jasperware formulas, primarily to take advantage of the hype around the material's mysterious associations.[41] Even though Wedgwood might have been interested in pursuing a regular supply of the Cherokee clay, he was certainly put off by the high cost as well as the outbreak of the American Revolution. He lost all interest in American kaolin after gaining access to a supply from Cornwall on the southeast coast of England, a figurative stone's throw from Wedgwood's factory in Staffordshire.

Just before John Bartlam began producing creamware and his version of a phosphatic soft-paste porcelain in Cain Hoy and later Charleston, South Carolina, Andrew Duché again went back to England in 1763, moving from Norfolk to Bath.[42] The historical record is again frustratingly mute regarding Duché's extended time there. While he was in England, however, English potter William Cookworthy, after years of experimentation, received a patent for porcelain manufacture on March 17, 1768.[43] This discovery became the basis for the establishment of the Plymouth and Bristol porcelain factories, which produced the first successful true porcelain in England.[44] Although there are no records to connect Duché to Cookworthy's patent application, the timing is probably not sheer coincidence.

Duché returned to America and his home in Norfolk in early 1769, and almost immediately announced that he was leaving the Virginia colony to return to his hometown of Philadelphia. His relocation to Philadelphia—the epicenter of American porcelain production in 1769—coinciding with the opening of Bonnin and Morris's American China Manufactory in 1770 has, of course, been noted by researchers.[45]

The arrival of Duché in Philadelphia also coincides with Henry Laurens's shipment of Cherokee clay from Charleston. Laurens's letter of November 28, 1771, to William Williamson of London states that he "put a Keg of Clay with one of my own into the hands of Mr George Morris, the Manager or one of the principal Managers of the China Manufactory in Philadelphia."[46] The arrival of Carolina clay at the Bonnin and Morris works is confirmed by the *Pennsylvania Packet* on August 3, 1772.[47] With that notice, the historical evidence places together in Philadelphia the necessary ingredient of porcelain making (Cherokee *unaker*), the expertise of Andrew Duché, and the desire of the proprietors of the American China Manufactory to produce "China superior to any brought from the East Indies."

The Archaeological and Historical Context of the Bowl

The Old City section of Philadelphia is its earliest area of settlement, dating back to the city's founding in 1682. The city, which was laid out in a standard grid, developed distinct neighborhoods (fig. 3). The Old City

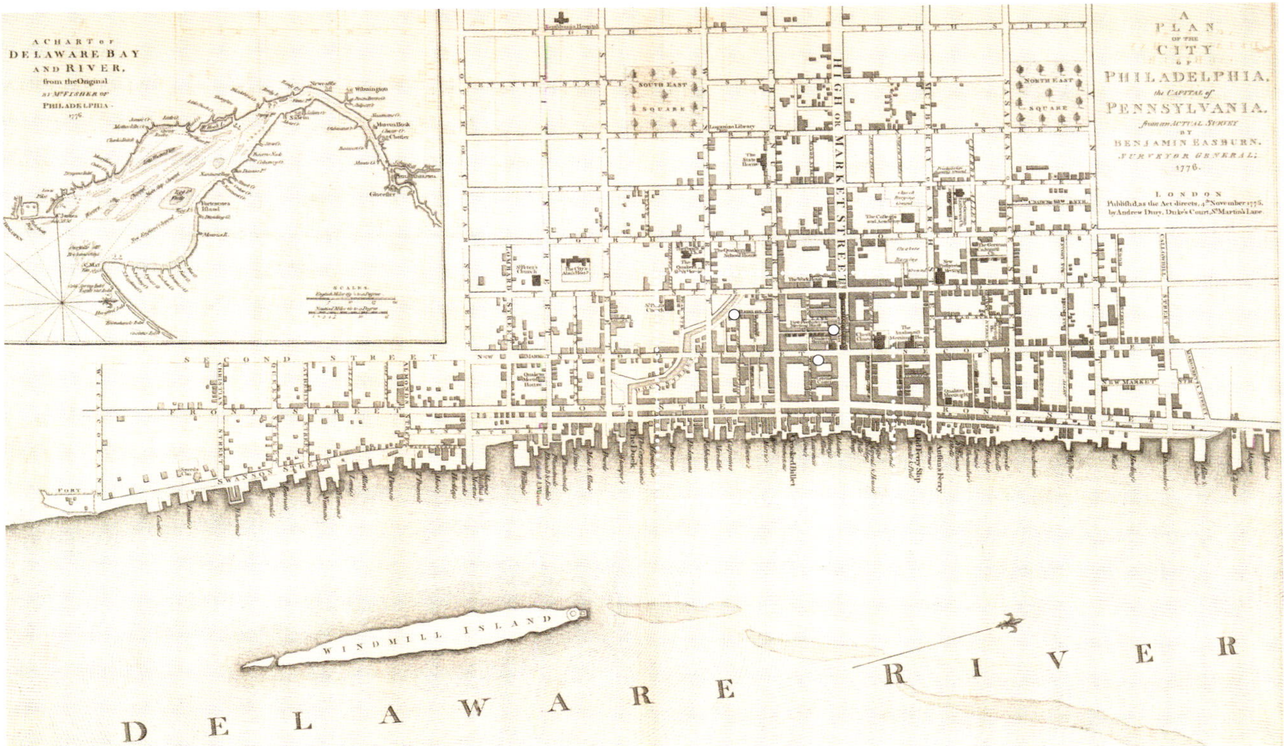

Figure 3 Plan of the City of Philadelphia, the Capital of Pennsylvania, from an Actual Survey by Benjamin Eastburn, Surveyor-General 1776. 19" x 26". Peter Andre (sculpt.). Andrew Drury, London, England, 1776. (The Huntington Library, Art Collections, and Botanical Gardens, Rare Books Department.) The area under discussion is highlighted and detailed in fig. 14.

became Philadelphia's mercantile and political center during the eighteenth century and remains one of the most historic areas in America. Adjacent to Philadelphia's wharfs, the surrounding blocks were home to merchants, tradesmen, shipbuilders, and private residents. Taverns were among the most active businesses. Benjamin Franklin had his print shop and residence on the main road known as High Street, later renamed Market Street. Notable Philadelphia craftsmen, among them chairmaker William Savery, were located there.

Since the 1960s, large-scale archaeological excavations have been conducted in advance of block-wide redevelopment projects in downtown Philadelphia.[48] Most recently archaeological investigations by crews from Commonwealth Heritage Group, Inc. spent months recovering artifacts from the site of the proposed Museum of the American Revolution (MoAr), located at 3rd and Chestnut Streets (see fig. 3). That investigation identified a total of fifty features, including foundations and brick-lined shafts. Of particular interest were four privies that dated to various parts of the eighteenth and early nineteenth centuries. These features contained an enormous quantity of ceramics including locally made Philadelphia objects and imported English, German, and Chinese wares, among other artifacts.

The punch bowl was recovered from deposits in Feature 16, a brick-lined privy shaft (fig. 4). The privy was associated with a house that had been built on the south side of Carter's Alley in 1753 by Nathaniel Goforth. The digging of that privy may date to that period of use, but when it was filled in the 1780s, the house belonged to Mary and Benjamin Humphreys, a Quaker couple who lived there from 1776 to 1813. A deed for the property

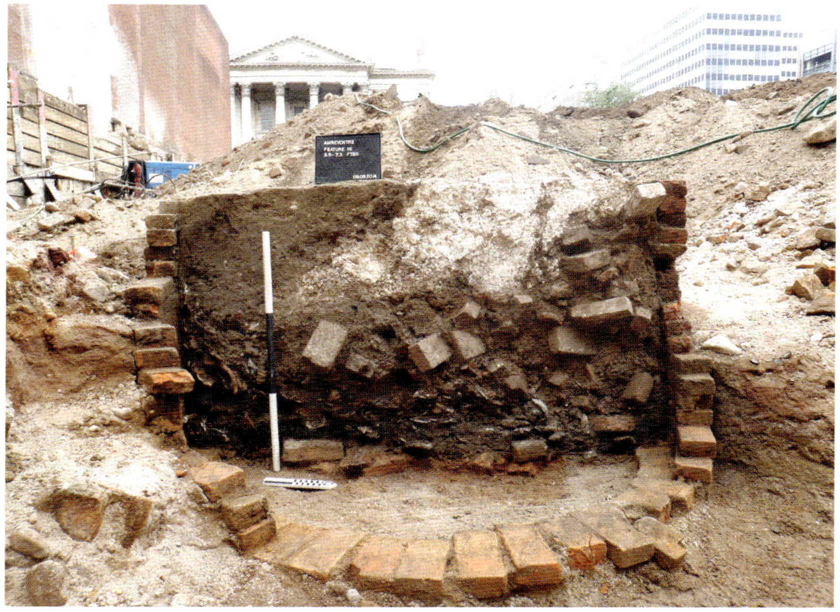

Figure 4 Section of brick-lined privy designated Feature 16. (Photo, courtesy Commonwealth Heritage Group, Inc.)

shows that the shaft was "now filled up" as of July 13, 1789, by which time an adjacent privy, Feature 12, was constructed.[49]

The Humphreys appear to have operated an unlicensed tavern from the house, which certainly accounts for the incredible quantities of ceramic and glassware found in the privy fill.[50] A minimum of 364 ceramic and 220 glass vessels were mended from 23,202 artifacts recovered in Feature 16. The artifacts include 18 tankards in different sizes (pint, half-pint, quarter-pint), 38 drinking glasses of one kind or another, 7 punch bowls, 8 jugs or bottles, 3 decanters, and 103 bottles, most of them for alcohol. Besides the drinking vessels and bottles, other artifacts found in the Humphreys's privy—29 slip-decorated pie dishes and chargers, 13 teapots, 6 porringers, 11 bowls, and 23 chamber pots—resemble artifacts found elsewhere in association with Philadelphia taverns. In addition, 73 tobacco pipe fragments and 107 buttons were recovered, quantities typically associated with tavern sites. The large number of chamber pots, in particular, suggests that the tavern may also have functioned as an inn. Figures 5–7 illustrate the variety

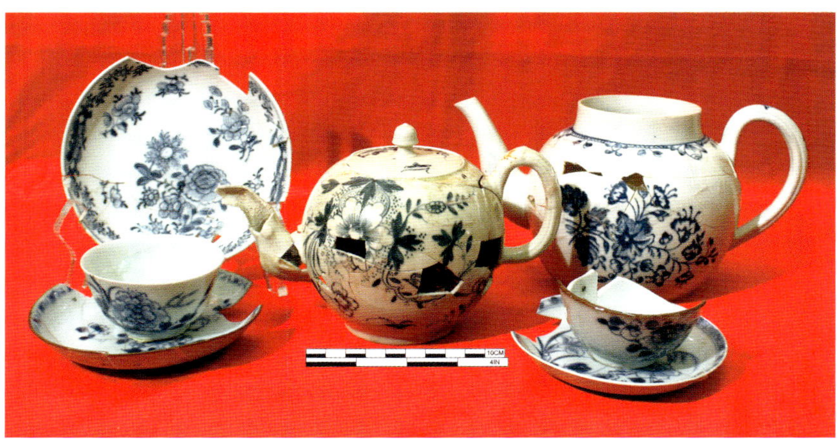

Figure 5 Porcelain teawares recovered from Feature 16. (Photo, Juliette Gerhardt.) The teabowls and saucers are Chinese porcelain and date to the 1760–1770 period. The center teapot is currently being analyzed.

of refined teawares and drinking vessels found among the mix of large quantities of redware, creamware, and Chinese export porcelain.

The porcelain bowl was initially cataloged as circa 1760 white salt-glazed stoneware assumed to be of British manufacture (fig. 8). An unusual matte finish to the glaze was noted but the surface lacked the characteristic orange peel appearance of salt-glazed stoneware. Subsequent examination confirmed that it was not stoneware but a previously unseen porcelain type in Philadelphia's archaeological contexts. It is more than 90 percent complete, having shattered into nineteen pieces when it was discarded into the privy. With a trimmed foot ring, the profile of the bowl conforms to the Chinese export porcelain shapes adopted by British potters in the second half of the eighteenth century (fig. 9).

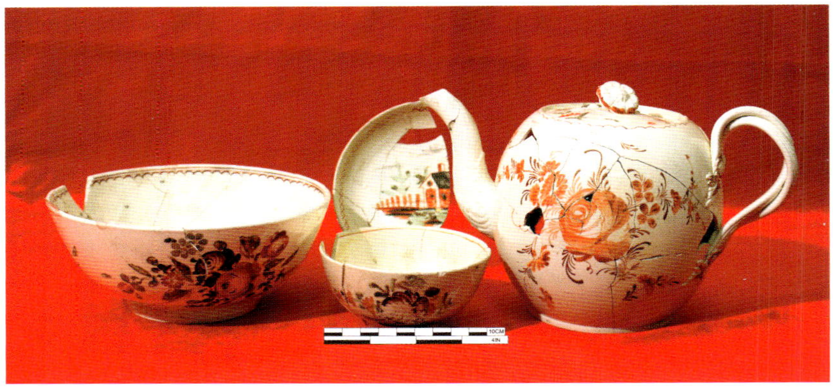

Figure 6 English enameled creamware teawares recovered from Feature 16. (Courtesy, The Museum of the American Revolution; photo, Juliette Gerhardt.) These date to the 1770s.

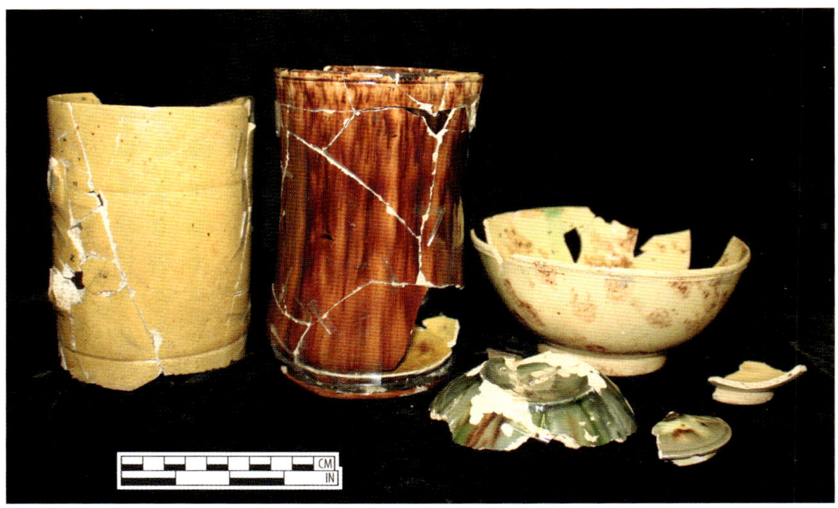

Figure 7 Ceramic drinking vessels recovered from Feature 16. (Courtesy, The Museum of the American Revolution; photo, Juliette Gerhardt.) The lead-glazed tankards, bowl, and teapot fragments with the clear and black manganese glaze, which are of Philadelphia manufacture and date to the last quarter of the nineteenth century, represent the efforts of Philadelphia potters to compete with English refined wares.

After the initial visual determination that the bowl was an unidentified porcelain body rather than stoneware, a small sample was taken from one of the nineteen fragments and submitted for analysis to J. Victor Owen of the Department of Geology at St. Mary's University, Halifax, Nova Scotia. His results showed that the composition of the bowl was a true, hard-paste porcelain with a non-lead glaze; he and his team have also determined that the bowl was once-fired, having been glazed in the raw state. Although the

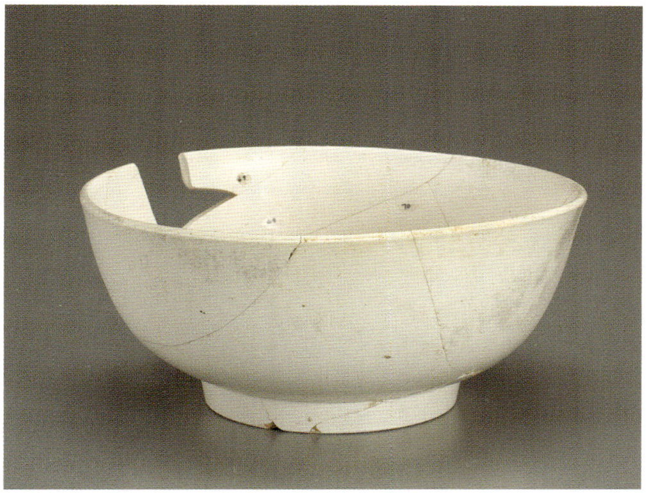

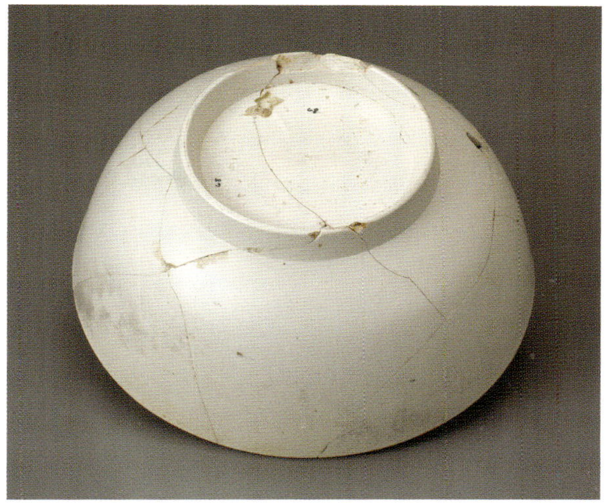

Figure 8 Punch bowl, attributed to the American China Manufactory, Philadelphia, Pennsylvania, ca. 1772. Hard-paste porcelain. D. 5½". (Courtesy, The Museum of the American Revolution; photo, Robert Hunter.)

Figure 9 Detail of the foot ring of the bowl illustrated in fig. 8.

Figure 10 Light passing through the bowl illustrated in fig. 8.

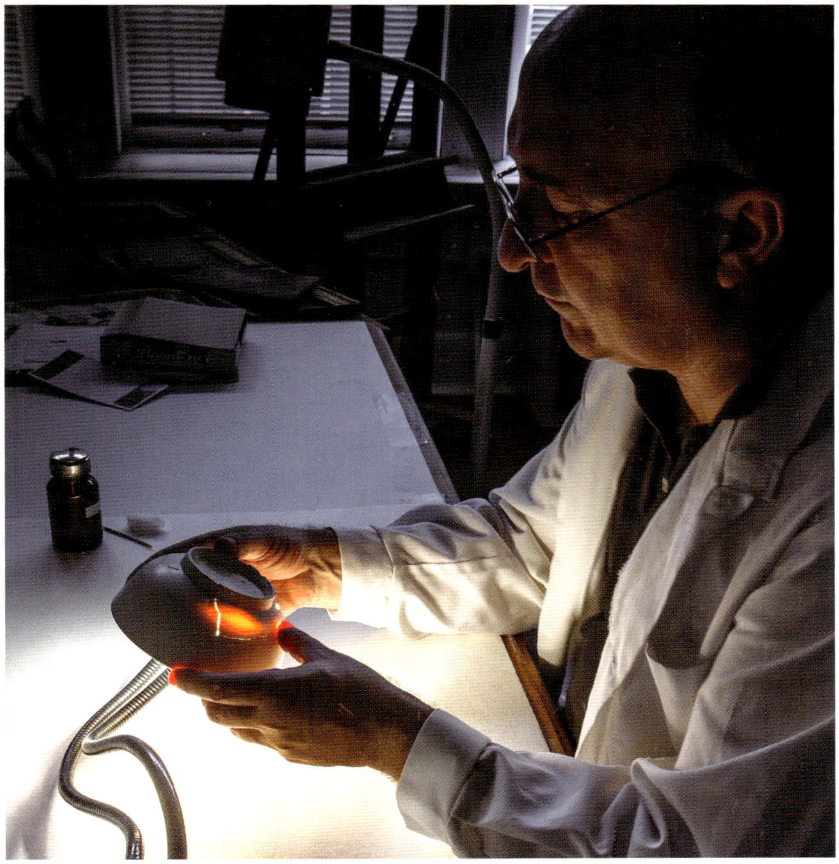

bowl clearly lacked any underglazed decoration, conservator Scott Nolley subsequently examined it for possible remnants of overglazed enamels, which might have become fugitive in its buried context. No such evidence was found.[51] Nolley reassembled the bowl to contemporary museum standards and further confirmed the translucency of the body with directed light sources (fig. 10).

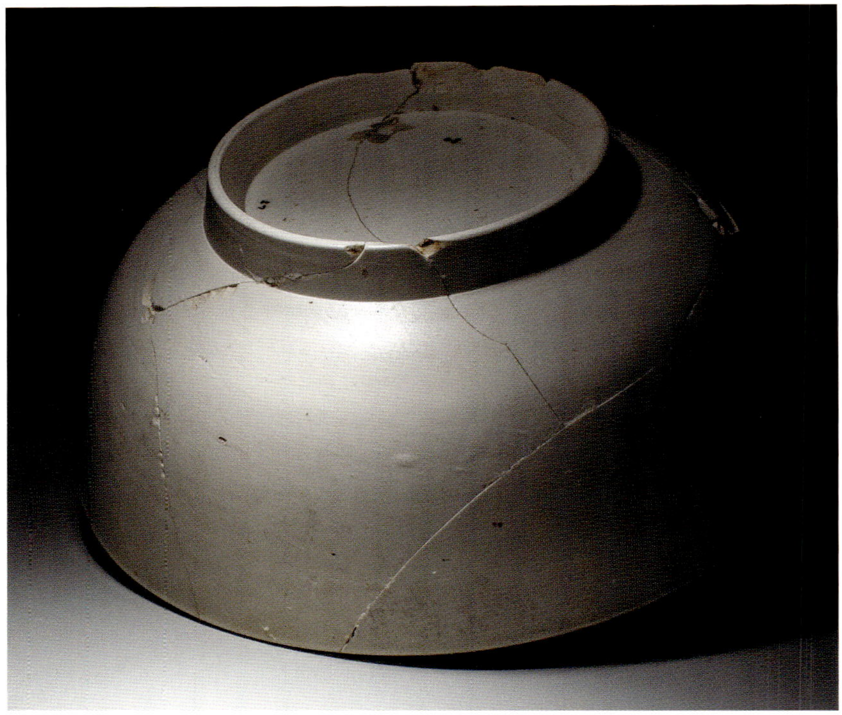

Figure 11 Detail of the surface of the bowl illustrated in fig. 8.

Discussion

The discovery of a true porcelain bowl in a brick-lined privy in the Old City section of Philadelphia is a thrilling addition to the mounting evidence for the history of porcelain making in eighteenth-century America. The undecorated porcelain bowl, found among numerous other Philadelphia ceramics, appears to be of local Philadelphia manufacture. The bowl retains visible turning or production marks and has numerous imperfections in the clay body (fig. 11). The sheen of the glaze is barely noticeable and the materials analysis suggests that it is under-fired. It cannot be mistaken for Chinese porcelain or fully finished examples of eighteenth-century Plymouth or Bristol porcelain.

The American China Manufactory in Philadelphia was a short-lived venture that opened in 1770 and was closed by 1773. Despite attempts to keep the operation afloat, the property was sold in 1774. It was a period of considerable experimentation and adaptation to both material and economic factors. For years, various scholars have suggested that Bonnin and Morris were attempting to create a true porcelain body in addition to their phosphatic and glassy frit formula. Evidence for their phosphatic soft-paste body is found in the various advertisements for the factory seeking large quantities of cattle and beef bones as well as physical confirmation by Victor Owen's analysis.[52] The initial workers hired for the Philadelphia venture were brought in from the Bow factory in London, which used a very similar soft-paste formula. Owen's examination of an openwork basket in the collection of Colonial Williamsburg with the inscription "23 April 1773" has provided proof positive of the Philadelphia factory's change in formula to produce a glassy frit, lead-bearing, non-phosphatic paste.[53]

Remarkably, the initial cataloging of the bowl as English white salt-glaze corresponds nearly identically with the 1741 observation made by the Georgia trustee secretary William Stephens. On August 17 of that year Stephens watched Duché turn some "useful cups . . . about the size of large Tea-Cups." He further noted that their "present Appearance differs very little (if any Thing) from some of our finest Earthen-Ware made in England."[54] The finest English wares of that period would have been the white salt-glazed stoneware being produced in Staffordshire and exported throughout the Western world.

The lack of decoration on the bowl, which might simply reflect the experimental nature of the object, also corresponds with another of Stephens's commentaries. Stephens noted that Duché himself admitted that "Glazing, and Colouring, is a peculiar Work, to be done by other Hands, who are Artists in that way."[55] So if Duché was not making decorated porcelain in Savannah in the 1740s, either lacking the knowledge or the interest, this trend may have continued if he was once again at work in Philadelphia in the 1770s.

If the bowl is not a product of Andrew Duché or the Bonnin and Morris factory, what are alternative explanations? Could the bowl be an example from a post-1774 attempt at making American porcelain? The privy context in which the bowl was found contains numerous other examples of tea- and tablewares that date to the early to mid-1770s, and the privy was known to have been filled by 1789. One putative attempt to produce hard-paste porcelain in eighteenth-century Philadelphia was recorded in August 1783: that an unidentified "French regimental surgeon" was attempting to make a "porcelain fabrick" using clays obtained in Maryland and that "some small samples had been fused very successfully . . . but many difficulties are yet to be overcome."[56] No further evidence of this apparently failed attempt is known. Another eighteenth-century mention comes from the Philadelphia city directories in 1795 and 1796: the entry of a Samuel Faulkner as a china manufacturer at 91 North Sixth Street—the operative word "china" equating with porcelain—but nothing further is known.[57]

Given the current evidence for dating the age of the punch bowl based on its archaeological context, its size and shape, and the absence of another likely American porcelain maker, we believe that the porcelain bowl is a creation of the 1772 experimentation with the Cherokee clay sent by Henry Laurens from Charleston, South Carolina. There is little to suggest that the production of hard-paste porcelain in Philadelphia went beyond limited production. The analysis by Owen and his team shows that the bowl is relatively under-fired and did not reach the temperature necessary to fully vitrify the hard-paste admixture.[58] While Bonnin and Morris may have acquired the appropriate materials and knowledge for their preparation, they might not have possessed the kiln technology to consistently reach the 1400°C level needed for hard-paste porcelain.

Another explanation for not pursuing production is that the partners did not have a steady source of the Cherokee clay to supply the quantity needed. With the financial failure of the factory looming in 1773, Gousse Bonnin abandoned Philadelphia and sailed to England. George Morris had already left Philadelphia, in 1772, to travel to North Carolina, where he died on October 5, 1773, from a "bilious fever." Diana Stradling has suggested that Morris could have been in North Carolina to make arrangements for acquiring a regular supply of the Cherokee clay—an intriguing speculation, but no other information has been found to support such an endeavor.[59]

We have to consider how an experimental object like the punch bowl could have ended up on the sales counter of the Bonnin and Morris warehouse. Examples of their soft-paste wares have been found in similar contexts in the Old City area of Philadelphia. Many of those objects have significant firing flaws and surely were considered "seconds" (figs. 12, 13). Of course the selling of seconds has always been part of the pottery business.

Figure 12 Punch bowl fragment, American China Manufactory, Philadelphia, Pennsylvania, 1770–1772. Soft-paste porcelain. D. approximately 5½". (Courtesy, Atwater Kent Museum of Philadelphia History; photo, Gavin Ashworth.)

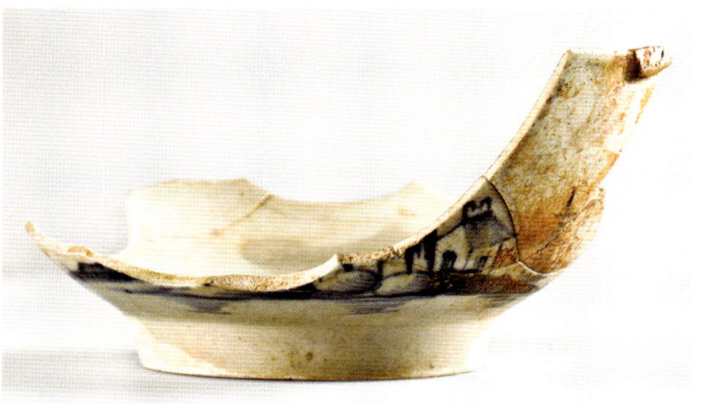

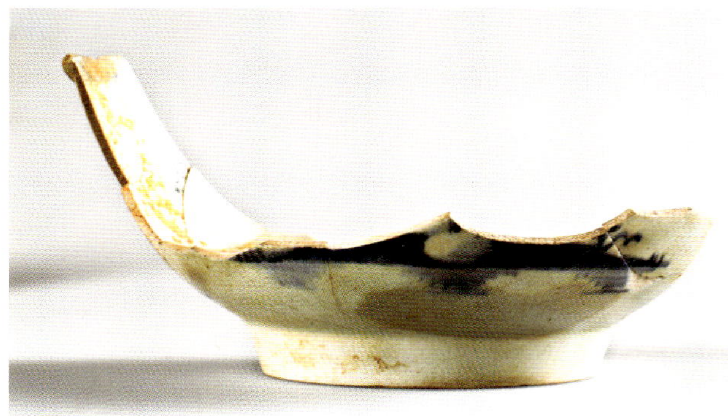

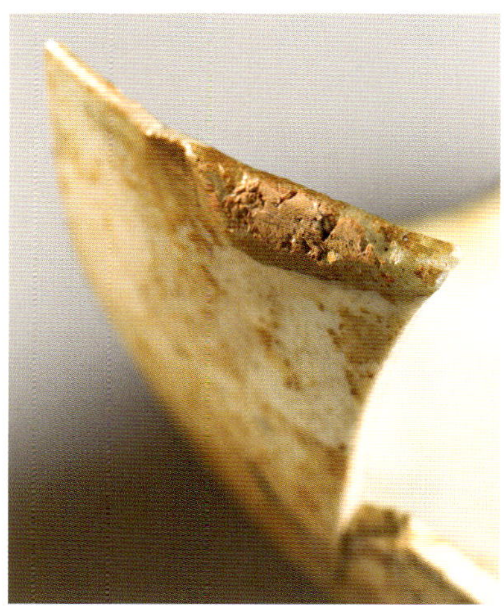

Figure 13 Detail of a glassy glaze flaw on the rim of the bowl fragment illustrated in fig. 12.

In a 1768 letter, English porcelain manufacturer William Cookworthy wrote that even the failed "Product of this kiln together with that of our former Experiments, hath been sold for upwards of Twenty two Pounds!"[60] Such was the demand for even imperfect specimens of his newly founded porcelain works in Plymouth. In another letter discussing the status of Cookworthy's porcelain, John Mudge noted: "Mr Cookworthy has inform'd you that some of the imperfect ware has sold for upwards of £20 & there now is a larger quantity of the same kind in stock."[61] These letters suggest that the selling of experimental and imperfect objects was standard practice for these fledgling porcelain enterprises.

Andrew Duché's role in the Bonnin and Morris enterprise has always been a subject of speculation. Researchers have noted that the timing of his move to Philadelphia would have placed him in a position of knowledge of, if not active participation in, the venture. Just how close was Duché to the Bonnin and Morris operation? According to their newspaper advertisements, Bonnin and Morris operated, in addition to their factory in the Southwark section of Philadelphia, a "Ware-House, the east side Second-street, a few doors below Market Street."[62] The inclusion of "below Market Street" puts its location on the south side. When Duché moved to Philadelphia in 1769, he acquired for £130 the rights to receive £9 Pennsylvania

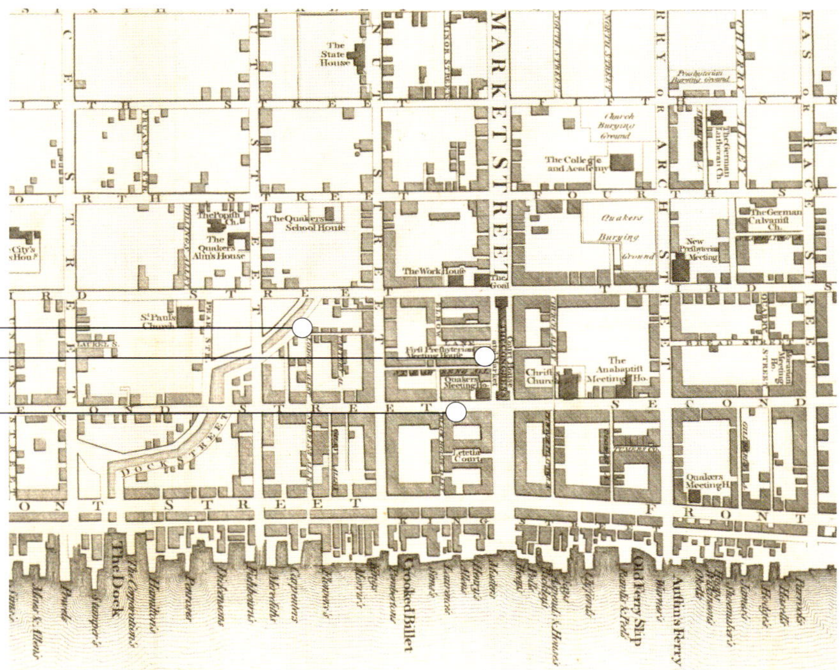

Figure 14 Detail of *Plan of the City of Philadelphia* illustrated in fig. 3, showing the location of Feature 16 at the site of the Museum of the American Revolution. The location of Andrew Duché's rental property on Market Street is shown in relation to the "Ware-House" of Bonnin and Morris.

currency rent for a lot, 14 feet by 30 feet, on High Street and Strawberry Alley. As figure 14 illustrates, these two locations are less than 100 yards from each other. It is inconceivable that the leading porcelain potters in America were so close and did not communicate or collaborate.

Other tangential evidence of the Duché connection to the products of Bonnin and Morris relates to their molded rococo-style sauceboats. As previously mentioned, British potter William Cookworthy purportedly met with Andrew Duché in 1745. Cookworthy went on to find a domestic source for kaolin and china stone in Cornwall. Many of the designs made by Bonnin and Morris, although similar to Bow porcelain prototypes,

Figure 15 Sauceboat, American China Manufactory, Philadelphia, Pennsylvania, 1770–1772. Soft-paste porcelain. H. 4", W. 3½", L. 7⅞". (Courtesy, Brooklyn Museum, gift of Daniel Berry Austin; photo, Gavin Ashworth.)

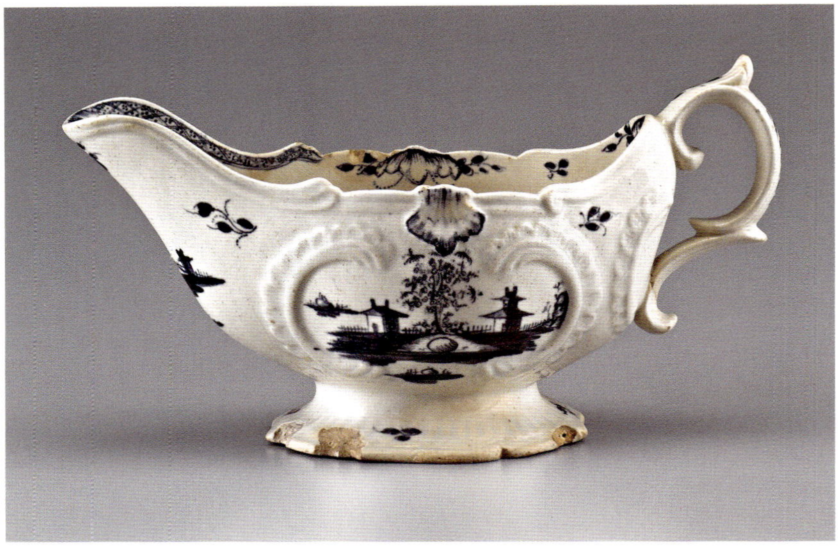

Figure 16 Sauceboat, William Cookworthy, Plymouth Porcelain Works, Devon, England, ca. 1768–1770. Hard-paste porcelain. L. 6⅛". (Courtesy, High Museum of Art, Frances and Emory Cocke Collection.)

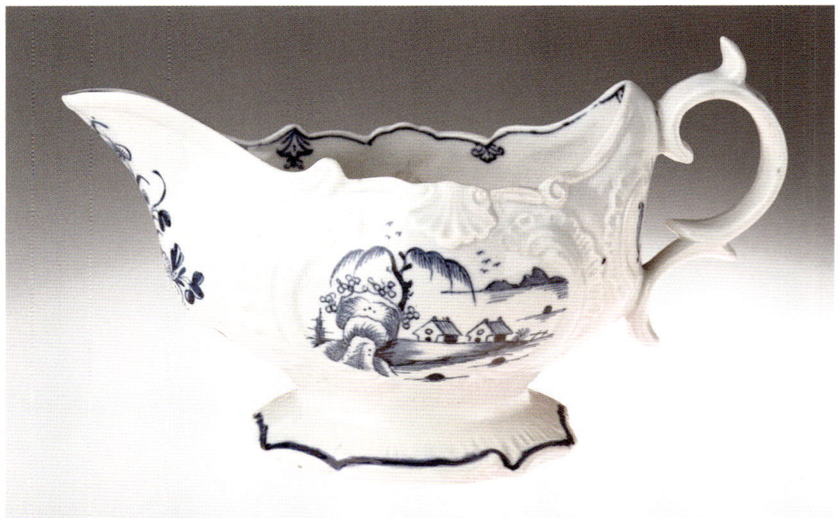

appear to have elements unique to their factory. Yet one of their molded sauceboats, now in the collection of the Brooklyn Museum (fig. 15), appears to be an exact copy of a Plymouth example (fig. 16).[63] In his 1980 article on Bonnin and Morris, Michael K. Brown addressed this similarity and speculated that the Plymouth mold may have found its way to Philadelphia via an "an émigré craftsman—perhaps one who was leaving England in 1770 when Cookworthy moved his manufactory from Plymouth to Bristol, which was also about the same time that Bonnin and Morris were recruiting workmen to come to America."[64] A more tantalizing explanation is that these sauceboats were given to Duché personally by Cookworthy for his use in America and were subsequently incorporated into the American China Manufactory's repertoire.

Mention must be made of another possible candidate for the maker of the hard-paste porcelain bowl. A mysterious "young German, completely skilled in the process of making porcelain" appears in a May 1773 advertisement of the pending auction of the china works.[65] The possible identity of this individual has been discussed by both Graham Hood and Diana Stradling without much success.[66] He was touted as "someone [who] would readily undertake the management, upon reasonable terms, either in partnership or otherwise" as part of the sale agreement, but the property was sold without further mention of him. While it is notable that the Meissen porcelain works in Germany is the site of the first successful European hard-paste porcelain effort, there is nothing more to suggest a connection between the bowl and this unknown man. If indeed he was "completely skilled in the process of making porcelain in the German hard-paste tradition," those abilities are not reflected in the poorly finished, undecorated, and underfired bowl illustrated in figures 8 and 11.

Conclusion

We believe that the newly discovered hard-paste porcelain bowl is an artifact from a short period of experimentation conducted at the American China Manufactory. It most likely was made during the trials announced in August 3, 1772. At the very least, the bowl carries elements of Andrew Duché's invisible hand, given his role in the discovery of the Cherokee clay and his experiments with porcelain manufacture both in his Savannah potteries and in one or more English porcelain works. Duché's physical and intellectual proximity to the Philadelphia porcelain works suggests he was directly involved with the creation of the bowl within the Bonnin and Morris factory setting.

Future research may corroborate Duché's role in Philadelphia's porcelain incubator. For now, we are confident that the bowl is the first physical proof of an American effort to produce hard-paste porcelain in the eighteenth century.[67] Its existence gives more confidence to the belief that Duché did succeed in making true porcelain much earlier, during his time in Savannah in the 1740s. And there is the outside chance that the bowl may be a relic from those earlier experiments and traveled with Duché when he relocated to his hometown. Certainly, scholars and archaeologists should pay closer attention to potential finds in or near his Savannah properties where he operated his kilns. Although the Earl of Egmont's 1739 boast of Duché's ability was pure exaggeration and historically inaccurate, perhaps Duché can truly claim the title as the first man in America to find the "true material and manner of making porcelain."

ACKNOWLEDGMENTS First and foremost we thank the Museum of the American Revolution for permitting this study. Scott Stephenson, vice president of collections, exhibitions, and programming; Neal Hurst, assistant curator; and Michelle Presnall, collections manager and registrar, were instrumental in allowing access to the finds recovered from the excavations and facilitating the study and analysis of this bowl. Staff members from Commonwealth Heritage Group, Inc., in particular Rebecca Yamin and Tod Benedict, were instrumental in the interpretation of Feature 16, and Sarah Ruch provided invaluable help with report photographs and graphics. The Chipstone Foundation is acknowledged for its support of the analysis and conservation of the bowl. Jon Prown and Luke Beckerdite are thanked for their enthusiastic understanding of the importance of this find. Michelle Erickson, drawing on her experience with Bonnin and Morris porcelain, was immensely helpful with her comments on the production aspects the bowl. Scott Nolley of Fine Art Conservation of Virginia contributed his expertise and encouragement in the conservation and mending of the bowl. Debbie Miller provided much-needed assistance with historical information from Philadelphia, and we are grateful for her stewardship of the larger Philadelphia archaeological resource. Ron Fuchs, curator of the Reeves Collection at Washington and Lee University, kindly commented on an earlier version of the manuscript. Heartfelt thanks to Graham Hood, who made the time to examine the bowl and has long encouraged and inspired the pursuit of the story of early American porcelain. Without the analysis and expertise of geologist Victor Owen, this article would not have been possible. Victor's enthusiastic pursuit of the secrets of eighteenth-century Anglo-American porcelain is unsurpassed. He would have a lot to talk about with Andrew Duché.

1. Rose Kerr and Nigel Wood suggest that porcelain was being developed at the end of the sixth or the beginning of the seventh century in northern China. See Kerr and Wood, *Science and Civilisation in China: Volume 5, Chemistry and Chemical Technology, Part 12, Ceramic Technology* (Cambridge: Cambridge University Press, 2004), p. 143.

2. Janet Gleeson, *The Arcanum: The Extraordinary True Story* (New York: Grand Central Publishing, 1998); Edmund De Waal, *The White Road: Journey into an Obsession*, 1st American ed. (New York: Farrar, Straus and Giroux, 2015).

3. Julie Emerson, Jennifer Chen, Mimi Gardner Gates, *Porcelain Stories: From China to Europe* (Seattle: Seattle Art Museum in association with University of Washington Press, 2000).

4. J. Victor Owen, "A New Classification Scheme for Eighteenth-Century American and British Soft-Paste Porcelains," in *Ceramics in America*, edited by Robert Hunter (Hanover, N.H.: University Press of New England for the Chipstone Foundation, 2007), pp. 120–40.

5. As this article was nearing completion, a new publication was released that chronicles the British search for the secrets of hard-paste porcelain. See Brian Adams, *True Porcelain Pioneers: The Cookworthy Chimera, 1670–1782* (Devon, U.K.: Printed for the author, 2016).

6. Stanley South, "John Bartlam's Porcelain at Cain Hoy, 1765–1770," in *Ceramics in America*, edited by Robert Hunter (Hanover, N.H.: University Press of New England for the Chipstone Foundation, 2007), pp. 196–203; Lisa R. Hudgins, "John Bartlam's Porcelain at Cain Hoy: A Closer Look," in *Ceramics in America*, edited by Robert Hunter (Hanover, N.H.: University Press of New England for the Chipstone Foundation, 2007), pp. 203–8.

7. Robert Hunter and Alexandra A. Kirtley, "Recent Scholarship on the American China Manufactory of Bonnin and Morris," *Antiques* 173, no. 2 (January 2008): 66–71.

8. Graham Hood, *Bonnin and Morris of Philadelphia: The First American Porcelain Factory, 1770–1772* (Chapel Hill: Published for the Institute of Early American History and Culture at Williamsburg, Va., by the University of North Carolina Press, 1972), a book reprinted in its entirety in *Ceramics in America*, edited by Robert Hunter (Hanover, N.H.: University Press of New England for the Chipstone Foundation, 2007), pp. 2–75. See also Michael K. Brown, "Piecing Together the Past: Recent Research on the American China Manufactory, 1769–1772," *Proceedings of the American Philosophical Society* 133, no. 4 (1989): 557–59, reprinted in the 2007 issue of *Ceramics in America*; Robert L. Giannini III, "Ceramics and Glass from Home and Abroad," in John C. Milley, *Treasures of Independence: Independence National Historical Park and Its Collections* (New York: Mayflower Books, 1980); Robert Hunter, "John Bartlam: America's First Porcelain Manufacturer," in *Ceramics in America*, edited by Hunter (Hanover, N.H.: University Press of New England for the Chipstone Foundation, 2007), pp. 193–96; Diana Stradling, "Bonnin & Morris: Reading between the Lines," *Transactions of the English*

Ceramic Circle 20, pt. 3 (2009): 613–31; and Diana Stradling, "Bonnin and Morris: Reading between the Lines," in *Ceramics in America*, edited by Robert Hunter (Hanover, N.H.: University Press of New England for the Chipstone Foundation, 2013), pp. 179–99.

9. Graham Hood, "The Career of Andrew Duché," *Art Quarterly* 31 (1968): 168–84. Bradford L. Rauschenberg, "Andrew Duche: A Potter 'a Little Too Much Addicted to Politicks,'" *Journal of Early Southern Decorative Arts* 17, no. 1 (May 1991): 1–101.

10. Rebecca Yamin, Principal Investigator, with contributions by Tod L. Benedict, Meagan Ratini, Kevin C. Bradley, Leslie E. Raymer, Juliette Gerhardt, Kathryn Wood, Timothy Mancl, and Claudia L. Milne, *Archaeology of the City—The Museum of the American Revolution Site Archaeological Data Recovery, Third and Chestnut Streets, Philadelphia, Pennsylvania*, 2 vols. (West Chester, Pa.: Commonwealth Heritage Group, 2016), vol. 1.

11. See J. Victor Owen, Joe Petrus, and Xiang Yang, "The Geochemistry of a True Porcelain Punch Bowl Excavated in Philadelphia," in this volume.

12. Robert L. Giannini III, "Anthony Duche Sr., Potter and Merchant of Philadelphia," *Antiques* 119 (January 1981): 198–203.

13. The earliest comprehensive treatment of Duché is found in a series of articles by G.A.R. Goyle [Rudolph P. Hommel], "First Porcelain Making in America," *Chronicle of the Association of Early American Industries* 1 (1934–35): 8–11; Rauschenberg, "Andrew Duche: A Potter."

14. *South Carolina Gazette*, April 5, 1735.

15. Jean-Baptiste du Halde, *The General History of China*, translated by R. Brookes (London: J. Watts, 1736).

16. Rauschenberg, "Andrew Duche: A Potter," p. 20.

17. Allen Candler, ed., *The Colonial Records of Georgia*, 26 vols. (Atlanta, Ga.: Franklin Printing and Publishing Co., 1904), 22:168–69.

18. Earthenware and stoneware fragments attributed to Andrew Duché's Savannah pottery have been excavated in the colonial settlement of Frederica, Georgia, and were reported on in Nicholas Honercamp, "The Material Culture of Fort Frederica—The Thomas Hird Lot" (master's thesis, University of Florida, 1975), pp. 40–42; and Nicholas Honerkamp, "Frontier Process in Eighteenth Century Colonial Georgia: An Archaeological Approach" (Ph.D. diss., University of Florida, 1980), pp. 45–50.

19. Candler, ed., *Colonial Records of Georgia*, 22:168–69.

20. Ibid.

21. Ibid., 5:141–42.

22. Ibid., 4 (supp.): 46; Rauschenberg, "Andrew Duche: A Potter."

23. Candler, ed., *Colonial Records of Georgia*, 5:657. E. Merton Coulter, ed., *The Journal of William Stephens, 1741–1743* (Athens: University of Georgia Press, 1958), pp. 140–42.

24. Candler, ed., *Colonial Records of Georgia*, 2:289, 5:196–97, 5:657.

25. Llewellynn Jewitt, *The Ceramic Art of Great Britain* (London, 1883), pp. 84–85.

26. W. R. H. Ramsay, Anton Gabszewicz, and E. G. Ramsay, "'Unaker' or Cherokee Clay and Its Relationship to the 'Bow' Porcelain Manufactory," *English Ceramic Circle Transactions* 17, pt. 3 (2001): 473–99.

27. William R. H. Ramsay, Frank A. Davenport, and Elizabeth G. Ramsay, "The 1744 Ceramic Patent of Heylyn and Frye: 'unworkable *unaker* formula' or Landmark Document in the History of English Ceramics?," *Proceedings of the Royal Society of Victoria* 118 (1): 11–34.

28. J. V. G. Mallet, "The 'A' Marked Porcelains Revisited," *English Ceramic Circle Transactions* 15 (1994): 240–57. Recent scientific analysis by W. Ross Ramsay, Judith A. Hansen, and E. Gael Ramsay has suggested that this group was made using the Cherokee *unaker*. They strongly believe that historical and scientific evidence shows that it was Andrew Duché who brought the clay and the results of his Savannah experimentation to the proprietors of the Bow manufactory.

29. Hugh Owen, *Two Centuries of Ceramic Art in Bristol* (London, 1873), pp. 7–8.

30. F. Severne Mackenna, *Cookworthy's Plymouth and Bristol Porcelain* (Leigh-on-Sea, Eng.: F. Lewis Publishers, 1946), pp. 26–27.

31. Adams, *True Porcelain Pioneers*, pp. 101–38.

32. A. S. Salley, ed., *Journal of the Commons House of Assembly of S.C. 19 Jan. 1748–29 June 1748* (Columbia, S.C., 1947), p. 171.

33. Bradford L. Rauschenberg, "'A Clay White as Lime . . . of Which There is a Design Formed by some Gentlemen to Make China': The American and English Search for Cherokee

Clay in South Carolina, 1745–75," *Journal of Early Southern Decorative Arts* 17, no. 2 (November 1991): 67–79.

34. Mabel L. Webber, ed., "Journal of Robert Pringle, 1746–1747," *South Carolina Historical and Genealogical Magazine* 26 (1925): 102.

35. David Ramsay, *The History of South Carolina, from Its First Settlement in 1670 to the Year 1808*, 2 vols. (Newberry, S.C.: W. J. Duffie, 1809), 2:470.

36. Philip M. Hamer and George C. Rogers Jr., eds., *The Papers of Henry Laurens* (Columbia: University of South Carolina Press, 1970), 2:431.

37. Edmund Berkeley and Dorothy Smith Berkeley, *Dr. Alexander Garden of Charles Town* (Chapel Hill: University of North Carolina Press, 1969), pp. 93–94.

38. Ibid., p. 486.

39. Hugh Owen, *Two Centuries of Ceramic Art in Bristol, Being a History of the Manufacture of "The True Porcelain" by Richard Champion* (London, 1873), p. 8; J.V.G. Mallet, "Cookworthy's First Bristol Factory of 1765," *English Ceramic Circle Transactions*, pt. 2 (1974): 212–20.

40. William L. Anderson, "Cherokee Clay, from Duché to Wedgwood: The Journal of Thomas Griffiths, 1767–1768," *North Carolina Historical Review* 63, no. 4 (October 1986): 477–510.

41. Robin Reilly, *Josiah Wedgwood 1730–1795* (London: Macmillan London, 1992), pp. 159–60.

42. For a fuller discussion of Duché's time in England and his residence in Bath, see Rauschenberg, "Andrew Duche: A Potter," pp. 76–77.

43. Mackenna, *Cookworthy's Plymouth and Bristol Porcelain*.

44. In his assessment of Andrew Duché, Brian Adams concludes that he was indeed successful in producing an American true porcelain in the 1740s. Adams, *True Porcelain Pioneers*, p. 24.

45. Graham Hood tries to connect this with Duché, "who was in Philadelphia about this time, has been named as an entrepreneur previously shipping clays from South Carolina to London for the use of the Bow factory" but acknowledges that "there is no known connection between him and Bonnin and Morris." Hood, "The Career of Andrew Duché," p. 180. Brad Rauschenberg tries harder to not dismiss it as coincidence. He writes, "It is perhaps ironic that Duche returned to the city of his birth in time to witness the success of the first documented American porcelain factory. Although proof of any association between Duche and the Bonnin and Morris manufactory has not been found, it might not be illogical to suppose that there was a connection, at least through proximity." He allows that "it is therefore possible that Duché contributed to the factory." Rauschenberg, "Andrew Duche: A Potter," pp. 77–78.

46. George C. Rogers Jr. and David R. Chestnutt, eds., *The Papers of Henry Laurens* (Columbia: University of South Carolina, 1980), 8:55.

47. Alfred Coxe Prime, comp., *The Arts & Crafts in Philadelphia, Maryland and South Carolina, Gleanings from Newspapers*, 2 vols. ([Topsfield, Mass.]: Walpole Society, 1929–32), vol. 1, *1721–1785* (1929), p. 120.

48. John L. Cotter, Daniel G. Roberts, and Michael Parrington, *The Buried Past: An Archaeological History of Philadelphia* (Philadelphia: University of Pennsylvania Press, 1993).

49. Philadelphia County Recorder of Deeds 1789, pp. 182–84.

50. "Archeology of the City—The Museum of the American Revolution Site Archeological Data Recovery," pp. 69–80.

51. Fine Art Conservation of Virginia, available online at http://www.facva.com.

52. J. Victor Owen, "Geochemical Distinctions between Bonnin and Morris Porcelain (Philadelphia, 1770–72) and Some Contemporary Phosphatic British Wares," *Geoarchaeology* 16 (2001): 785–802.

53. J Victor Owen, "'Too little, too late': The Geochemistry of a 1773 Philadelphia Porcelain Openwork Basket," *Journal of Archaeological Science* 36, no. 2 (January 2009): 333–42; Hunter and Kirtley, "Recent Scholarship on the American China Manufactory"; Nicholas Pane, "Mr Ball the English Potter and the American China Manufactory," *English Ceramic Circle Transactions* 20, pt. 3 (2009): 633–45.

54. Candler, ed., *Colonial Records of Georgia*, 4 (supp.), pp. 220–21.

55. Ibid.

56. Alice Cooney Frelinghuysen, *American Porcelain, 1770–1920* (New York: Metropolitan Museum of Art, 1989), p. 11.

57. Ibid.

58. Owen, Petrus, and Yang, "The Geochemistry of a True Porcelain Punch Bowl Excavated in Philadelphia," in this volume.

59. Stradling, "Bonnin and Morris: Reading between the Lines," p. 192.

60. Cornwall Record Office reference DDF(4) 80/21, as cited in Adams, *True Porcelain Pioneers*, p. 411.

61. Cornwall Record Office reference DDF(4) 80/23, as cited in ibid., p. 413.

62. *Pennsylvania Journal*, September 5, 1771, and *Pennsyslvania Gazette*, January 13, 1772.

63. Nicolas Panes, *British Porcelain Sauceboats of the 18th Century* (Wales: Gomer Press, 2009). Two recently recorded archaeological examples also appear to represent copies of the Cookworthy sauceboat design and are illustrated in J. Victor Owen, John D. Greenough, and Nick Panes, "Statistical Evaluation of Analytical Data for Eighteenth-Century American and British Sulphurous Phosphatic Porcelains," in this volume.

64. Brown, "Piecing Together the Past" (*Ceramics in America*), p. 86.

65. *Pennsylvania Chronicle*, May 24–31, 1773. Hood, Bonnin and Morris of Philadelphia (reprint in Ceramics in America), p. 16.

66. Stradling, "Bonnin and Morris: Reading between the Lines," pp. 191–92.

67. A Chinese-shaped porcelain punch bowl with underglaze blue decoration, postulated as being of Duché manufacture in Savannah, was illustrated and discussed in Ruth Monroe Gilmer, "Andrew Duché and His China, 1738–1743," *Apollo* 45 (1947): 128–30. This claim was challenged by Arthur W. Clement in a subsequent article, but still defended by Ruth Monroe Gilmer in "Andrew Duché—America's First China Maker, born circa 1709," *Apollo* 48 (1948): 63–65. A color image of this bowl is reprinted in Adams, *True Porcelain Pioneers*, p. 22. The whereabouts of this bowl are not known.

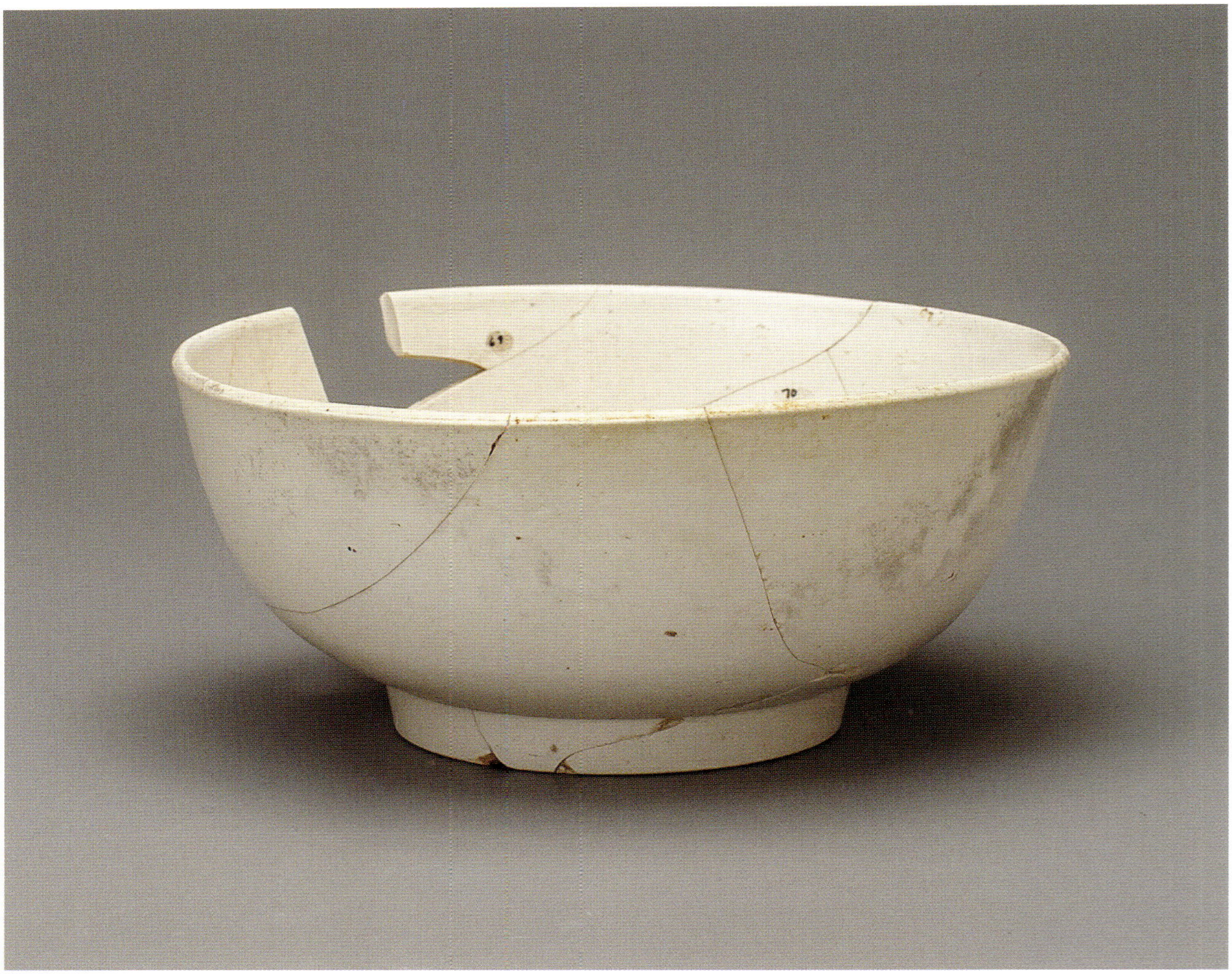

Figure 1 Punch bowl, attributed to American China Manufactory, Philadelphia, Pennsylvania, ca. 1772. Hard-paste porcelain. D. 5½". (Courtesy, The Museum of the American Revolution; photo, Robert Hunter.) Excavated from a privy" on the site of the proposed Museum of the American Revolution in Philadelphia.

J. Victor Owen, Joe Petrus, and Xiang Yang

Bonnin and Morris Revisited: The Geochemistry of a True-Porcelain Punch Bowl Excavated in Philadelphia

Introduction

It is not known with certainty just who first made porcelain in America, but there are several contenders for this honor: Gousse Bonnin and George Anthony Morris of the American China Manufactory (Philadelphia, ca. 1770–1773); John Bartlam (Cain Hoy, South Carolina, ca. 1760–1765); and Andrew Duché (Savannah, Georgia, ca. 1737–1738). At present, Bartlam has an edge on the other two candidates, since there is no firm evidence supporting Duché's overstated and oft-cited claims about being the first person outside of Asia who "ever found the true material and manner of making porcelain or China ware. . . ," and the historical record clearly shows that Bartlam had set up shop in South Carolina several years before the establishment of the American China Manufactory.[1] That said, it would be reassuring to confirm the presence of phosphatic (bone-ash) porcelain biscuit wasters at Cain Hoy to remove any niggling concerns that the porcelain fragments found there might have derived from ceramic objects produced elsewhere.

▼ CONSIDERATION OF what is known about the comings and goings of our trio outwardly suggests that there is little connection between them. However, in their discussion (also in this volume) of the same white, true-porcelain punch bowl that is the subject of this essay, Hunter and Gerhardt review the historical and archaeological record of the nascent American porcelain industry and show this not to be the case.[2] Duché's background in pot-making (and perhaps even porcelain-making), access to Carolina ("Cherokee") clay (kaolin), connections with leading British porcelain manufacturers, and familial links to Philadelphia, the city to which he returned just before the establishment of the Bonnin and Morris enterprise, reveal him to be a lynchpin in this story. Andrew Duché arrived in Philadelphia on November 14, 1769, six weeks before Bonnin and Morris publicly announced their success in using American clay to produce high-quality porcelain. Moreover, he took up residence close to the Southwark location of the American China Manufactory. Based on this evidence, Hunter and Gerhardt have concluded that this punch bowl constitutes tangible evidence for a link between Andrew Duché and Bonnin and Morris.

This paper presents compositional data for the punch bowl, which was found in an archaeological context in Philadelphia.[3] Its paste and glaze are shown here to differ from those of contemporary varieties of Anglo-American true porcelain for which analytical data (including high-precision trace elements) are available. Its glaze is vanishingly thin, and features of the microstructure of this bowl indicate that its paste was poorly prepared and relatively underfired, which in turn suggests that it was made by a comparative novice in the production of these wares. As such, the bowl is a plausible example of an experimental ware produced by Bonnin and Morris, perhaps under the influence of Andrew Duché and likely after they had abandoned use of their phosphatic paste. In acknowledgment of the location of its discovery, albeit at the risk of prejudging its origin, this object will henceforth be referred to as the "Philadelphia bowl." Should it eventually be shown to have been made elsewhere, this ersatz title will have to change, but for our purposes, it has a cachet evocative of the importance of this unusual and undoubtedly soon-to-be-controversial artifact.

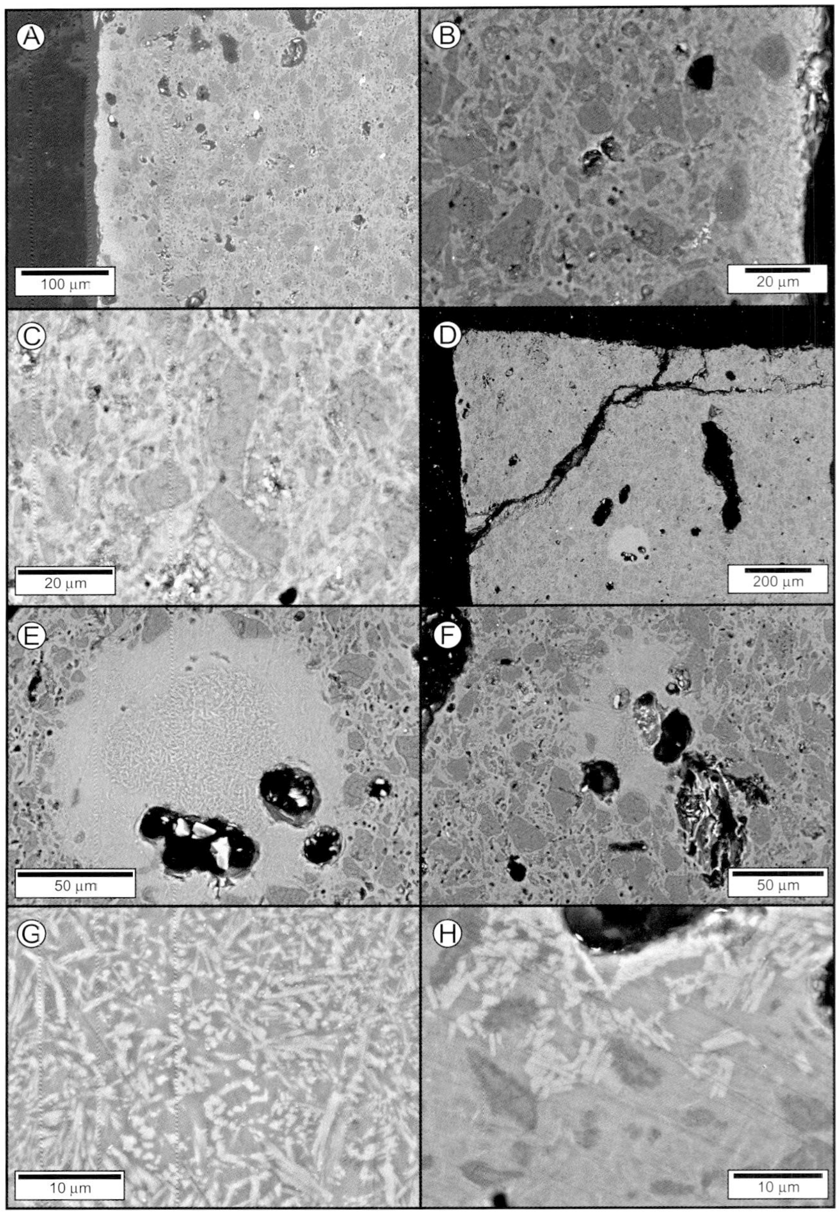

Figure 2 Backscattered-electron images of the Philadelphia bowl. (A) Thin glaze (left side of image) with an S-A-C composition coating the aluminous-silicic paste. Black patches in this and other images are holes (pores). (B) Close-up showing the glaze (right side of image, above scale) and its ceramic substrate. Note the calcic plagioclase microlites in the glaze, the integrated body-glaze layer, and the subangular silica polymorphs (dark gray grains, probably α-quartz) in the substrate. (C) Subangular silica polymorphs (probably α-quartz) in a mullite-laden melt phase. (D) Overview of the corner of a mounted sherd from the bowl, showing a circular melt patch (lower middle part of image). (E) Close-up of the circular melt patch showing the concentration of relatively coarse grained mullite crystallites in the center. (F) Irregular-shaped melt patch. Note the coarse mullite crystallites in the center. (G) Relatively coarse-grained mullite microlites in the core of the circular melt patch. (H) Calcic plagioclase microlites in the irregular-shaped melt patch. Finer-grained mullite microlites (indistinct, felty, medium-gray crystals) occur in the melt in the lower left part of the image.

Description of the Sample

The Philadelphia bowl (fig. 1) has been reconstructed from fragments. Ninety percent complete, it is small, white, undecorated, thinly potted, and very thinly glazed. Indeed, the glaze is only about 20 microns thick (figs. 2A, B). The dish was originally catalogued as a white, salt-glazed stoneware slop bowl with an unusual matte finish, the latter descriptor, of course, alluding to its rather cryptic glaze.[4]

Archaeological Context

The Philadelphia bowl was recovered from a privy ("Feature 16") on a property once owned by a family with the name Humphreys.[5] They lived on the southeast corner of the property, which is now part of the site of the proposed Museum of the American Revolution (MoAR). Most of the artifacts from this privy were objects once used in a tavern operated by the Humphreys—tankards, drinking glasses, punch bowls, bottles, and decanters—that date to the 1780s or earlier. The privy itself was sealed in 1789. The abundance of chamber pots in the privy suggest that some of the artifacts originated in an inn on this property, archival evidence for which is known. Besides the Philadelphia bowl, 326 other ceramic vessels were found. These are dominated by lead-glazed earthenware, English creamware, and Chinese Export porcelain. Approximately one third of the ceramics found are believed to have been locally made.[6] They consist of red/buff earthenware vessels and one stoneware butter pot. If the Philadelphia bowl proves to be a domestically produced artifact, then true porcelain will need to be added to this list.

Analytical Methods

The Philadelphia bowl was analyzed for major and minor elements using a scanning electron microscope (SEM) equipped with a silicon-drift, energy-dispersive detector (EDS) and a field emission gun.[7] So too were other samples of broadly contemporary porcelains: two figures made by William Cookworthy, proprietor of English porcelain factories in Bristol (ca. 1768–1770) and Plymouth (ca. 1770–1774, later operated by Richard Champion ca. 1774–1781); a true-porcelain sherd recovered from the site of John Bartlam's potworks (ca. 1765–1770) in Cain Hoy, South Carolina; a dated (1773) and marked ("Philadelfia") openwork basket attributed to the Bonnin and Morris factory; a first-patent Bow ("A-marked") porcelain teapot lid excavated in London; and unvitrified, aluminous-silicic biscuit (unglazed) wasters from an archaeological site in Hopewell, Virginia. Although apparently consisting of fine earthenware (possibly creamware), the average composition of five of the wasters was included for comparison with the porcelain artifacts because of the geographically intermediate position of the site of their discovery between Cain Hoy and Philadelphia. Data for the Philadelphia bowl are compared with these diverse samples in order to test the hypothesis that this bowl was made near where it was found, and is not an imported piece.

TABLE 1 Comparison of SEM/EDS and electron microprobe analyses of the bulk paste and glaze compositions of Cookworthy samples CK1 and CK2.

	SEM/EDS									MICROPROBE (OWEN ET AL., 2000)									
	SiO_2	TiO_2	Al_2O_3	FeO	MgO	CaO	Na_2O	K_2O	TOTAL	SiO_2	TiO_2	Al_2O_3	FeO	MgO	CaO	Na_2O	K_2O	P_2O_5	TOTAL
CK1 PASTE	66.1	0.0	27.7	0.4	0.0	0.7	1.5	3.6	100.0	64.2	0	28.0	0.6	0.5	0.8	1.4	4.1	0.4	100.0
CK1 GLAZE	82.0	0.0	10.4	0.4	0.9	0.7	1.5	4.1	100.0	80.0	0	11.1	0.8	1.3	0.8	1.4	4.6	0	100.0
CK2 PASTE	70.6	0.0	24.2	0.4	0.0	0.3	1.1	3.5	100.0	68.9	0	24.9	0.5	0.3	0.3	0.9	3.1	0.3	99.2
CK2 GLAZE	80.3	0.0	12.5	0.5	0.5	0.6	1.3	4.3	100.0	77.75	0	13.6	0.8	0.7	0.7	1.1	5.0	0.4	100.0

TABLE 2 Major, minor, and trace element contents of the Philadelphia bowl and contemporary aluminous-silicic and S-A-C-type wares.

	PHILADELPHIA BOWL		PHILADELPHIA BOWL MELT BLEB*		BARTLAM BART11		COOKWORTHY CK1		COOKWORTHY CK2		B&M 1773 BASKET		A-MARKED TEAPOT LID		AVG. OF 5 HOPEWELL, VA.	
	PASTE	GLAZE	CORE	CORONA	PASTE	GLAZE	PASTE	GLAZE	PASTE	GLAZE	PASTE	GLAZE	PASTE	GLAZE	BISCUIT WASTERS	STD. DEV.
SiO_2 (WT%)	76.20	65.08	54.15	70.33	68.70	69.30	66.11	81.99	70.56	80.31	68.38	47.86	61.55	64.68	76.86	0.32
TiO_2	0.60	0.56	0.59	0.63	0.10	0.03	0.00	0.00	0.00	0.00	0.52	0.16	0.07	0.08	0.19	0.03
Al_2O_3	19.13	20.52	35.94	19.73	23.10	14.27	27.68	10.45	24.21	12.48	5.12	0.88	22.40	14.02	20.03	0.35
FeO	0.52	0.69	3.86	1.83	1.00	0.52	0.41	0.40	0.42	0.54	0.48	0.49	0.31	0.52	0.39	0.02
MnO	0.00				0.10	0.04					0.02	0.07	0.03	0.09	0.00	
MgO	0.04	0.52	0.98	0.87	0.10	0.15	0.00	0.92	0.00	0.49	0.29	0.29	1.14	1.39	0.08	0.02
CaO	0.93	7.84	1.90	2.34	1.00	9.44	0.68	0.65	0.30	0.64	13.04	5.35	6.49	11.51	0.03	0.06
Na_2O	0.73	2.21	0.66	1.00	2.30	1.59	1.54	1.51	1.05	1.28	1.07	2.16	3.48	2.49	0.50	0.03
K_2O	1.90	2.58	1.94	3.27	3.00	4.31	3.57	4.08	3.46	4.27	5.51	4.66	3.15	2.96	1.91	0.05
PbO						0.01					5.29	37.02	0.48	0.43		
P_2O_5											0.20	0.11	0.23	0.33		
SO_3											0.07	0.21	0.00	0.00		
SnO_2												0.57	0.67	1.49		
Total	100.05	100.00	100.00	100.00	99.40	99.66	99.99	100.00	100.00	100.00	99.99	99.83	100.00	99.99	100.00	
Li(ppm)	97.9	321	151	205	114	37.9	478	516	341	500	8.1	35.9	54.2	48.7	69.8	6.7
P	212	208	117	164	209	201	894	809	693	1102	803	473	1083	907	545	398
Sc	10.4	8.7	17.9	15.6	3.41	3.01	3.93	3.60	6.88	4.02	6.51	3.71	2.25	2.13	8.05	0.90
V	73.0	71.6	91.3	104.4	5.07	2.48	0.77	2.38	1.19	5.85	25.30	12.27	6.38	4.04	28.3	7.3
Cr	63.0	46.6	55.3	40.8	5.89	7.52	5.17	3.30	4.74	4.04	25.32	7.00	7.56	7.21	25.8	6.3
Co	3.15	12.86	9.32	7.29	6.41	4.32	4.05	4.04	0.90	1.46	6.33	9.51	20.3	21.6	1.4	0.1
Ni	23.3	40.0	67.1	41.5	2.1	11.3	3.0	6.7	2.0	8.8	14.5	11.9	18.6	29.6	11.2	1.1
Cu	10.5	73.4	12.1	14.5	8.3	16.2	29.3	43.7	13.0	24.0	68.4	90.0	63.9	87.7	29.1	4.3
Zn	32.1	173.2	77.6	63.8	43.4	48.2	17.7	21.6	10.1	11.0	44.9	18.8	37.4	104	89.8	22.8
As	17.0	14.1	10.0	14.5	3.5	6.3	9.8	12.1	4.1	4.0	2034	3910	15.6	52.1	15.1	2.9
Rb	119	144	190	249	454	327	724	1040	680	927	44.2	37.6	41.0	30.9	196	18

TABLE 2 Continues on page 205.

TABLE 2 (Continued)

	Philadelphia bowl		Philadelphia bowl melt bleb*		Bartlam Barth		Cookworthy Ck1		Cookworthy Ck2		B&M 1773 basket		A-marked teapot lid		Avg. of 5 Hopewell, Va.	
	PASTE	GLAZE	CORE	CORONA	PASTE	GLAZE	PASTE	GLAZE	PASTE	GLAZE	PASTE	GLAZE	PASTE	PASTE	BISCUIT WASTERS	STD. DEV.
Sr	75.4	333	121	151	40.0	85.7	174	146	183	240	123	107	282	22C	83.C	14.3
Y	11.99	9.17	9.57	9.68	14.3	15.9	3.54	1.14	2.62	2.00	7.85	4.28	7.42	3.09	7.5	1.4
Zr	77.4	67.8	54.1	65.1	23.1	23.6	14.6	10.9	18.3	14.7	47.3	35.9	42.2	24.4	31.8	2.8
Nb	10.62	8.97	8.91	10.53	11.58	10.75	18.73	45.68	25.31	49.15	5.35	2.34	4.24	1.14	6.4	1.4
Sn	12.82	12.1	64.7	33.4	16.9	15.4	24.8	24.9	32.6	24.7	52.1	36.6	6349	580	20.4	1.4
Cs	39.15	24.1	50.3	67.5	53.2	17.6	26.0	26.3	22.4	26.8	1.35	0.62	2.77	1.16	24.7	2.2
Ba	246	491	338	426	61.2	70.0	86.9	62.9	97.4	110	67.2	53.3	118	148	297	66
La	30.86	30.91	29.87	31.45	6.25	14.66	3.88	1.36	3.24	2.74	12.0	5.76	8.43	14.6	13.5	3.7
Ce	53.60	55.77	55.43	57.55	13.98	28.54	7.89	2.75	7.02	5.45	25.8	12.1	11.4	16.0	30.4	9.2
Pr	6.20	6.06	6.08	6.55	1.50	3.02	0.90	0.30	0.83	0.61	2.87	1.33	1.39	1.82	3.4	1.0
Nd	24.3	22.3	22.7	24.3	6.24	12.1	3.40	1.26	3.47	2.45	11.10	5.28	5.20	5.80	13.2	4.2
Sm	4.54	4.26	4.43	4.69	2.30	3.41	0.77	0.28	0.79	0.53	2.11	1.03	1.19	0.99	2.9	0.9
Eu	0.86	1.62	1.06	1.14	0.27	0.47	0.14	0.08	0.15	0.23	0.41	0.20	0.28	0.49	0.5	0.1
Gd	3.32	2.63	2.85	3.05	2.45	3.62	0.69	0.23	0.64	0.48	1.77	0.91	1.29	0.73	2.1	0.6
Tb	0.43	0.32	0.37	0.46	0.50	0.57	0.10	0.04	0.10	0.07	0.22	0.12	0.23	0.11	0.3	0.1
Dy	2.56	1.87	1.96	2.00	2.78	3.11	0.58	0.19	0.53	0.36	1.27	0.77	1.34	0.61	1.7	0.4
Ho	0.50	0.32	0.42	0.48	0.44	0.53	0.10	0.04	0.09	0.07	0.24	0.15	0.23	0.11	0.3	0.1
Er	1.34	0.96	1.18	1.16	1.02	1.29	0.28	0.09	0.21	0.17	0.68	0.42	0.61	0.28	0.7	0.2
Tm	0.20	0.15	0.13	0.20	0.14	0.19	0.04	0.02	0.03	0.03	0.09	0.07	0.08	0.04	0.1	0.0
Yb	1.44	1.10	1.21	1.52	0.95	1.13	0.33	0.12	0.23	0.22	0.58	0.37	0.57	0.28	0.7	0.1
Lu	0.22	0.16	0.19	0.19	0.11	0.15	0.04	0.02	0.03	0.02	0.08	0.05	0.08	0.04	0.1	0.0
Hf	2.14	1.79	1.30	1.96	1.35	1.70	0.85	0.68	1.12	0.89	1.20	0.86	1.35	0.71	1.0	0.1
Ta	1.24	0.92	1.26	1.20	2.35	2.13	7.51	20.97	9.19	19.09	2.45	0.15	0.74	0.13	0.9	0.3
Pb**	136	193	237	254	92.5	377	79.5	18.2	17.0	64.4	151340	314400	3179	3053	94.0	41.4
Th	8.86	7.37	7.31	8.11	4.80	7.88	5.85	1.38	4.72	2.95	2.58	1.21	2.22	0.66	7.3	2.0
U	3.00	2.96	2.72	3.29	10.2	10.1	3.97	4.52	4.22	5.32	0.76	0.36	2.14	0.45	3.8	0.6
B	ns	ns	11570	7383	ns	ns	ns	ns	ns	ns	4190	8550	ns	ns	76.3	16.0

Some of the samples list above were previously analyzed for major and minor elements by electron microprobe and have been described elsewhere.[8] The two methods (SEM/EDS and microprobe) give comparable results, as shown by the replicate analyses of the paste and glaze of two Cookworthy porcelain figures (Table 1). Trace element concentrations for all samples were determined by laser ablation–inductively coupled plasma–mass spectrometry (LA-ICP-MS).[9] The major, minor, and trace element compositions of all samples described here are reported in Table 2.

TABLE 3 Composition of melt blebs in the body of the Philadelphia bowl.

	MELT BLEB 1 (FIGS. 2 D,E)		MELT BLEB 2 (FIG. 2F)	
	CORE	CORONA	CORE	CORONA
SiO_2	54.15	70.33	54.38	71.21
TiO_2	0.59	0.63	0.52	0.66
Al_2O_3	35.94	19.73	38.21	20.13
FeO	3.86	1.83	1.97	0.73
MgO	0.98	0.87	0.96	0.70
CaO	1.90	2.34	1.29	2.23
Na_2O	0.66	1.00	0.62	1.01
K_2O	1.94	3.27	2.06	3.37
TOTAL	100.00	100.00	100.01	100.02

Mineralogy and geochemistry

The paste of the Philadelphia bowl has an aluminous-silicic composition.[10] It contains about 76 wt.% silica (SiO_2) and 19% alumina (Al_2O_3), with subordinate potash (~2% K_2O) and minor (<1%) lime (CaO) and soda (Na_2O). The paste is vitrified, and in places the melt phase has coalesced to form large blebs (figs. 2D–F) with mullite-rich cores that mirror their alumina-rich composition (36% Al_2O_3). The bowl has a high-temperature glaze containing approximately 65 percent SiO_2, 21 percent Al_2O_3 and 8 percent CaO, with nearly 5 percent alkalis (Na_2O+K_2O), and only trace concentrations (193 ppm) of lead. The glaze is readily apparent on compositional maps that include the adjacent paste, since it is enriched in chromium (Cr), strontium (Sr), cobalt (Co), barium (Ba), lead (Pb), and the light Rare Earth elements (REE; fig. 3). It contains calcic plagioclase (andesine [An_{61-69}]) microlites (fig. 2B). The glaze has fused to the ceramic substrate, from which it is separated by a narrow, integrated body-

TABLE 4 Composition of selected minerals in the Philadelphia bowl.

	GLAZE PLAGIOCLASE (FIG. 2B)			MELT BLEB PLAGIOCLASE (FIG. 2H)			ALUMINO SILICATE	STD DEV (N=6)	META-KAOLIN*	MULLITE ***
SiO_2	60.28	58.51	61.91	51.49	52.28	57.93	60.64	4.18	53.38	71.79
TiO_2	0.33	0.32	0.52			0.37	0.70	0.10	0.59	
Al_2O_3	24.07	26.45	21.33	31.93	31.91	27.74	33.25	3.69	43.18	28.21
FeO_t**	0.6	0.4	1.04	0.84	0.76	0.36	0.80	0.11	1.16	
MgO	0.35		0.85				0.42	0.07	0.35	
CaO	10.21	10.61	11.08	15.24	14.12	11.36	1.49	0.11	0.05	
Na_2O	2.43	2.53	1.58	0.51	0.49	1.23	0.66	0.13		
K_2O	1.71	1.18	1.7		0.46	1.01	2.04	0.2		
Total	99.98	100	100.01	100.01	100.02	100	100.00			
%An	61	64	69	94	91	77				
%Ab	26	28	18	6	6	15				
%Or	12	8	13	0	4	8				

* MK metakaolin analysis is presented for comparison with the unidentified aluminosilicate. It source is given in endnote xiv.

** total iron as FeO_t

*** ideal composition based on an empirical formula of $Al_{4.5}Si_{1.5}O_{9.75}$

glaze layer (figs. 2A, B). Collectively, these features show that the body and glaze were subjected to a single, high-temperature firing. We conclude that the bowl consists of true porcelain.

The paste of the bowl consists of subangular (partly resorbed) silica polymorphs (fig. 2C) and subordinate, irregularly shaped (xenomorphic) grains of an unidentified aluminosilicate phase that is more silicious and less aluminous than metakaolin (Table 4) set in a matrix comprising a melt phase containing variable amounts of minute (micron-scale) mullite crystallites.[11] Two large (~150–200 micron diameter) melt patches (figs. 2D–F) were exposed on a polished grain mount prepared from the sample. The center of these patches contains relatively coarse mullite crystallites (fig. 2G). The coarse mullite-bearing core of one of these melt patches (fig. 2E) has calcic plagioclase (bytownite to anorthite) microlites ($An_{77\text{-}94}$) along part of its periphery (fig. 2H). Image analysis of the coarse interior of one of the melt patches (fig. 2G) shows that the two phases are present in the ratio

Figure 3 Compositional maps of part of the glaze and ceramic substrate of the Philadelphia bowl. "Hot" colors (yellows and reds) denote high concentrations of each component; "cool" colors (greens and blues) denote low concentrations. These images were generated by ICP-LA-MS using an 8-micron beam and 20 lines across the imaged areas.

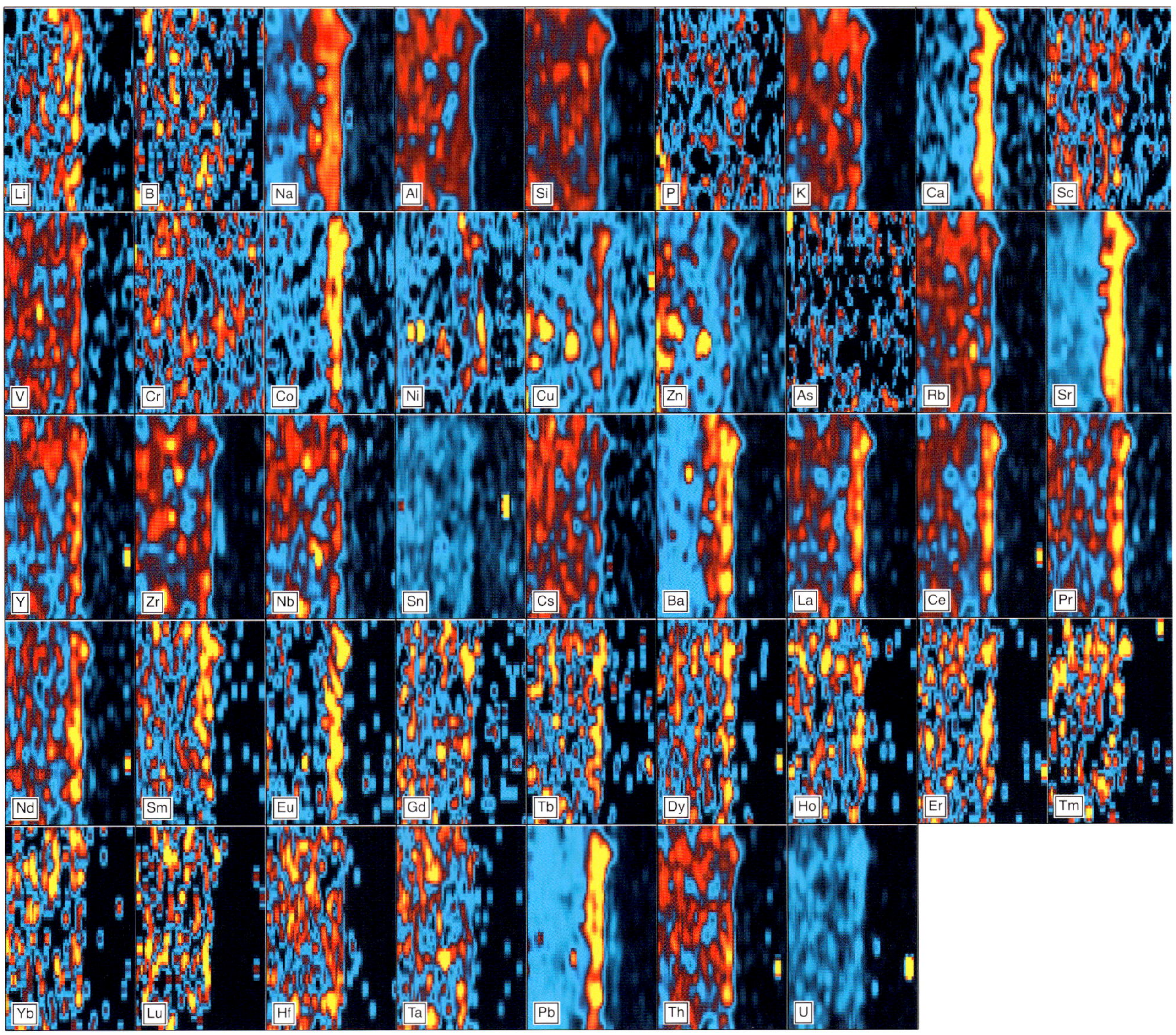

55:45 (melt:mullite); the finer-grained corona-like envelope enclosing it has a much higher proportion of melt (~85:~15). This accounts for the relatively silica-rich (~70% versus ~54% SiO_2) and alumina-poor (~20% versus ~36% Al_2O_3) composition of this corona.

The higher proportion of melt in this corona is also mirrored by its minor and trace element composition. Figure 4 shows compositional maps of this melt bleb based on the concentrations of forty-three components. The melt bleb is clearly visible on almost half of these images, and the corona is compositionally distinct from the core on many, particularly maps based on components preferentially concentrated in the melt phase (i.e., incompatible elements).[12] These include silica, the alkalis (Na, K), cesium (Cs), barium (Ba), yttrium (Y), the light Rare Earth elements (La, Ce, Pr, Nd), and lead (Pb). The most incompatible of these components are concentrated in this corona, owing to its higher proportion of melt. Calcium and a geochemically affiliated trace element, strontium (Sr)—ordinarily relatively compatible components that are preferentially partitioned into high-temperature minerals—are also concentrated in this corona, probably because there were no discrete calcic minerals in the paste to retain them during low degrees of partial melting. For obvious reasons, alumina is enriched in the mullite-rich core of this melt bleb, but so too are tin (Sn) and boron (B),

Figure 4 Compositional maps of a spherical melt bleb (see figs. 2D, E) in the body of the Philadelphia bowl. These images were generated by ICP-LA-MS using an 8-micron beam and 20 lines across the imaged area.

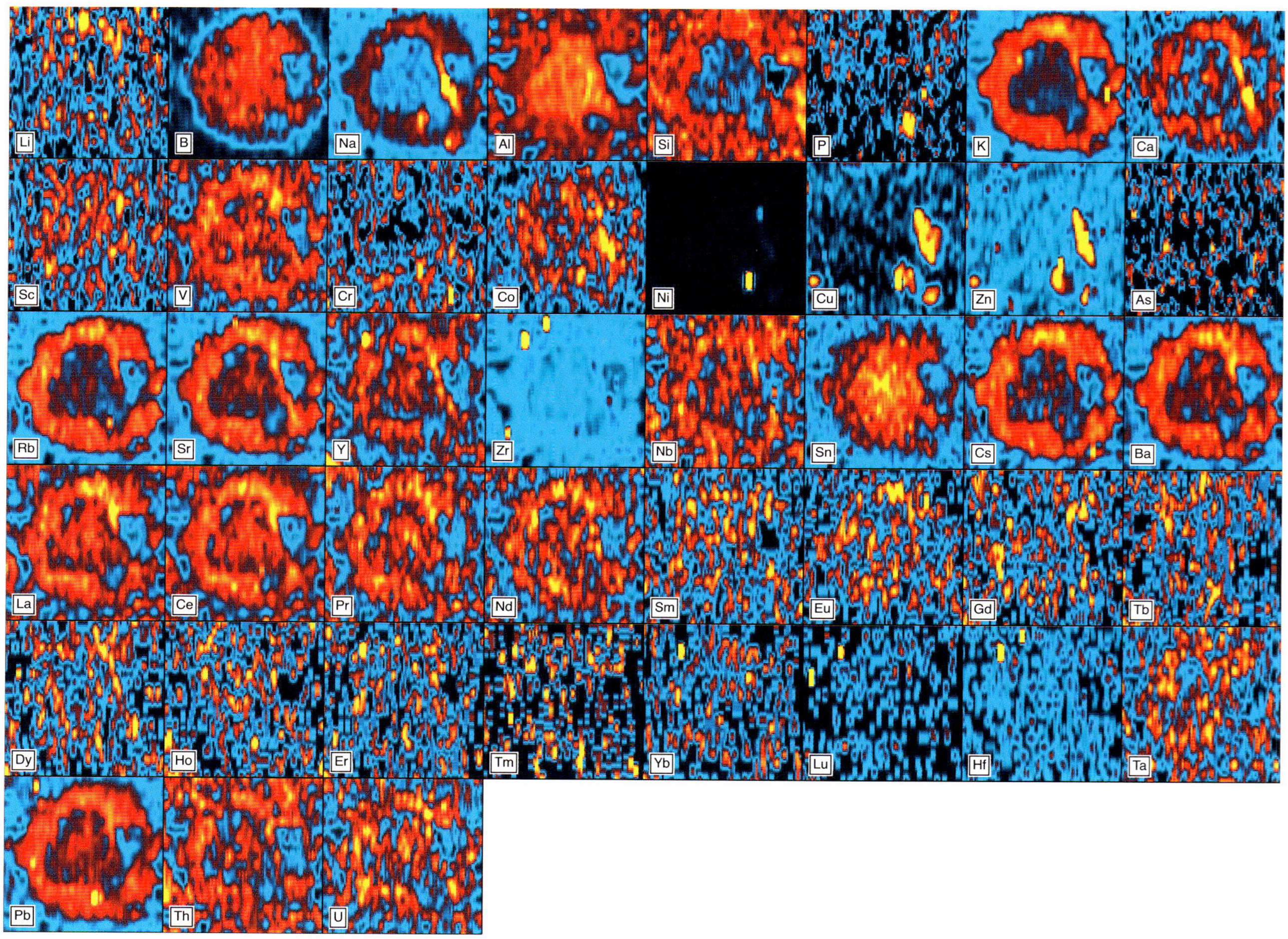

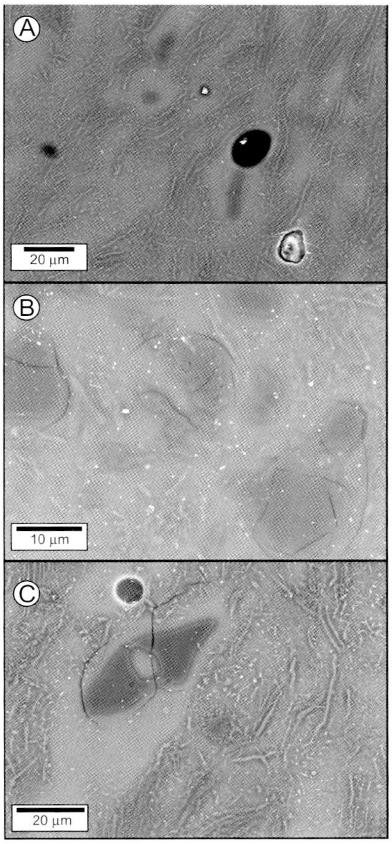

Figure 5 Backscattered-electron images of the aluminous-silicic paste of (A) a sherd (Bart11) from the Cain Hoy site of John Bartlam's potworks, and (B, C) a William Cookworthy figure (sample Ck1). Note the abundance in both samples of the matrix melt phase, the well-rounded character of relict quartz grains entrained in the melt, and the high aspect ratio of mullite. The melt patch enclosing the resorbed quartz grain in (C) contains 79 percent SiO_2.

which evidently are favored (i.e., preferentially partitioned) by this mineral. We are unaware of any studies on the tin content of this mineral, but its enrichment in boron is consistent with the existence of boron-bearing, mullite-type compounds.[13]

The aspect ratio (i.e., length/width ratio) of mullite in aluminous-silicic ceramic wares dramatically increases with temperature.[14] The aspect ratio of longitudinal sections of mullite in the Philadelphia bowl (fig. 2E) is lower (≤ 10) than in the other samples (≤ 40) containing this mineral (see figs. 5A–C), implying a lower firing temperature. Compositional data can be used to place constraints on kiln conditions.

Firing Temperature

The temperature at which the Philadelphia bowl was fired can be estimated by consulting an appropriate phase diagram on which the compositions of the melt phase in the paste have been plotted, and using software that uses glaze compositions to calculate melting temperature.

The composition of the bowl's paste and its melt phase can be modeled using the SiO_2-K_2O-Al_2O_3 phase diagram. Owing to the presence of entrained mullite microlites, melt compositions in the bowl—and in the two Cookworthy samples—plot as a linear array of points between the composition of mullite and the thermal valley ("cotectic") separating the silica polymorph and K feldspar fields (fig. 7). This array of points intersects the cotectic at a temperature of approximately 1300°C. Lines of equal temperature ("isotherms") are very closely spaced on this part of the diagram, however, so this temperature constraint is poor. Consequently, glaze compositions were used to calculate melting temperature.

The software used here to estimate firing temperatures, MELTS for silica-rich (rhyolitic) systems ("pMELTS"), not only calculates the temperature at which the analyzed material is fully melted (i.e., its liquidus temperature), but also can predict which minerals will form as the melt cools.[15] For minerals that exhibit solid solution, it determines their composition. This software was created for use by geologists. Mullite is not a common mineral in igneous rocks, so it is not included among the minerals currently modeled by this program. Consequently, we determined melting temperature for the glaze on the Philadelphia bowl, rather than its paste. This is justified by (1) the calcic nature of the glaze, which ensured that calcic plagioclase (andesine)—a mineral whose formation is modeled by MELTS—formed instead of mullite as the glaze cooled, (2) the Philadelphia bowl was fired but once, so the calculated melting temperature of the glaze corresponds to (or at least places a constraint on) the firing temperature of the paste as well, and (3) the glaze completely melted, so, by definition, its composition (once andesine microlites have been reintegrated back into the glaze) corresponds to a melt, and therefore the liquidus temperature calculated by this software should approximate actual firing temperature. The latter assumption presupposes minimal compositional interaction between the glaze and its ceramic substrate, and minimal loss of relatively volatile components (e.g., alkalis) to the kiln atmosphere.

MELTS gives a liquidus temperature of 1221°C for the glaze on the Philadelphia bowl.[16] The bowl could have been heated above the liquidus; once the last glaze ingredient melts, the melt is no longer confined to the liquidus surface.

The first mineral that would have crystallized during cooling of a glaze of this composition is calcic plagioclase. MELTS determined that its composition initially (i.e., at 1220°C) would correspond to bytownite (An_{76}) but at 1000°C, andesine would form (An_{63}). This falls within the range of compositions of plagioclase (An_{61-69}) of the plagioclase actually seen in the glaze on the Philadelphia bowl, indicating that the glaze was finally quenched at a temperature at least 200°C below that at which it melted.

Although we have no constraint on the extent to which this bowl might have been heated above its glaze's liquidus, unless this was substantial, it seems that it was fired at a much lower temperature (i.e., approaching 1400°C or even higher)[17] than expected for this type of ware, one that is more in line with the firing conditions used for many soft-paste porcelains.[18] The calculated liquidus temperature is unlikely to be an artifact of substantial loss of volatile components (e.g., alkalis) during firing, because these components usually act as fluxes. Such losses would therefore leave the glaze with a more *refractory* composition, with a higher liquidus temperature than it would have had if volatilization had not occurred.

Either the Philadelphia bowl was fired at "soft-paste" porcelain kiln temperatures, or MELTS has underestimated the liquidus temperature of the glaze because it falls outside the composition of the samples for which this software was designed, or the bowl was fired at an indeterminate temperature above the liquidus temperature of its glaze.[19] To put the 1221°C liquidus temperature calculated for the Philadelphia bowl in context, liquidus temperatures were determined for some of the other glazes, whose compositions are reported in Tables 1 and 2. MELTS predicts a relatively high liquidus temperature (1298°C) for the low-calcium glaze on Cookworthy sample Ck1. A silica polymorph would be the first mineral to crystallize from this glaze during cooling. Significantly lower liquidus temperatures were calculated for the glazes on samples Bart11 (1092°C) and the A-marked porcelain teapot lid (1174°C, calculated exclusive of its small [0.43% PbO] lead content[20]). Owing to their comparatively calcic compositions, these two glazes would first crystallize calcic plagioclase (An_{68} and An_{75}, respectively). The calculated liquidus temperature of the Philadelphia bowl glaze is thus higher than those determined for the other low-lead and high-calcium glazes reported here, but less than the low-lead and low-calcium glaze on Cookworthy porcelain. Again we emphasize that these calculated temperatures are minimum values, because any of these glazes could have been heated to temperatures well above the liquidus. Firing temperatures above the liquidus are indeterminate in these samples.

The comparatively stubby character of mullite in its paste suggests that the Philadelphia bowl was fired at a lower temperature than the true porcelain manufactured at well-established manufactories (e.g., Cookworthy's enterprise), consistent with the calculated temperatures. If this inference is

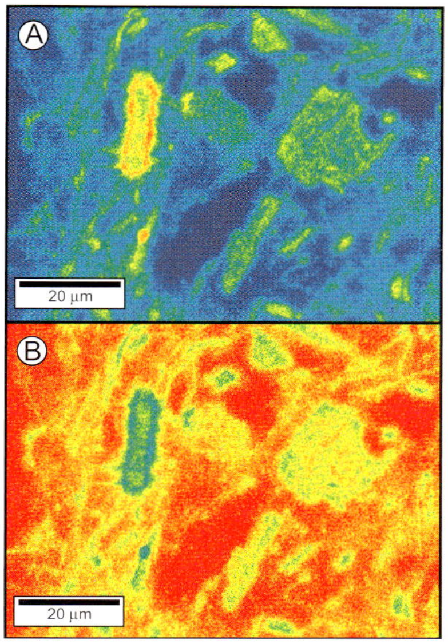

Figure 6 Compositional map for a small area of the body of a Cookworthy figure (sample Ck1). (A) Alumina; (B) silica. In (A), mullite appears as elongated, yellow and red domains, whereas quartz occupies blue patches, and the melt phase the milky-green areas, and in (B), quartz is red, mullite green, and the melt orange. The minute red speckles in (A) are possibly corundum (Al_2O_3), a high-temperature breakdown product of mullite.

correct, it seems that the manufacturer of this bowl had difficulty achieving the very high kiln temperatures attained by some of their competitors. In this regard, it is noteworthy that the 1773 S-A-C-type Bonnin and Morris openwork basket preserves relicts of the lead-rich ("flint") glass flux used in its manufacture, so it too has been interpreted to have been underfired.[21]

Comparison with Contemporary British and American Aluminous-Silicic and S-A-C Porcelains

High-fired aluminous-silicic wares include both true porcelain and stoneware. Fine earthenwares (e.g., creamware) can be compositionally similar.[22] They, of course, are not vitrified. Stoneware is vitrified, but Anglo-American stoneware is typically made from less pure clay than porcelain, so has higher concentrations of femic components (e.g., iron, titanium) than true porcelain. Some true porcelains are calcic; these include the S-A-C porcelains, which can be fired using a traditional, "hard-paste" porcelain firing schedule (in which case they will have a low-lead glaze, e.g., first-patent Bow ["A-marked"] porcelain), or, as in the case of the lead-bearing-variant described here—the Bonnin and Morris basket—a "soft-paste" firing schedule (in which case they will have a lead-rich glaze).[23] Mineralogically, high-fired aluminous-silicic porcelaneous wares are characterized by the presence of mullite, an aluminosilicate mineral. Owing to its fine grain size, crystals of this mineral can be difficult to analyze by microbeam methods (e.g., SEM/EDS and electron microprobe), but compositional maps of mullite-bearing parts of true porcelain (figs. 6A, B) and the displacement of melt compositions on phase diagrams (fig. 7) reveal its presence. Owing to their higher calcium content, S-A-C wares can contain pseudowollastonite ($CaSiO_3$) instead of mullite. Pseudowollastonite tends to be coarser grained than the mullite in high-fired porcelains, so it can be analyzed by microbeam methods.

Although having an aluminous-silicic paste, the suspected Philadelphia bowl is compositionally dissimilar to some contemporary British (Cook-

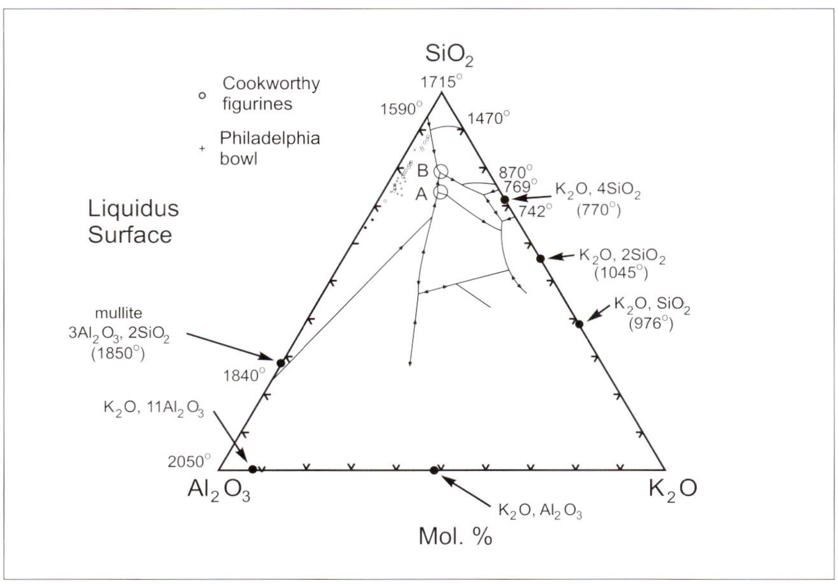

Figure 7 Silica-alumina-potash diagram showing the displacement of melt compositions by mullite microlites. Modified after E. F. and A. Muan, *Phase Equilibrium Diagrams of Oxide Systems: 5, K_2O-Al_2O_3-SiO_2* (Columbus, Ohio: American Ceramic Society, 1960).

worthy) and American (putative Bartlam) counterparts. Compared with these three samples, the Philadelphia bowl is relatively silicious and less aluminous (SiO_2/Al_2O_3 = 4.0 versus 2.4, 3.0), and it has lower alkali contents (Na_2O+K_2O = 2.6 versus 4.5–5.3). Its glaze is even more distinct. In contrast to Cookworthy's glaze, which has a low lime content (0.6% CaO), it contains almost 8 percent CaO; in this regard, these glazes mirror the compositions of the pastes of these artifacts. Its alumina content is approximately twice that of the Cookworthy glaze (20.5 versus ~11% Al_2O_3), with correspondingly lower silica content (65 versus ~81% SiO_2), and it contains a higher concentration of soda (2.2 versus ~1.4% Na_2O), but less potash (2.6 versus ~4.2% K_2O). In this respect, the glaze on the Philadelphia bowl resembles the paste of S-A-C porcelains.[24] So too does the glaze on the aluminous-silicic porcelain sherd recovered from Bartlam's potworks site. Although both have similar lime contents (7.8 versus 7.0% CaO), the Philadelphia bowl glaze has a significantly higher alumina content (~21 versus 14% Al_2O_3), lower silica (65 versus 72% SiO_2), and lower potash (2.6 versus 4.5% K_2O).

The Philadelphia bowl also differs from contemporary British and American true porcelain in terms of its trace element content. Its paste and glaze have order of magnitude higher concentrations of vanadium (V) and Cr than the other samples described in Table 2. Barium is also high, even

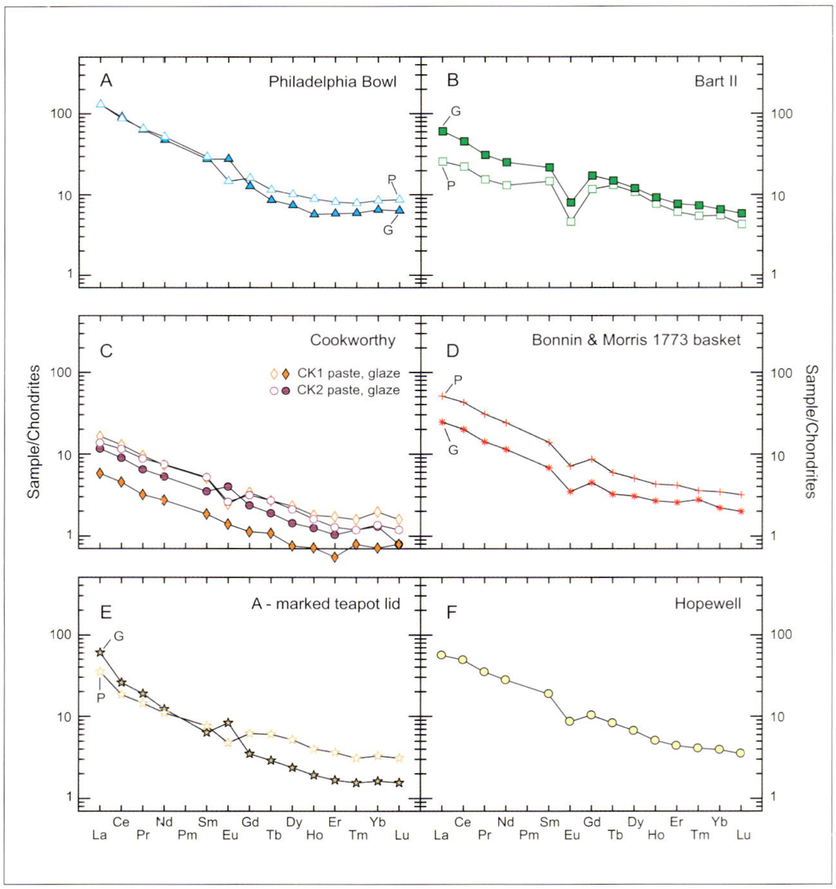

Figure 8 Chondrite-normalized REE plots for the pastes (P) and glazes (G) of (A) the Philadelphia bowl, (B) sample Bart11, (C) two Cookworthy figures, (D) the 1773 Bonnin & Morris openwork basket, (E) the A-marked porcelain teapot lid, and (F) the average of five unvitrified aluminous-silicic biscuit wasters from Hopewell, Virginia.

though its potassium content is comparatively low. It is similarly enriched in the Rare Earth elements compared with both Cookworthy figures, but less so compared with the other samples. On a chondrite-normalized REE diagram, the profiles of the paste and glaze of the Philadelphia bowl closely track one another (fig. 8A), but they have negative and positive europium (Eu) anomalies, respectively.[25] The same, of course, holds true on a multi-element ("spider") diagram. Apart from comparative enrichment of the glaze in strontium (Sr), consistent with its relatively calcic composition, and lower barium (Ba), the compositional profiles of the paste and glaze of this sample closely resemble one another (fig. 9A).

The origin of sample Bart11 remains uncertain. Its major/minor element composition (other than its higher Na_2O/K_2O ratio, 0.8 versus 0.4) resembles that of Cookworthy sample Ck2 (see Tables 1, 2), but it differs from both in the lower concentrations of Li, P, Rb, Sr, and Ta, and higher concentrations of V, U, Y, and REEs in its paste. The composition of Bart11 is therefore inconsistent with available data for Cookworthy porcelain. It also differs from the Philadelphia bowl, having much lower V, Cr, Ni, As, and REEs, and higher Rb and U contents.

The other samples have their own trace element signatures. The paste and glaze of the Bartlam sherd, for example, is enriched in Y and U, whereas the two Cookworthy samples (particularly their pastes) have high Li and Ta

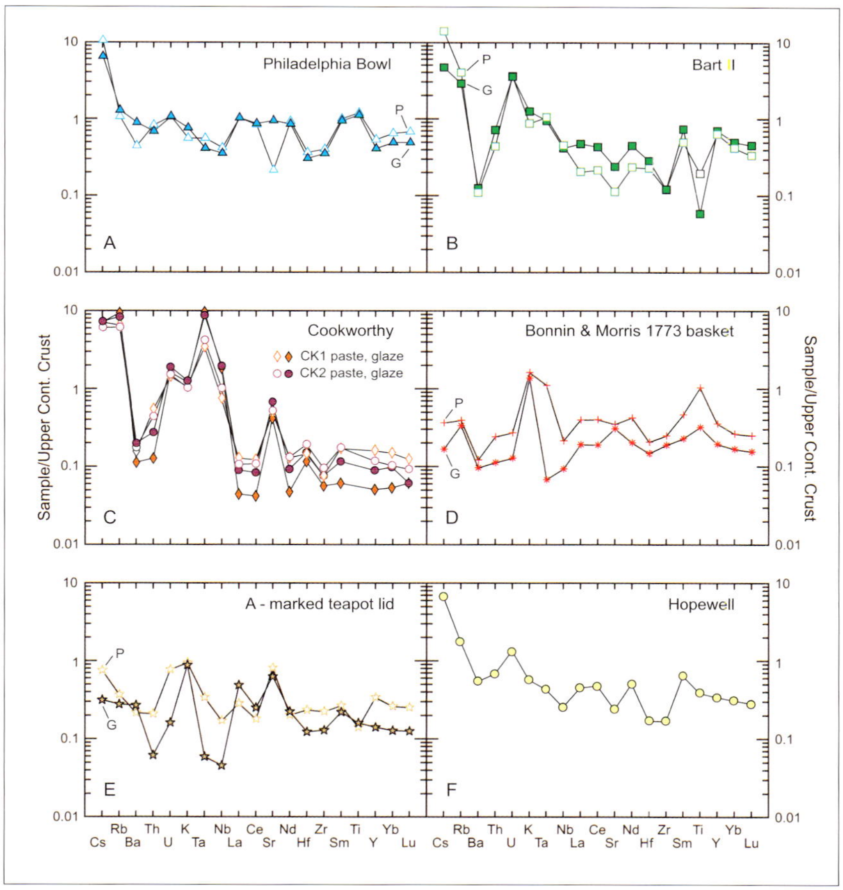

Figure 9 Mean-continental crust-normalized spider diagrams for the pastes (P) and glazes (G) of (A) the Philadelphia bowl, (B) sample Bart11, (C) two Cookworthy figures, (D) the 1773 Bonnin & Morris openwork basket, (E) the A-marked porcelain teapot lid, and (F) the average of four unglazed and unvitrified aluminous-silicic sherds from Hopewell, Virginia.

contents and low REEs. The REE profiles of the pastes of these samples are nearly coincident (fig. 8B). The glazes on these samples have lower and contrasting concentrations of the REEs, but both lack the negative Eu anomaly shown by their substrate. Similarly, the Cookworthy pastes and glazes have comparable profiles on the spider diagram shown in figure 8C. As with the other samples, the paste and glaze of the Bartlam-site sample show coherent profiles on the REE (see fig. 8B) and spider (fig. 9B) diagrams.

The Cookworthy samples cluster together, separate from other samples on triangular plots comparing their Na_2O-CaO-K_2O (fig. 10A) and chromium-scandium-yttrium (Cr-Sc-Y) (fig. 10B) concentrations. So too do the pastes and glazes of the Philadelphia bowl and the sample from the Bartlam site, but only on the latter diagram. The remaining samples are inherently different from the aluminous-silicic porcelains. The Bonnin and Morris basket is a lead-bearing S-A-C porcelain, whereas the A-marked porcelain teapot lid is a lead-poor variant of the same type of ware. Moreover, their glazes are distinct. The basket has a low-temperature, lead-rich (37% PbO) glaze, whereas the teapot lid has a high-temperature, low-lead (0.43% PbO), calcium-rich (11.5% CaO) glaze. The Bonnin and Morris basket was therefore fired using a soft-paste porcelain firing schedule involving an earlier, high-temperature firing of the body followed by glazing and firing at a lower temperature in a glost kiln. The REE signature of the body of the Bonnin and Morris basket (fig. 8D) closely resembles the REE profile of the aver-

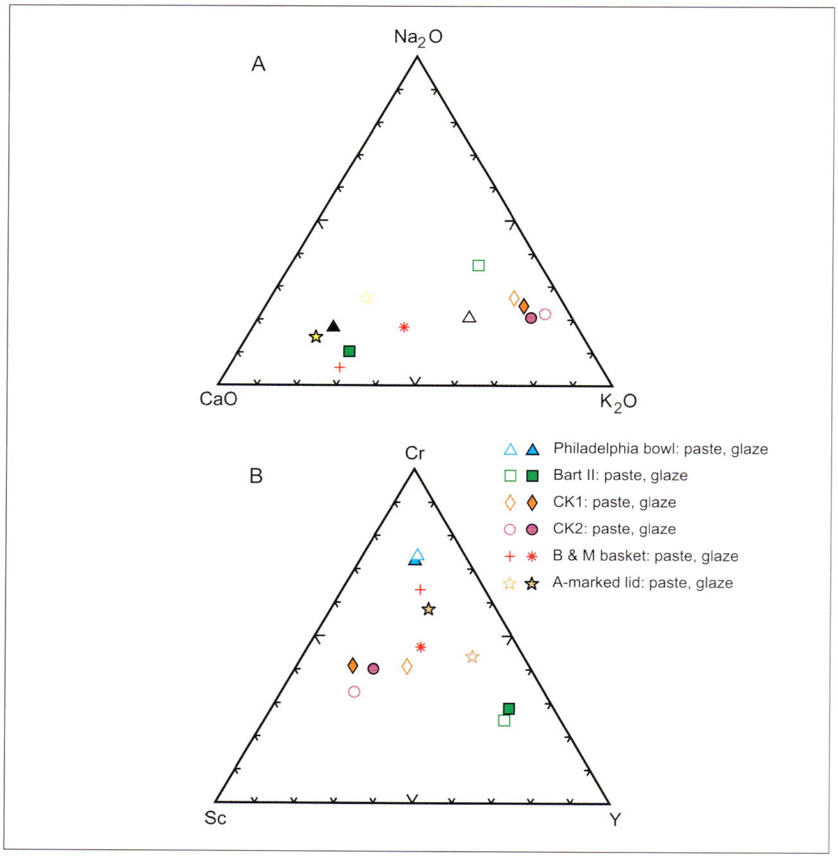

Figure 10 Comparison of the pastes and glazes of the samples described here plotted on (A) Na_2O-CaO-K_2O and (B) Cr-Sc-Y diagrams.

age paste composition of five aluminous-silicic biscuit sherds from Hopewell (fig. 8F). The spider diagram profile for the basket, however, is distinct (figs. 9D, F), as is, of course, its major element composition (see Table 2).

Not surprisingly, the glaze on the Bonnin and Morris basket is very distinct. In addition to being lead-rich, it shows a two to three orders of magnitude enrichment in arsenic (As). The REE profiles of the paste and glaze are nearly parallel, but the glaze has much lower concentrations of these components. It is also depleted in most other trace elements compared with its substrate, but, overall, their spider-diagram profiles resemble one another (fig. 9D).

Both the paste and glaze of the teapot lid are enriched in Sn and Pb compared with the other high-temperature glazes, and depleted in Cs. There is a crossover of their REE patterns on either side of Eu, and the paste has a negative Eu anomaly, whereas the glaze presents a positive Eu anomaly (fig. 8E). Their profiles are less concordant than those of the other samples, with divergent concentrations of Y and the heavy REEs (fig. 9E). Indeed, the body of this sample has more in common with the glaze on the Philadelphia bowl than with any of the other high-temperature glazes. In terms of microstructure, it is most unusual insofar as it preserves the corona structures that record features related to its vitrification and cooling history.[26]

In summary, a perusal of the compositional profiles (see figs. 8, 9) shows that samples from different sites/factories are compositionally distinct in terms of their trace element signatures, but the pastes and glazes of individual samples have a similar compositional fingerprint, differing mainly in the overall concentration of various components (e.g., glazes tend to have lower trace element contents than their ceramic substrate). Of course, the very limited number of analyzed British and American aluminous-silicic and S-A-C porcelain samples, particularly with respect to high-precision trace elements, allows only general comparisons to be made.[27] The compositional differences between the samples described here, however, are accompanied by contrasts in the microstructure of their pastes.

The BSE images of the Philadelphia bowl suggest that the bowl is relatively underfired compared to its aluminous-silicic counterparts, the Bartlam and Cookworthy samples. Unlike the generally subangular shape of quartz grains in the Philadelphia bowl, quartz grains in the latter samples are relatively well-rounded. They contain a higher proportion of melt, which accounts for the relatively high degree of resorption of the quartz in these samples. This, in turn, accounts for the relatively silicious composition of melt rinds that entrain relict quartz grains (see fig. 4C). It has been shown elsewhere that the composition of the melt phase in high-fired ceramics can vary with the type of adjacent or entrained minerals that have been partly resorbed.[28] Finally, the Bartlam and Cookworthy samples contain mullite with relatively high aspect ratios, suggesting higher firing temperatures.

The distinct firing history of the Philadelphia bowl might not have been restricted to its heating trajectory. Unlike the other samples, its glaze has partly crystallized; it contains plagioclase microlites that lack a quenched morphology (i.e., swallowtail crystal terminations), which indicates relatively slow cooling from peak kiln temperature. This, too, points to the inexperience of its manufacturer.

Discussion

Who in Philadelphia in the 1760s or 1770s knew how to make true porcelain? Bonnin and Morris brought in immigrants with a background in the production of phosphatic wares, notably those produced at the Bow factory in London. Until the discovery of the Philadelphia bowl, it has been thought that after cessation of its phosphatic production line, the American China Manufactory only produced a lead-bearing variety of S-A-C porcelain. This inference is based on an advertisement in which Bonnin and Morris announced that they no longer were in the market for bone ash, but sought glass (used as a flux in frit porcelains) instead. The new type of porcelain so made is now represented by a single intact exemplar, the 1773 openwork basket. This important artifact evidently was produced as a last-gasp effort to forestall bankruptcy during the last few months of the factory's existence. The discovery of the Philadelphia bowl, however, introduces the possibility that around this time Bonnin and Morris also produced aluminous-silicic porcelain.

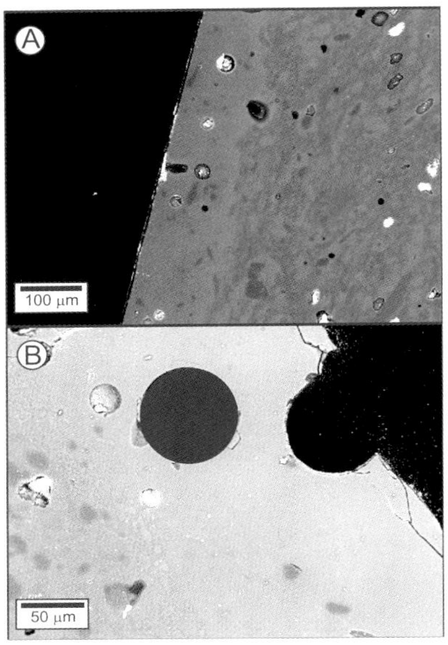

Figure 11 Backscattered-electron images of the glazes on (A) Cookworthy sample Ck1 and (B) Bartlam sample Bart11. Note the comparative thickness of these glazes (vis-à-vis the glaze on the Philadelphia bowl, see figs. 2A, B), and the absence of microlites.

The Philadelphia bowl described here does not resemble Chinese export porcelain. A continental origin is another possibility but it cannot be evaluated here, as we have no samples of late-eighteenth-century European porcelain to analyze and we are unaware of any published, high-precision trace element data for these wares. Then, there are rumors of a German resident of Philadelphia who was described in 1773 as being familiar with the porcelain trade.[29] Too little about him is known for us to evaluate his possible role in producing true porcelain in the City of Brotherly Love.

In general, several lines of evidence suggest the bowl has characteristics of being American-made. First, it is undecorated, a feature that is highly unusual among the products of known eighteenth-century British porcelain tableware. Moreover, its very thin glaze, about one fifth the thickness of the glaze on the Cookworthy (fig. 11A) and Bartlam (fig. 11B) samples, points to production by someone inexperienced in the manufacture of this medium. It is so thin that it is difficult to determine with the naked eye if the bowl is actually glazed.[30] Despite being so thin, it has partly crystallized, indicating insufficient dumping of heat from the glost kiln during production. The comparatively stubby shape of mullite microlites in its body points to a relatively low peak temperature (or short soaking interval) during kiln firing. Although circumstantial, these features point to the inexperience of the manufacturer.

The large melt blebs (see figs. 2D–F) in this artifact are particularly distinctive and unusual—we have not seen these in any other porcelain samples that we have analyzed over the past two decades. One explanation for them centers on the observation that fluxing compounds act as a point source of the fluxing agents they contain.[31] These fluxes infiltrate the surrounding clayey matrix, causing a melt phase to form in these domains at or near the peak kiln temperature. The fluxes used in historical porcelains include glass and/or alkali carbonate minerals. Some of the glass fluxes were specially made, and included lime-bearing alkali glass and alkali-lead glass. We postulate that the large melt patches in the body of the Philadelphia bowl formed in parts of the paste that were enriched in flux particles, probably alkali carbonate. This implies that paste ingredients were insufficiently ground (powdered) and/or mixed, allowing large melt blebs to form in domains enriched in fluxing components. Collectively, these diverse features suggest that the Philadelphia bowl was not made as the commercial product of a well-established producer of true porcelain. Indeed, it is more akin to an experimental ware made by a novice to these wares. We conclude that it is not an import, but rather was produced domestically.

Although we cannot exclude the possibility that the German gentleman previously mentioned might have had a role in its creation, until more is known about him and his activities in Philadelphia, it would be ill-advised to consider this hypothesis further. A more plausible possibility is that it originated at the Bonnin and Morris factory sometime after they ceased making phosphatic porcelain. As succinctly pointed out by Hunter and Gerhardt in their companion paper to the present contribution, Andrew Duché was in the right place at the right time to have influenced the proprietors of the American China Manufactory. Moreover, it is known that he had experience in the production of very high-fired wares (stoneware) and, of course, he claimed to have made true porcelain while in Georgia in the late 1730s, although this remains unsubstantiated.

The hypothesis that the Philadelphia bowl originated at the American China Manufactory could be tested by first scanning with a UV lamp all of the porcelain sherds recovered from the site, and analyzing those that fluoresce differently from the phosphatic samples. If any such sherds have compositions matching those of the Philadelphia bowl, then a Bonnin and Morris provenance is, for all intents and purposes, confirmed. Even if none match, however, the possibility that this artifact originated at this site still cannot be discounted on the basis of the available analytical database.

Are we prepared to attribute the Philadelphia bowl to Bonnin and Morris solely on the basis of analytical data? Not yet, but the differences in the trace element signatures of true porcelains from different areas reassure us that no obvious link can made to the products of other contemporary porcelain factories. Only future archaeological digs, archival research, and the discovery of matching sherds or a documentary piece, along with the acquisition of high-precision trace element data for historical true porcelains from a variety of producers, will confirm this. And what of Andrew Duché's contribution? Given his prominence and connections in the fine ceramics industry, it is hard to imagine that he had no influence on Bonnin and Morris. He may have been prone to hyperbole, as his 1737–1738 claim suggests, but with the Philadelphia bowl firmly in hand, we are tempted to say "touché, Duché!"

ACKNOWLEDGMENTS We thank Rob Hunter for inviting us to investigate this important artifact, for making available a fragment from it, for sending us an early draft of his companion paper, which helped direct our interpretation of the Philadelphia bowl, and for his incisive comments on an early draft of this contribution. He also arranged funding of the acquisition of the trace element analyses reported here. The acquisition of other data and of BSE images and drafting expenses were funded by a grant from Saint Mary's University Faculty of Graduate Studies and Research to the first author. Randolph Corney drafted the figures.

1. See John A. Burrison, *Brothers in Clay: The Story of Georgia Folk Pottery* (Athens: University of Georgia Press, 2008).

2. See Robert Hunter and Juliette Gerhardt, "An Eighteenth-Century American True-Porcelain Punch Bowl," in this volume.

3. Rebecca Yamin, Principal Investigator, with contributions by Tod L. Benedict, Meagan Ratini, Kevin C. Bradley, Leslie E. Raymer, Juliette Gerhardt, Kathryn Wood, Timothy Mancl, Claudia L. Milne, *Archeology of the City—The Museum of the American Revolution Site Archeological Data Recovery, Third and Chestnut Streets, Philadelphia, Pennsylvania*, 2 vols. (Jackson, Mich.: Commonwealth Heritage Group, 2016), vol. 1.

4. See ibid.

5. Ibid.

6. Ibid.

7. The SEM used was a LEO 1450VP operated with a beam current of 20kV and equipped with an Oxford Instrument INCA X-max 80 mm2 silicon drift EDS detector and a field emission gun. Samples were rastered over broad areas, allowing the bulk compositions of both pastes and glazes to be determined.

8. The microprobe used was a JEOL 1750 Superprobe equipped with three wavelength-dispersive spectrometers and one energy-dispersive spectrometer. For Cookworthy figures Ck1 and Ck2. see J. V. Owen, B. Adams, and R. Stephenson, "Nicholas Crisp's 'Porcellien': A Petrological Comparison of Sherds from the Vauxhall (London; c. 1751–1764) and Indeo Pottery (Bovey Tracey, Devonshire; c. 1767–1774) Factory Sites," *Geoarchaeology* 15, no. (2000): 43–78. For Bartlam sample Bart11, see J. Victor Owen, "Geochemistry of High-fired Bartlam Ceramics," in *Ceramics in America*, edited by Robert Hunter (Hanover, N.H.: University Press of New England for the Chipstone Foundation, 2007), pp. 209–18. For the dated and marked openwork basket, see J. Victor Owen and Robert Hunter, "Too Little, Too Late: The Geochemistry of a 1773 Philadelphia Openwork Porcelain Basket," *Journal of Archaeological Sciences* 36, no. 2 (February 2009): 333–42. For the first-patent Bow teapot lid, see J. Victor Owen, "Double Corona Structures in 18th Century Porcelain (1st Patent Bow, London, mid-1740s): A Record of Partial Melting and Subsolidus Reactions," *The Canadian Mineralogist* 50, no. 5 (October 2012): 1255–64.

9. The LA-ICP-MS equipment used was a Resonetics RESOlution M-50 193 nm ArF excimer laser ablation system coupled to a Thermo X Series II quadrupole mass spectrometer. Ten analyses consisting of 30 s of washout/background followed by 30 s of ablation were made with a spot size of about 40 μm (for pastes) and 20 μm (for glazes), a laser repetition rate of 6 Hz, and a fluence of 6 J/cm2. Data were processed with Iolite software (C. Paton, J. Hellstrom, B. Paul, J. Woodhead, and J. Hergt, "Iolite: Freeware for the visualisation and Processing of Mass Spectrometric Data," *Journal of Analytical Atomic Spectrometry* 26 [2011]: 2508–18) using NIST 612 glass as an external reference and the Si concentrations determined by EDS as an internal reference. Samples were rastered on lines to determined average trace element concentrations of both pastes and glazes.

10. J. Victor Owen, "A New Classification Scheme for Eighteenth-Century American and British Soft-Paste Porcelains," *Ceramics in America*, edited by Robert Hunter (Hanover, N.H.: University Press of New England for the Chipstone Foundation, 2007), pp. 121–40.

11. These stubby crystals are more silicious (60% versus 51–53% SiO_2) and less aluminous (33% versus 42–44% Al_2O_3) than metakaolin. Mullite is much more aluminous than either of these minerals (71.8% Al_2O_3, 28.2% SiO_2). The metakaolin analysis in Table 4 is from N. Torres Castellanos, J. Torres Agredo, and R. Majía de Gutiérrez, "Performance under Sulfate

Attack of Concrete Additioned with Fluid Catalytic Cracking Catalyst Residue (FCC) and Metakaolin (MK)," *Ingeniería e Investigación* 33, no. 1 (2013): 18–22.

12. Incompatible elements (or components) are those that are preferentially concentrated in the melt phase during partial melting (and crystallization); the converse is true for compatible elements, which are partitioned into minerals as they crystallize from a melt (and are retained by minerals during low degrees of partial melting).

13. Martin Fisch, "Crystal Chemistry of Boron-Bearing Mullite-Type Compounds," Ph.D. diss., Universität Bern, 2011.

14. Jorge Martín-Márquez, Jesús Ma. Rincón, and Maximina Romero, "Mullite Development on Firing in Porcelain Stoneware Bodies," *Journal of the European Ceramic Society* 30, no. 7 (May 2010): 1599–1607.

15. Guilherme A. R. Gualda and Mark S. Ghiorso, "MELTS_Excel: A Microsoft Excel-Based MELTS Interface for Research and Reaching of Magma Properties and Evolution," *Geochemistry, Geophysics, Geosystems* 16, no. 1 (January 2015): 315–24. Rocks and ceramic materials melt over a range of temperatures, starting at a eutectic (thermal minimum). The liquidus is the temperature above which the sample is entirely molten. On the liquidus, crystals start to form from the liquid as it cools, and tracks the liquidus surface down-temperature toward the eutectic (unless, by happenstance, the sample already has a minimum melt composition).

16. This temperature was calculated for a pressure of 1 bar (as would be the case in a kiln). We assumed relatively oxygenating conditions, so used the HM buffer. MELTS indicated an oxygen fugacity of -4.97 at the calculated liquidus temperature for the glaze.

17. For example, it is commonly stated that true ("Chinese type") porcelain is fired at a temperature of about 1450°C. References include historical reviews in refereed journals, such as Keven J. Anderson, "Porcelain," *MRS Bulletin* 15, no. 7 (July 1990): 71–72, and Lydia H. Liu, "Robinson Crusoe's Earthenware Pot," *Critical Inquiry* 25, no. 4 (Summer 1999), 728–57; and in reference books, e.g., Lesley Acton and Paul McAuley, *Repairing Pottery & Porcelain: A Practical Guide*, 2nd ed. (London: A & C Black, 2003). It also appears in *Encyclopaedia Britannica* online, at http://www.britannica.com/art/porcelain.

18. See J. Victor Owen and Jacob J. Hanley, "Bartlam Porcelain Re-created," *American Ceramic Circle Journal* 19 (forthcoming).

19. Other than its lower sodium content, this glaze is not radically different from some igneous rocks (e.g., some anorthosites).

20. Lead is not included among the components modeled by current versions of MELTS.

21. Owen and Hunter, "Too Little, Too Late," p. 339.

22. J. Victor Owen, "What Analytical Data Can (and Cannot) Tell Us about 18th Century Fine Ceramics," *English Ceramic Circle Transactions* 22 (2010): 215–30.

23. E.g., W.R.H. Ramsay and E. G. Ramsay, "A Classification of Bow Porcelain from First Patent to Closure: c. 1743–1774," *Proceedings of the Royal Society of Victoria* 119, no. 1 (September 2007): 1–68. Owing to its lead-rich glaze and lead-bearing S-A-C paste, many workers might classify this artifact as a type of "soft-paste" porcelain. This distinction is not important to us. We include this sample in the present investigation because this dated basket was made in the same city where the Philadelphia bowl was found, and so it begs comparison.

24. Owen, "A New Classification Scheme," p. 132.

25. Chondrites are a type of stony meteorite. They have a primitive composition compared with crustal materials. Normalizing data to chondritic values indicates how compositionally evolved are the samples (generally rocks, but in this case, ceramics) being investigated.

26. See Owen, "Double Corona Structures," p. 1258.

27. In marked contrast, the trace-element contents of ancient Chinese porcelains have been reported in the literature for more than a decade.

28. See Y. Iqbal, P. F. Messer, and W. E. Lee, "Microstructural Evolution in Bone China," *British Ceramic Transactions* 99, no. 5 (2000): 193–99; and J. Victor Owen, Robert Hunter, Rod Jellicoe, and Martha Zierden, "Microstructures of Phosphatic Porcelain from Sintering to Vitrification: Evidence from Sherds Excavated in Charleston, South Carolina," *Geoarchaeology* 26, no. 2 (March–April 2011): 292–313.

29. Diana Stradling, "Bonnin and Morris: Reading between the Lines," in *Ceramics in America*, edited by Robert Hunter (Hanover, N.H.: University Press of New England for the Chipstone Foundation, 2013), pp. 191–92.

30. Rob Hunter, email communication, December 21, 2016.

31. See Owen, "Double Corona Structures," p. 1262.

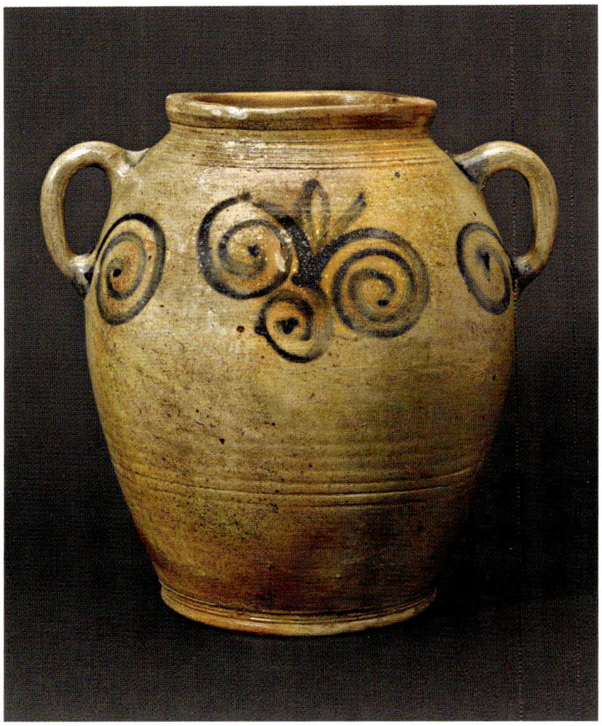

Figure 1 Jar, attributed to Morgan pottery, Cheesequake, New Jersey, 1775–1784. Salt-glazed stoneware. H. 12¼". (Private collection; photo, Gavin Ashworth.)

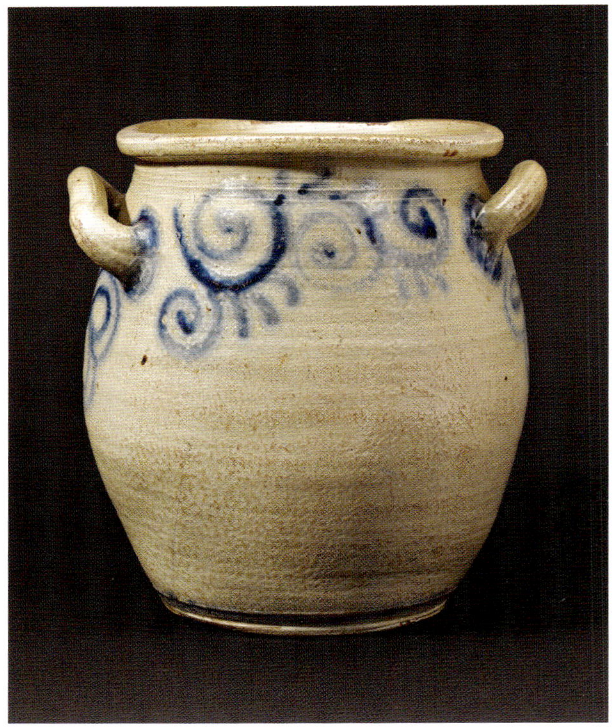

Figure 2 Jar, Kemple pottery, Ringoes, Hunterdon County, New Jersey, ca. 1746–1790s. Salt-glazed stoneware. H. 11". (Courtesy, Monmouth County Historical Association, Gift of Mrs. J. Amory Haskell; photo, Gavin Ashworth.)

Johanna R. Bernstein, Arthur F. Goldberg, and Jennifer Mass

A Comparative Scientific Study of James Morgan and the Kemple Family Stoneware

▼ THIS STUDY IS a scientific comparison of stoneware fragments found at the New Jersey archaeological sites of the pottery of Captain James Morgan of Cheesequake (ca. 1775–1784) and the Kemple family pottery in Ringoes (1746–1778) (figs. 1, 2). The study was undertaken to help resolve unanswered questions discussed in "The Eighteenth Century New Jersey Stoneware of Captain James Morgan and the Kemple Family," published in this journal in 2008.[1] Those questions focused in part on the source of the clay used by the Kemple family pottery. A scientific examination of the stoneware fragments would, it was hoped, provide an elemental signature for the clay that was used at these potteries, allowing identification of intact vessels currently attributed to these potteries but without documented provenance. Important to the history of New Jersey manufacturing, this research was the first attempt to identify the materials used to make some of the earliest stoneware in America.

The importance of this investigation was based in part on the findings of archaeologist Meta Janowitz's study of sherds from the African Burial Ground that were considered to be from 1730/40–1760, when the area was a dump site for the potteries of William Crolius and John Remmey.[2] Some of these sherds exhibited the watch-spring design typical of James Morgan, others showed the cogged watch spring used by the Kemple family pottery. Janowitz's findings raised doubts about attributions of Morgan and Kemple vessels that were based solely on their designs. For this reason, an examination of differences in the elemental composition of the clay among all the potteries was undertaken to help determine where the pots were made.

There is no known historical documentation about the clays that the Crolius and Remmey families used circa 1730–1760 to make their stoneware. William C. Ketchum Jr. states that, according to Adriaen van der Donck in his *Description of New Netherland*, one finds that "New York has hills of fuller's earth and some fine clay such as white, yellow, red and black suitable for pots, dishes, plates tobacco pipes and like wares."[3] Ketchum notes that this is most likely where John William Crolius first worked and possibly he could have been using that clay.[4] It is possible, however, that clay was purchased from the clay bank located in the vicinity of Cheesequake and owned by the Morgan family since 1710. Captain James Morgan inherited the banks in 1764 and started to make stoneware in the 1770s.[5] While there is no evidence of any sale of clay during that time, the quality of clay from the Morgan bank was widely known and in the 1790s was shipped as far away as New England by Captain Morgan's son (also James), General Morgan.[5]

Unlike the clay banks owned by the Morgan family, the clay source for the Kemple family pottery has not been documented and is the focus of this scientific study. Three possible clay sources for this pottery have been considered. First, the Kemples might have purchased their clay from the Morgan clay banks located about thirty miles away from their operation. Second, they might have been mixing higher quality clay from the Morgan banks with local clay of a different/lesser quality.[7] Third, perhaps there was another, undocumented source of stoneware clay, although this is considered improbable based on the geology of the surrounding area. Any differences in the clays would show up in trace element analysis of stoneware from each pottery. The presence or absence of trace elements in the composition of the fired clay has been commonly employed as an indicator of differences in the clay source.

The primary method of analysis employed was energy dispersive X-ray fluorescence (XRF), a standard method of nondestructive elemental analysis for cultural materials. XRF has been used to examine the trace elements in archaeological pottery to determine clay provenience.[8] Elemental analysis is used routinely in art conservation to identify materials such as overpaint on paintings and furniture, and to identify anomalous materials that could call into question the authenticity of an object. In the case of ceramic materials, elemental analysis using X-ray–based techniques such as XRF are among the most useful. That XRF is nondestructive gives it an advantage over other elemental analysis methods.

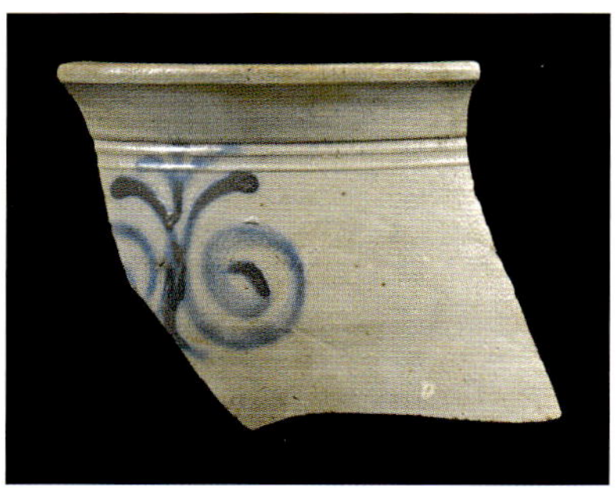

Figure 3 Jar fragment, Morgan pottery, Cheesequake, New Jersey, 1775–1784. Salt-glazed stoneware. (Courtesy, Robert J. Sim Collection, MCHA; photo, Gavin Ashworth.) This fragment displays a typical example of Morgan's cobalt-blue watch-spring motif.

Summary of Analysis

The unglazed edges of thirty-five stoneware fragments—thirteen from the Morgan pottery (fig. 3) and twenty-two from the Kemple pottery (fig. 4)—were examined at the Winterthur Museum Scientific Research and Analysis Laboratory using an energy dispersive X-ray spectrometer (Roentec/Bruker ArtTAX spectrometer with the molybdenum X-ray tube measurement head; measurement spot size is ~75 m). The fragments came from the collections of the New Jersey State Museum (NJSM), the Monmouth County Historical Association (MCHA), and Brenda Springsted, a private collector.

Each specimen was analyzed three times on an unglazed edge at three different spots using 300s exposures at 50 kV (600 A). An average of the three scans from each fragment was performed to account for the variations in signals from the clay bodies due to variations in the physical and chemical structure of the clay body and the variations in the smoothness of the test surface. Variations in the chemical signature within a single sherd are likely due to the heterogeneous nature of the fired clay body as compared to the spot size of the X-ray beam.

The area under selected peaks in the spectra was measured, concentrating on the following elements: calcium (Ca), chromium (Cr), copper (Cu), iron (Fe), manganese (Mn), nickel (Ni), potassium (K), rubidium (Rb), strontium (Sr), titanium (Ti), yttrium (Y), zinc (Zn), and zirconium (Zr). Data analysis focused on Cu, Fe, Rb, Sr, Y, Zn, and Zr concentrations. The mea-

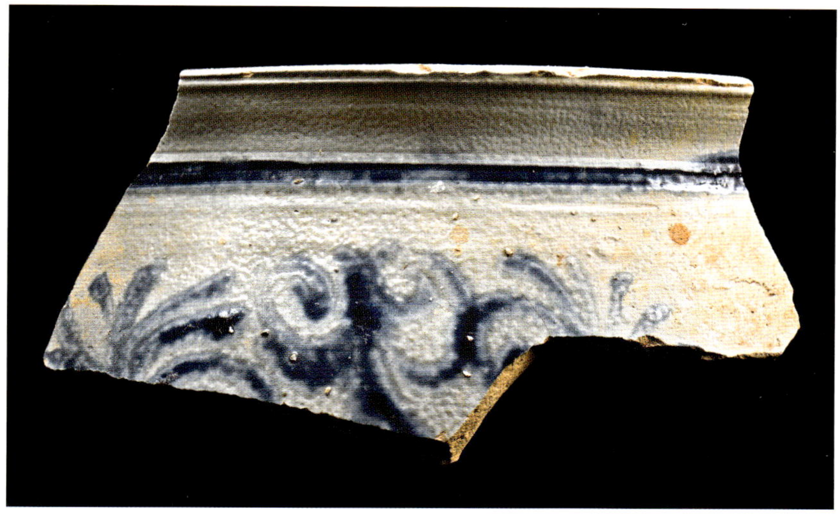

Figure 4 Jar fragment, Kemple pottery site, Hunterdon County, New Jersey, ca. 1746–1790s. Salt-glazed stoneware. (Courtesy, Robert J. Sim Collection, MCHA; photo, Gavin Ashworth.)

sured values for each peak were averaged for each sherd and were plotted as bivariate component elemental plots. Each averaged measurement was divided by the corresponding averaged measurement for Fe for each sherd in order to minimize the effect of intensity variations in the average signals.

Discussion

Examination of the plot comparing the Zn/Fe and Y/Fe contents of the Morgan and Kemple sherds (fig. 5) shows that there might be a weak correlation for yttrium, suggesting that the clay used at the Kemple pottery contains slightly more yttrium than that of the Morgan pottery. The plot for Rb/Fe versus Sr/Fe (fig. 6), however, does not show significant variation in the Rb/Sr ratio for each pottery.

If both potteries were using the same clay source, one would expect little difference in the chemical content of the clay bodies and no correlation of elemental composition to pottery for any of the selected elements. It is possible that the yttrium concentration difference can be explained if the Kemple pottery mixed local clay with clay from the Morgan bank, as has been previously suggested.

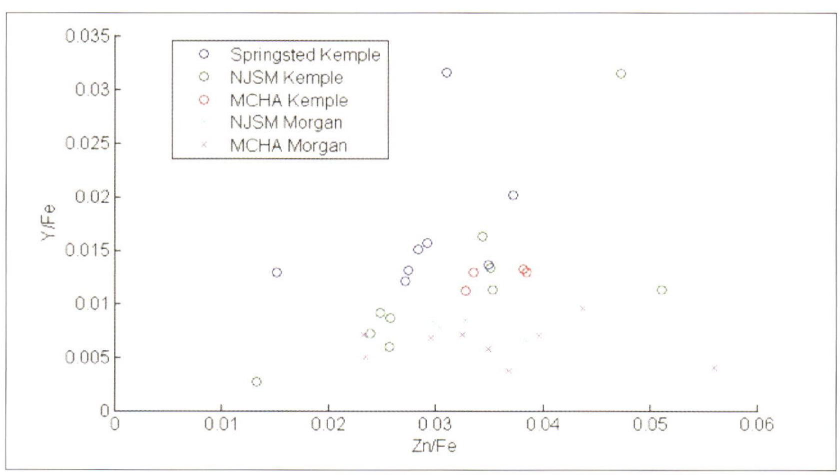

Figure 5 Bivariate elemental plot comparing Zn/Fe and Y/Fe content for the Morgan and Kemple sherds.

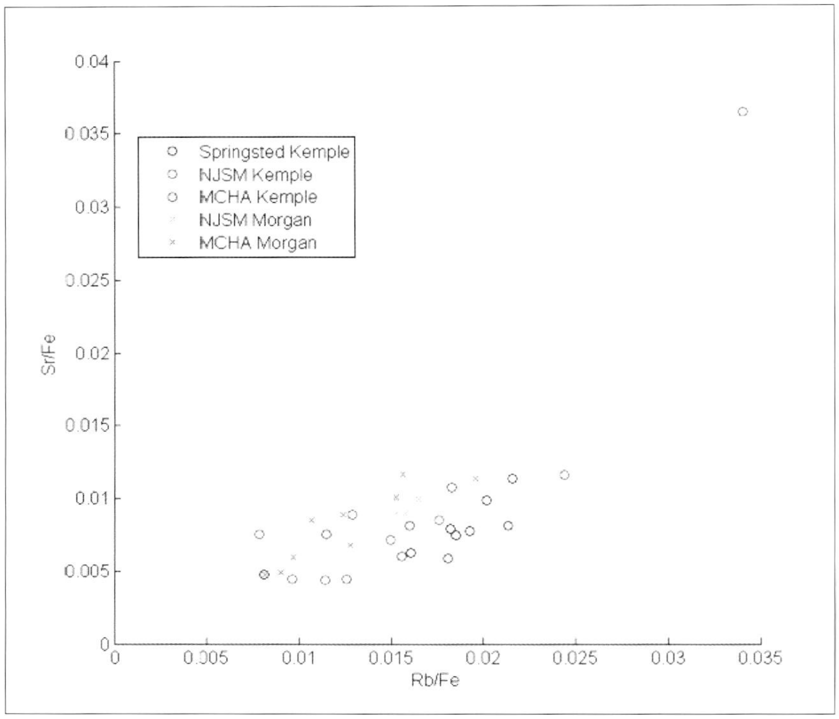

Figure 6 Bivariate elemental plot comparing Sr/Fe and Rb/Fe content for the Morgan and Kemple sherds.

Further examination of the plot for yttrium shows greater scatter for the Kemple stoneware, which is consistent with the clay-mixing theory. Possibly the Kemple pottery had another unknown clay source, with similar but not identical composition. Given the geology of the area, however, this last possibility may be the least likely.

A 1977 report written by James P. Owens, N. F. Sohl, and J. P. Minard indicates that the ratios of the clay minerals vary from one part of the Morgan clay bank to another.[9] While XRF is generally not able to show these differences, the slight variation in the yttrium content of the clays from the two potteries indicating differences in the ceramic bodies could be explained by variations in the Morgan clay bank itself. However, closer examination of figures 5 and 6 show that there is generally more scatter associated with the Kemple data. If the variations in the Morgan clay bank were significant, we would likely also see similar scatter associated with the Morgan stoneware results. In fact, there is actually less scatter associated with the Morgan samples.

Ultimately, the results of this XRF study show that subtle differences in the clay bodies from these two potteries are observable by examination of their trace element contents. XRF is an ideal nondestructive tool for identifying trace elements in clays.

Future research will concentrate on sampling clays from the historic Morgan clay bank, and examination of fragments using sampling techniques that are able to quantify bulk trace element content (rather than the surface trace element content that is observed with XRF). Such techniques may include ICP-MS (inductively coupled plasma analysis with mass spectrometry), XRD (X-ray diffraction), and petrographic analysis.

Other areas for examination include analysis of the cobalt blue decoration on the Morgan and Kemple sherds and their attributed vessels. Inclusion of the Crolius/Remmey sherds from the African Burial Ground site will also help to establish differences between unmarked eighteenth-century New York and New Jersey stoneware. Further analysis of Morgan and Kemple sherds will reveal whether the yttrium effect is statistically significant. Originally, this research was to have included fragments from the Crolius/Remmey potteries that were excavated at the African Burial Ground site. Unfortunately, the Government Services Administration, which houses these collections, changed its lending policies and the fragments became unavailable for study. The date for their release and transfer is not known.

ACKNOWLEDGMENTS The authors are grateful for the New Jersey Historical Commission grant (2007–8) that provided the funding to perform the initial research for this study, as well as to the New Jersey State Museum in Trenton, the Monmouth County Historical Association in Freehold, New Jersey, and archaeologist Brenda Springsted for lending us their examples.

1. Arthur F. Goldberg, Peter Warwick, and Leslie Warwick, "The Eighteenth-Century New Jersey Stoneware Potteries of Captain James Morgan and the Kemple Family," in *Ceramics in America*, edited by Robert Hunter (Hanover, N.H.: University Press of New England for the Chipstone Foundation, 2008), pp. 3–40.

2. Meta F. Janowitz, "New York City Stonewares from the African Burial Ground," in *Ceramics in America*, edited by Robert Hunter (Hanover, N.H.: University Press of New England for the Chipstone Foundation, 2008), pp. 41–66.

3. William C. Ketchum Jr., *Potters and Potteries of New York State, 1650–1900*, 2nd ed. (Syracuse, N.Y.: Syracuse University Press, 1987), pp. 39–40, quoting Adriaen van der Donck, Esq., *Description of New Netherland*, first English edition published in 1841 (widely considered a bad translation), Collection of the New-York Historical Society.

4. Ketchum, *Potters and Potteries of New York State*, p. 40.

5. Goldberg, Warwick, and Warwick, "Eighteenth-Century New Jersey Stoneware Potteries of Captain James Morgan and the Kemple Family," p. 32.

6. Lura Woodside Watkins, *Early New England Potters and Their Wares* (Cambridge, Mass.: Harvard University Press, 1950), pp. 82–83.

7. Goldberg, Warwick, and Warwick, "Eighteenth-Century New Jersey Stoneware Potteries of Captain James Morgan and the Kemple Family," p. 25.

8. E. H. Balffaji, "Application of Multivariate Sstatistical Methods to Classify Archeological Pottery from Tel-Alramad site, Syria, Based on X-ray Fluorescence Analysis," *X-ray Spectometry* 35, no. 3 (2006): 190–94. F. P. Romano, G. Pappalardo, L. Pappalardo. S. Garraffo, R. Gigli, and A. Pautasso, "Quantitative Nondestructive Determination of Trace Elements in Archeological Pottery Using a Portable Beam Stability-Controlled XRF Spectrometer," *X-ray Spectometry* 35, no. 1 (2006): 1–7.

9. J. P. Owens, N. F. Sohl, and J. P. Minard, "A Field Guide to Cretaceous and Lower Tertiary Beds of the Raritan and Salisbury Embayments, New Jersey, Delaware and Maryland," presented at the Annual AAPG/SEPM Convention, Washington, D.C., June 12–16, 1977. Excerpts of this report were provided by Peter Sugarman, New Jersey Geological Survey.

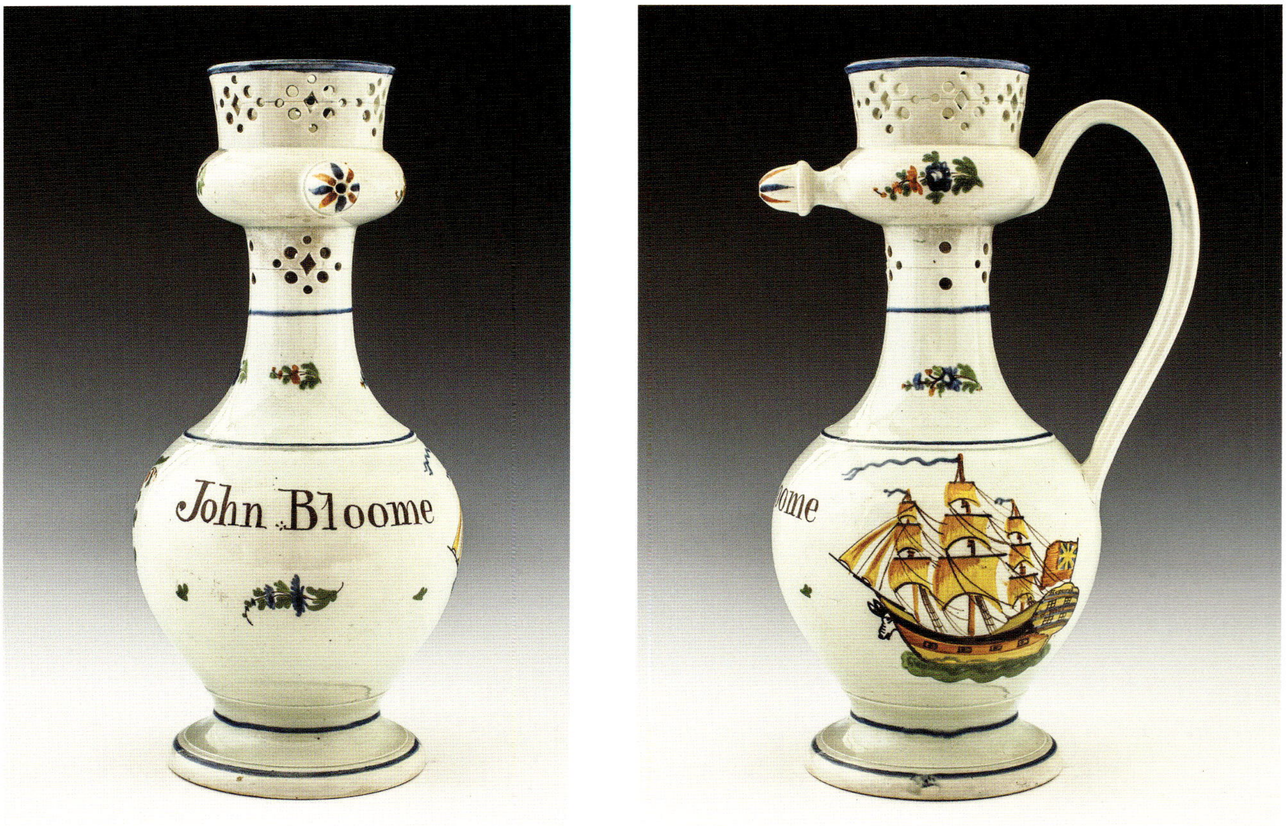

Figure 1 Puzzle jug, probably South Yorkshire, England, ca. 1798. Pearlware. H. 11". (Author's collection; unless otherwise noted, all photos by Robert Hunter.) This undated pearlware puzzle jug was made for shipmaster John Bloome and includes polychrome-painted decoration depicting his ship *Hopewell*.

Ivor Noël Hume

A New Bloome

▼ "PUBLISH OR PERISH" is one of the oldest clichés in the academic lexicon, but in ceramic research publishing can do more harm than good. If we see it on paper we are inclined to believe it and to pass it on as truth. As with a bad brick, any wall built with it will eventually fall down, but not until the brick maker has left town and no one can be sure which brick was the culprit. In my own work I have taken the view that it is better to publish what one erroneously thinks is right than not to publish at all—even if other students and collectors are allowed the joy of finding the errors.

That mea culpa–ish introduction brings me to rewriting the story of John Bloome's puzzle jug, which I had believed to be hiding the evidence of a secret romance (fig. 1). For forgetful readers of the 2014 issue of *Ceramics in America*, it goes like this: along with Bloome's name, the pearlware jug is decorated with a well-painted portrait of a late-eighteenth-century merchant ship.[1] Across her stern is an identifying *Hopewell* and, below, the name "Wells" (fig. 2). To the left of "Wells" is what I took to be a heart-pierced arrow, suggesting that Mr. Wells was declaring his love for someone. Or was someone recording her love for Mr. Wells? And where did John Bloome fit into that scenario?

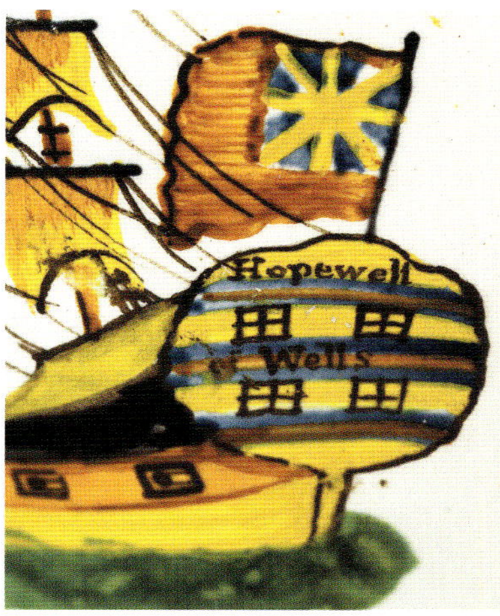

Figure 2 Detail of the ship stern on the puzzle jug illustrated in fig. 1. The image shows the ship's name as "Hopewell" and below it "Wells" and what at first looked like an arrow-pierced heart but turned out to be "of"—i.e., *Hopewell of Wells*.

That was not the jug's only secret. Puzzle jugs had been a tavern joke since the Elizabethan era, first in lead-glazed earthenware, then in delft and later in other wares from white salt glaze to brown stoneware. As figure 3 shows, they continued to be made into the mid-nineteenth century. This dated

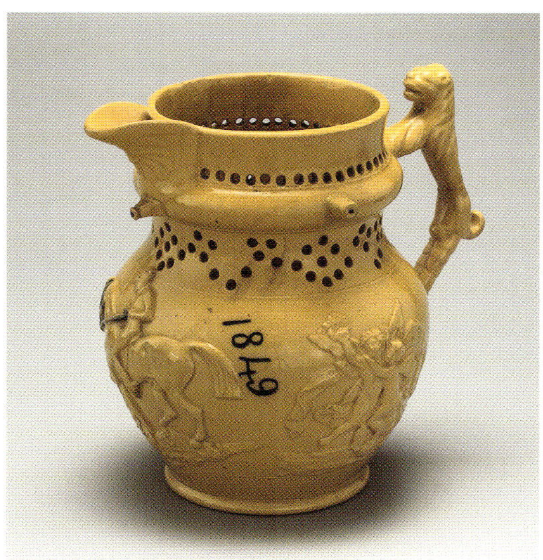

Figure 3 Puzzle jug, attributed to Ferrybridge Pottery, Yorkshire, England, 1849. Yellow ware. H. 7½". Dated 1849 and inscribed with the initials MHS in blue slip. (Author's collection.) The form is an example of declining elegance in puzzle jug production.

Figure 4 The ship *Hopewell* in a detail of the puzzle jug illustrated in fig. 1. The ship is under sail and flying the flag (red ensign) of the British Merchant Navy.

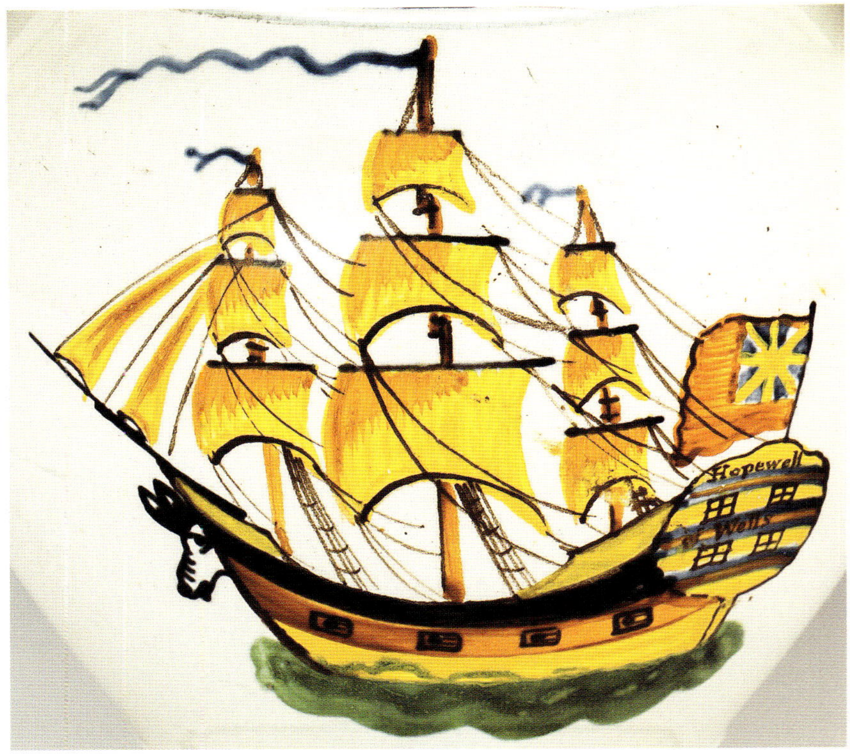

1849 example in yellow-glazed earthenware is attributable to Yorkshire's Ferrybridge Pottery. The joke centered on the number of spouts that the drinker had to cover to control the flow. The majority had three, though some had as many as six, theoretically making imbibing even more difficult. Why novice drinkers should continue to fall for the same trick across three centuries is hard to explain—as is John Bloome's jug that has but one spout. Was he a simple fellow who could not cope with more?

Support for that assumption was offered by the jug's painter. Below the bowsprit, where most ships braved the waves behind figureheads carved with mermaids, pulchritudinous maidens or heraldic beasts, the *Hopewell* displayed the head of a jackass (fig. 4). One immediately asks, "Did the ship really sail under an animal renowned for its stupidity?" Much more reasonable can be the conclusion that she only acquired that figurehead from the hand of the pearlware painter. In putting the single spout and the jackass together on a jug made for a specific individual, someone wanted to tell John Bloome that he was an idiot. So who could that be?

My answer claimed it to be one of two people, either a failed business partner or his wife. Had Bloome invested her money to buy or lade the *Hopewell* for a disastrous voyage, her annoyance might have been expressed in ceramic terms. Or, one might read a tavern mug as a gift from a man. When I published these speculations, British local historian Michael Welland made no bones about saying that unless they can be established as fact, "I don't see how it can be written into an article."[2] He also reminded me that "isolated records do not necessarily illustrate what was happening." That, of course, is true of any assemblage of historical documents, be they state papers or an inscription on a pearlware jug. Much more pertinent was Welland's blast that blew away my straw-built Ziggurat. My amorous Wells turned out to be the name of the Norfolk town where John Bloome lived, and the *Hopewell* was one of several trading vessels he owned to carry Yorkshire goods from Hull to Wells and on to Holland and Germany (fig. 5).

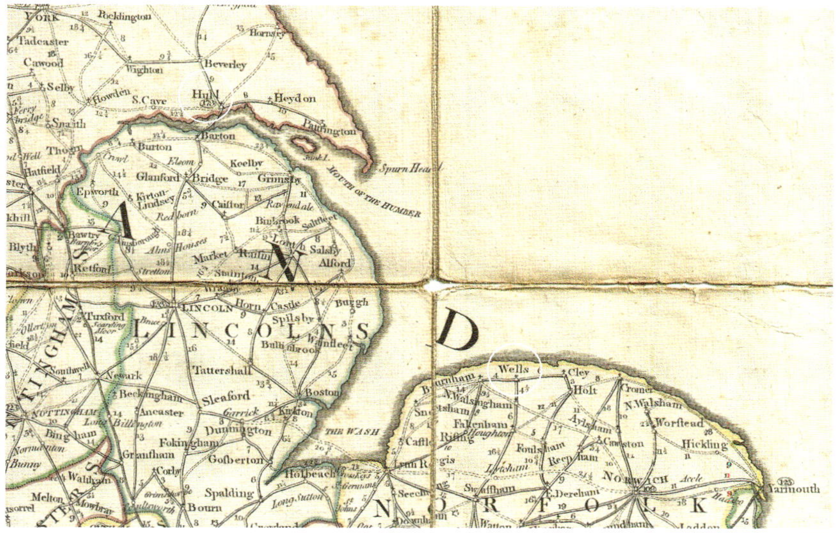

Figure 5 Detail of William Faden's *The Roads of Great Britain. Initneraire de la Grande Bretagne*, 1790. (Courtesy, Geographicus Rare Antique Maps.) The map shows a portion of the east coast of England, with the ports of Wells and Hull indicated.

The scope of the *Hopewell*'s cargoes was laid bare on March 8, 1782, when she ran aground and broke up off the Norfolk coast. The *Norfolk Chronicle* reported that "On Wednesday morning last the Hopewell of Wells, John BLOOM Master, bound from Hull to the above port, laden with wheat, iron, cheese, earthenware etc. was drove onto Overstrand Beach, three miles from Cromer, in a hard gale of wind. The crew and part of the cargo was saved, but the vessel is entirely lost."[3] The Norfolk coast with its tourist-appealing sands was a trap for wind-blown vessels, which, once grounded, were soon battered to pieces by the wind and sea that stranded them (fig. 6).

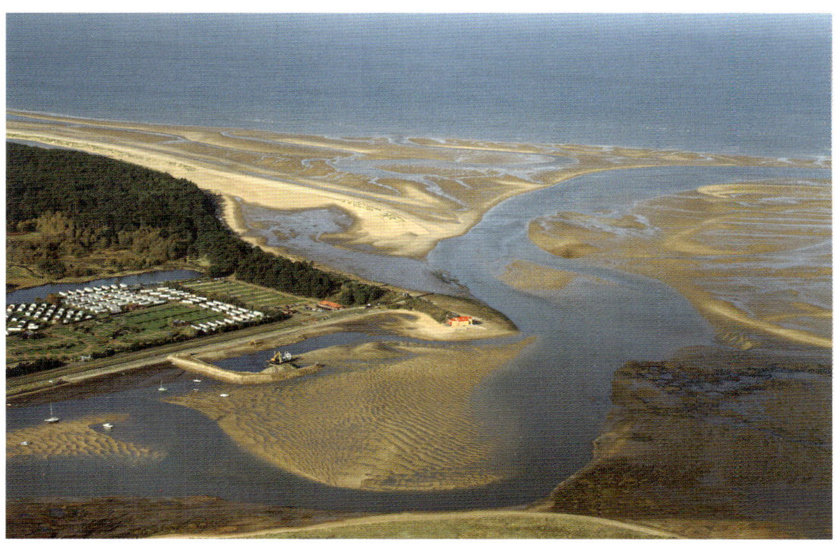

Figure 6 The Wells estuary at low tide. The *Hopewell* ran aground about three miles to the east. (Photo, Mike Page Aerial Photography, www.mike-page.co.uk.)

Was this the disaster that prompted John Bloome's puzzle jug to be made? It seemed reasonable until one went beyond the painting to the pearlware body to be reminded that it was not in general production until two years later; nonetheless, the jug seems to be an example of pearlware at its best. The answer had to be that Bloome christened another *Hopewell*, perhaps dating as late as the 1790s. Under the heading of Hull traders *The Universal British Directory* of 1793 listed "*Hopewell*, John Bloom, *King's Coffeehouse, Hull*," thus proving beyond doubt that the puzzle jug related to a later *Hopewell* and not to the wreck of 1782.[4]

With that established one asks, what can we learn from the painted ship? Was the artist working from a drawing of the *Hopewell* or was he painting a generic merchantman and adding only the figurehead? Naval historian Sean Kingsley has described the rendering as "quite a large ship by the size of the aftcastle with four cabin windows." She is shown with four of eight guns plus a stern-mounted salute swivel in the process of being discharged. The presence of the armament is a reminder that eighteenth-century commerce had more to face than boisterous seas and treacherous sands.

The North Sea (which Europeans called the "German Ocean") was renowned for its storms and the presence of French pirates who preyed on merchantmen plying between England and the Continent. On June 2, 1744, the *Charlotte* sailing with four other Wells vessels was attacked by a Frenchman mounting ten cannon, eight swivels, and a crew of seventy-five. The *Charlotte* carried only four three-pounder cannon, two swivels, and a crew of seven, three of them boys. Repelling two attempts to board her, the *Charlotte*'s gunner scored a lucky hit on the Frenchman's powder store with explosive results.[5] Had the boarding been successful, the pirates would have discovered that the *Charlotte* was carrying only a cargo of wheat.

That engagement was but one of the many that plagued the North Sea to the end of the century, when Hull merchants appealed to the Lords of Admiralty to provide a thirty-two-gun frigate to escort their convoys. The Dutch already had two to protect their ships carrying exports to England. John Bloome could not afford to put to sea unarmed. Even so, England's short-trip export trade remained a hazardous undertaking—as old newspaper reports and modern scuba divers can attest.

The inclusion of earthenwares in the *Hopewell's* lost cargo opens a new avenue of research, another of those curious coincidences that litter the rocks and eddies of maritime and ceramic history.

The first question to ask is, "What did John Bloome mean by 'earthenware'?" Through most of the eighteenth century the term was used in two words, "earthen" and "ware." In an account of the Low Countries published in 1672, the town of Delft was said to produce "Earthen Ware as stone-jugs, Pots, etc." In 1792, however, a notice explained that "when earthen ware is mentioned in this paper the cream-coloured or queen's ware is meant." Another reference defined "earthen ware" as being products of Staffordshire, Nottingham, and Kent. The assumption, therefore, is that the term meant virtually any mid- to late-eighteenth-century ceramics being exported from Hull via Wells. Regardless of the 1672 reference to Delft potters making stoneware, it is reasonable to deduce that English delftware was not part of John Bloome's cargo but was, instead, the English creamware often found in Dutch excavations. The 1793 *Directory* confirmed that "the town of Wells had a tolerable good trade with the Dutch in pottery."

The next question relates to the source of Bloome's ceramic cargo. We know it was being shipped from Hull, but was it the work of Hull potters? The 1793 *Directory* listed "no manufactories in the town" other than those directly associated with the shipping industry. However, the sage of English ceramic history, Llewellyn Jewitt, was confident that potters were working in the vicinity of Hull in the late seventeenth century. In a deed of 1802, Shelton potter Job Ridgway went into partnership with Hull potters James and Jeremiah Smith. Unfortunately, no examples of the Smiths' wares are known. However, Jewitt cited a dinner service made for the owner of the Baltic trader *Crow Isle* to commemorate that ship's encounter with the noted American pirate John Paul Jones off the Yorkshire coast in 1779. Jewitt noted that only one plate from the service was known to have survived and it was in the Hull museum.[6] Although it was likely to be of transfer-printed creamware, there was a remote possibility that it was of hand-painted pearlware or, as it was then called, "china glazed." The museum reported that the plate had been destroyed in World War II, though a poor-quality photograph published in 1916 proves that the plate was, in fact, creamware (fig. 7).

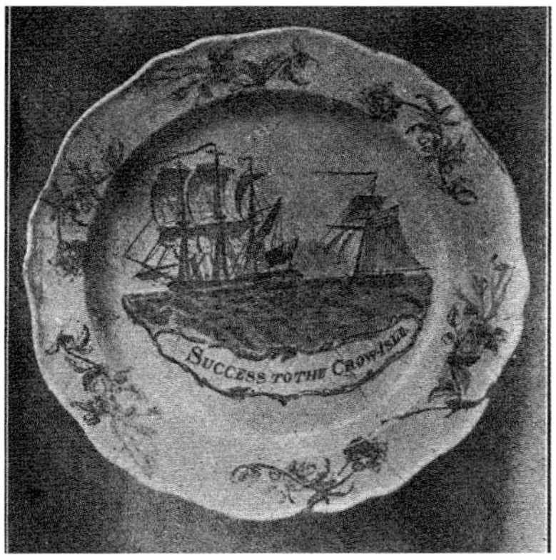

Figure 7 Photo of a transfer-printed creamware plate showing the encounter between the Baltic trader *Crow Isle* and John Paul Jones in 1779. Reproduced from Oxley Grabham, *Yorkshire Potteries, Pots and Potters* (1916; reprint, Wakefield: S. R. Publishers, 1971). The plate, now lost, is allegedly the product of a Hull potter.

Bloome may have purchased his earthenware in Hull. The port's export trade was extensive and, thanks to the canal system and improved roads, goods intended for Europe could be carried overland from as far away as Bristol. In the words of the 1793 *Directory*, "most of the trade of Leeds, Wakefield, Huddersfield, and Halifax is negotiated here." Of those, only Leeds earned fame as a potting center and for manufacturing a wide range of pierced creamwares. By the time that Jewitt was writing in 1872, the whitened "china glaze" was dismissed as "white ware." Consequently, there are no confirming references to pearlware in the early Leeds records. With that caveat aside, it was reasonable to attribute John Bloome's puzzle jug to Leeds. The distance from Hull to Leeds is approximately fifty miles, in modern terms a short trip but in the eighteenth century hardly worth the travel time to order one jug. It is probable, therefore, that somewhere in this mystery there was an agent selling Leeds products in Hull and that it was he who carried the Bloome request to the factory.

Building on that shaky assumption, one returns to the primary questions: Who placed the order and why? In 1792 John Bloome married his second wife, Ann Richardson, who was said by Michael Welland "to be an extremely wealthy woman, a multi-millionaire in modern terms."[7] It seems unlikely, therefore, that she would have stooped to puzzle jug imagery over a financial setback. It is equally improbable that she would have misspelled her husband's name.

Throughout his adult life John's name was printed as BLOOM, as were those of other family members. Welland considers the misspelling no more than the kind of slip common in handwriting in the eighteenth century, but

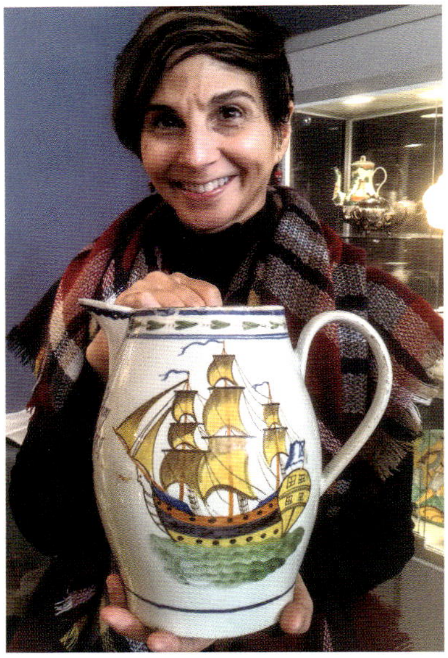

Figure 8 Ceramic artist Michelle Erickson holds the Admiral Duncan commemorative pitcher discovered at the 2016 *New York Ceramics Fair* in the booth of London dealer Garry Atkins.

the consistent absence of an "e" can be significant. It suggests that whoever placed the order knew Bloome in person but not in writing. This might well have been either a business partner or a rival who suffered a loss due to Bloome's stupidity and enjoyed direct or indirect contact with the Leeds pottery. If that be true, it is unlikely that the artist's drawing accurately portrayed the lines of the *Hopewell*. The trick was to prove it.

In 2004 the Bloome jug was offered for sale by the late Jonathan Horne at the annual New York ceramics fair. He failed to find a buyer, nobody having been intrigued by the *Hopewell*'s enigmatic figurehead. Twelve years later Jonathan's successor at the same fair, Garry Atkins, brought a ship-painted pearlware pitcher that undoubtedly was the work of the same artist (fig. 8). Even more valuable was its very precise date: October 11, 1797. The quality of the ware and the palette used to paint the *Hopewell* left little doubt that both jug and pitcher were of roughly the same age (fig. 9). Above the date on the front of the latter was a drawing of an anchor below "Duncan's Glory," which I first assumed to be the name of the ship (fig. 10). Had I been smarter (and my memory better.) I would have known that the date commemorated Admiral Adam Duncan's victory over the Dutch at the Battle of Camperdown and that the painted vessel represented his flagship, the *Venerable*.

Bloome's lightly armed *Hopewell* was clearly a merchantman, but on the pitcher Duncan's man-of-war mounted thirty-four guns. More helpful was

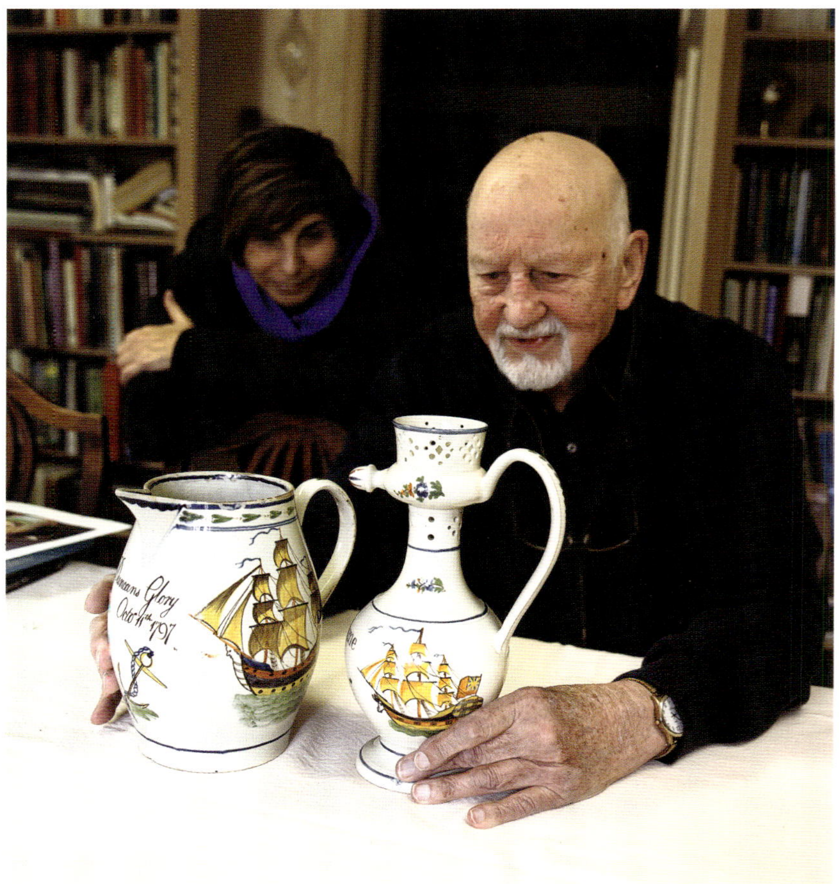

Figure 9 The *Venerable* pitcher and *Hopewell* jug in hand-painted pearlware, the former dated 1797 and a memorial to victory at the Battle of Camperdown. Both are believed to be the product of a single Leeds painter.

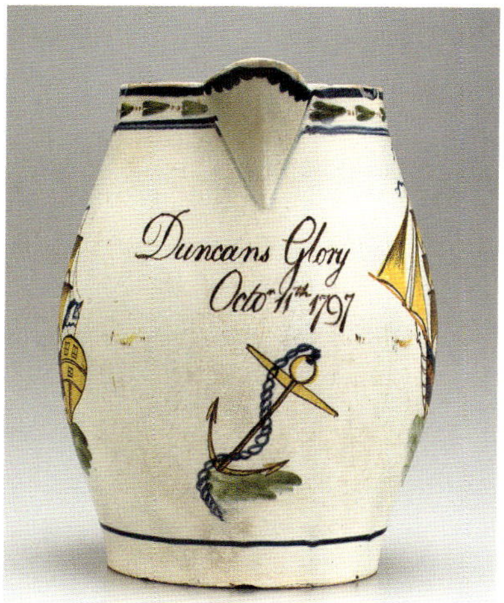

Figure 10 Pitcher, probably South Yorkshire, England, ca. 1797. Pearlware. H. 8½". Inscribed on front: "Duncans Glory / Octo 11th 1797" over a blue-chained anchor.

the artist's decision to paint the ship twice, once with three masts and four cabin windows, and again with two masts and three windows (fig. 11). Therefore, Duncan's painted warship bore no more resemblance to the real *Venerable* than Bloome's did to the *Hopewell*. Duncan's jug had yet more to tell us about Bloome's jackass figurehead. In one version, the warship had no figurehead, and in the other it showed an appropriately scaled outline of a male head, this in contrast to Bloome's jackass, which was drawn sufficiently out of proportion not to pass unnoticed (fig. 12). The red ensign flown by British merchantmen is fairly carefully drawn (the Union flag in

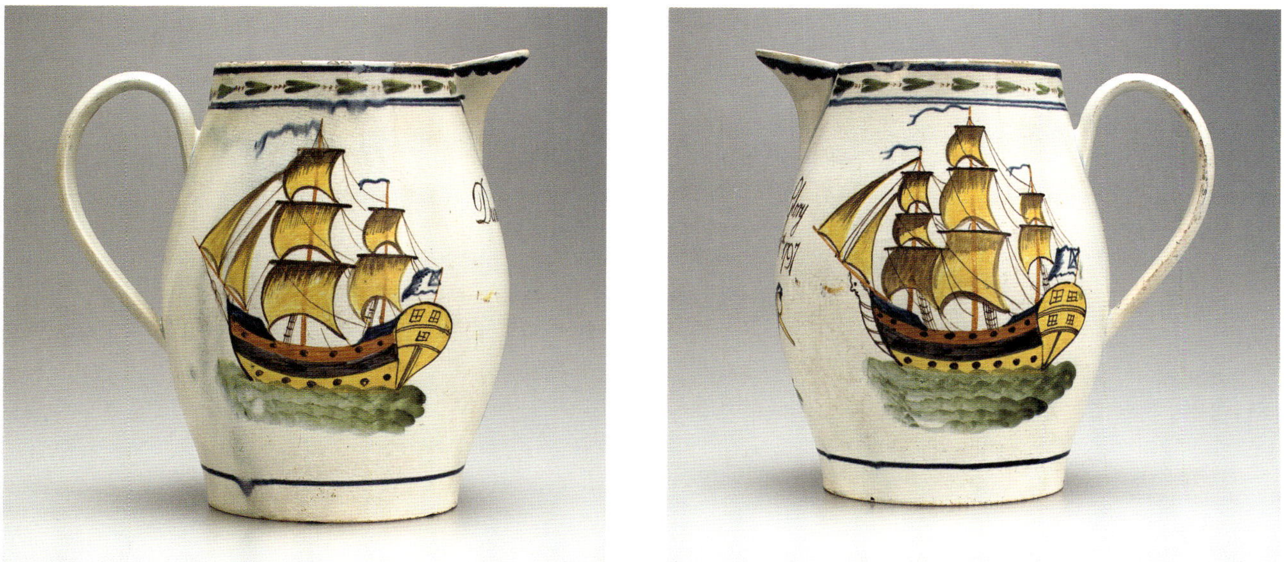

Figure 11 Side views of the pitcher illustrated in fig. 10. Note the marked differences between the two renderings of the same ship.

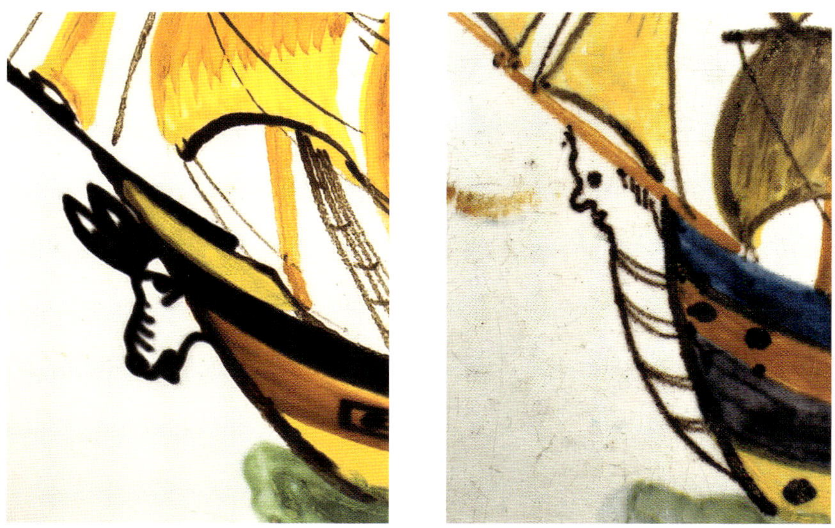

Figure 12 Figurehead details of the *Hopewell* (left) and the *Venerable* (right).

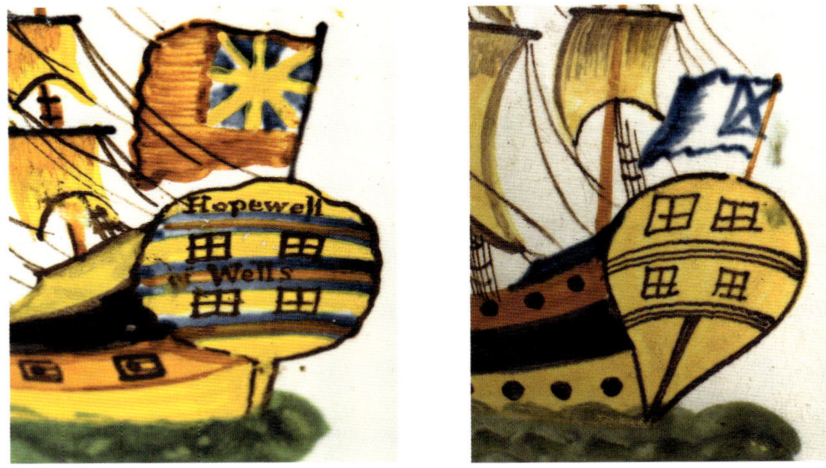

Figure 13 English and Scottish flags at the sterns of the *Hopewell* and *Venerable*.

its canon corner) at the stern of the *Hopewell*, but Admiral Duncan's *Venerable* flew his own flag, namely, the St. Andrew's cross on a white and blue ground (fig. 13). That might be read as showing that in the Battle of Camperdown the Scottish Duncan commanded both white and blue squadrons. He had been made admiral of the blue squadron in 1793 and of the white a year later.

A contemporary engraving shows twenty-five Dutch and French warships in a barricading line along the Camperdown shore against two prongs of the British attack (fig. 14). Led by the *Venerable*, Duncan's naval forces broke through the Dutch line and attacked it from the rear. The water was shallow, prompting him to tell his officers that "when the *Venerable* goes down, my flag will still be flying." His reference to "my" flag may have been to the Scottish flag on his pitcher, though it cannot be identified in the several paintings of the battle (fig. 15).

Duncan's commemorative ceramics were never as plentiful as those made for Admirals George Rodney and Horatio Nelson. All but one of the Dun-

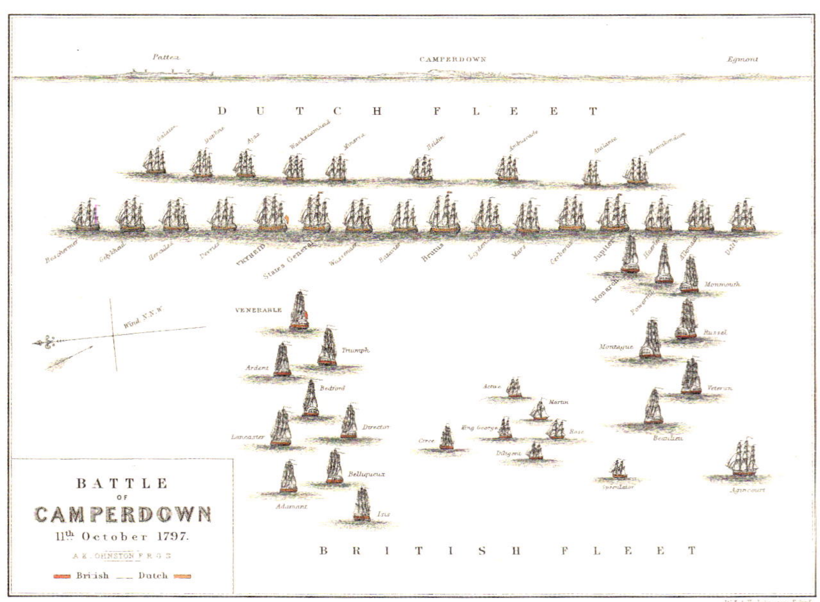

Figure 14 *Battle of Camperdown, 11th October 1797*, hand-colored engraving by A. K. Johnston F.R.G.S., published by William Blackwood and Sons, Edinburgh and London, 1848.

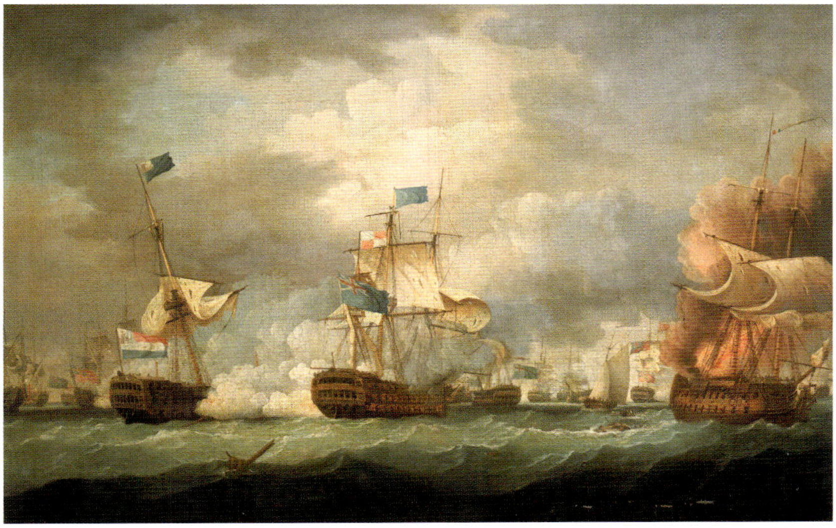

Figure 15 Thomas Witcombe, *Battle of Camperdown*, 1798. Oil on canvas, 35" x 53". (Courtesy, National Maritime Museum, Greenwich, London.) This is one of several paintings of the Battle of Camperdown, showing the blue-flagged *Venerable* at the center of the action alongside Admiral Van de Winter's flagship.

can souvenirs are of creamware, one hand-decorated and the others transfer-printed with portraits and battle scenes. The odd man out was described by ceramic historians John and Jennifer May: "Another painted piece is a pearlware jug . . . decorated on the one side with a ship in sail, on the other with a purely floral decoration but with, on the front, inside an open cartouche of laurels, the inscription 'Admiral Duncan for ever.'"[8] That this small jug is only 5¼ inches in height makes it likely that its ship's size is similar to that on the Bloome puzzle jug. Supportive, too, may be the floral sprig on the opposite side (fig. 16). Unfortunately, the whereabouts of the jug described by the Mays in 1972 are unknown and no illustration graces his excellent book. One likes to believe that when one's treasures are given to a museum, they will be there in perpetuity, but changing times and new curators provide no such assurance. While seeking proof that the Bloome puzzle jug was a Leeds product, I found a book by Oxley Grabham, retired keeper of the York Museum, which illustrated a puzzle jug whose foot appears to parallel that of Bloome's jug (fig. 17). Its floral decoration embraced the painted inscription "A Trifle / Shews Respect / E.S /1799."[9] Attributing the dated

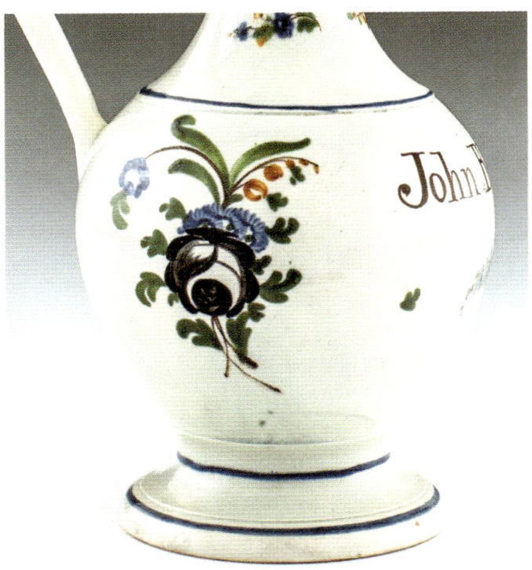

Figure 16 Detail of the floral decoration painting on the puzzle jug illustrated in fig. 1.

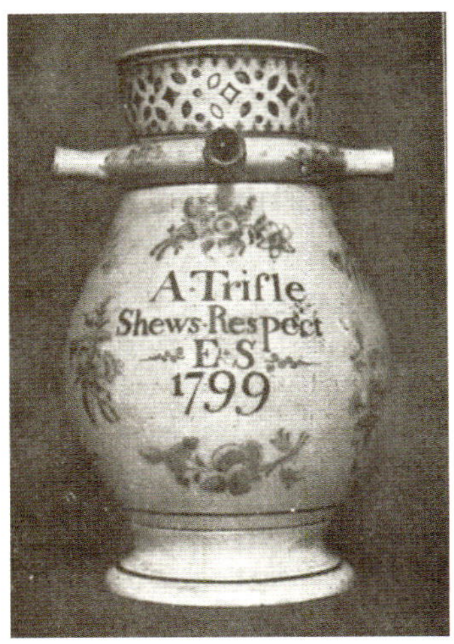

Figure 17 Photograph of a pearlware puzzle jug attributed to Leeds, dated 1799 and with a foot similar to that of the Bloome jug. Reproduced from Oxley Grabham, *Yorkshire Potteries, Pots and Potters* (1916; reprint, Wakefield: S. R. Publishers, 1971). Once in the York Castle Museum, the jug is now missing.

jug to Leeds would provide Bloome's jug with the evidence I needed. The author failed to confirm that the "E.S" jug was of pearlware or to provide any measurements. Nevertheless, he did say it was in the York Museum, but when I asked to see it I was told that no such jug was in its collection.

In spite of that dead end, the pursuit of John Bloome has atoned for the previous errors, if only to prove that an amorous Mr. Wells was not in love with Miss "X." In any research we have a tendency to see the first discovered nugget to be the foundation for everything that follows. Only belatedly do we realize that that may not be so. My assumption that the Bloome jug was made in the 1780s had meant that it preceded the Duncan commemoratives. Wrong again!

A Yorkshire pottery factory (probably Leeds) had perfected the making of pearlware by the mid-1790s and was including the hard-to-make puzzle jugs when the Battle of Camperdown provided the impetus to create a line of Duncan-related souvenirs. When the victory fervor faded, the factory went on making name-identified puzzle jugs. In 1797 or soon after, someone in the Hull shipping business saw one of the Duncan ship jugs and applied to the factory to make another for John Bloome.

The lucrative export trade of the 1780s faded with the outbreak of the Napoleonic Wars and one by one the shippers of Wells and Hull went bankrupt. As long as John Bloome's wife, Ann, remained alive, he had a financial buttress against his losses. But the good times had not returned when she died in 1805. Her will had been written prior to her marriage to John in 1795 and had not been updated. In it, she had left her assets to cousins and grandchildren. Of her two trustees, one had died and the other excused himself. With the will unproved, John went on spending her money until halted by extended legal action. In 1827 the court ordered him to return the money. Being unable to do so, his assets were put up for auction as "The Property of a Gentleman retiring from Business." He died two years later.

Were it not for an accident of survival, the career of earthenware trader John Bloome of Wells-next-the-Sea would be forgotten, as would nearly everything else that happened in the year 1797. Collectors who only aspire to own or admire ceramic treasures are liable never to look beyond the object to the hand that held the cup or the yokel who made a fool of himself by spilling the puzzle jug. Yet, in truth, we are within a hand's touch or a lip's sip of them both. If, in our imaginations, we see a happy group of tavern revelers or a cheering crowd welcoming the return of the *Venerable*, we might do well to remember the underlying mood of the British people in 1797. Embroiled in a seemingly endless war with France and the Netherlands, humiliated by the loss of her American colonies, "the aspect of public affairs afforded nothing to cheer the minds of the people. On the contrary, the ill success of our negotiations seemed to preclude a hope of terminating the present most destructive war with honour and advantage." The French are "our inveterate foe, breathing the most rancorous malice, and increasingly in strength from the desertion of our weak or perfidious allies."[10] The message on Duncan's jug reminds us that in the twenty-first century glory is as fleeting as the sight of a jackass on a puzzle jug.

ACKNOWLEDGMENTS The author is deeply indebted to Norfolk historian Michael Welland for much valuable informationt; also to maritime historian Sean Kingsley and stoneware authority Philip Mernick, who helped keep the *Hopewell* afloat.

1. Ivor Noël Hume, "The Ten Commandments," in *Ceramics in America*, edited by Robert Hunter (Hanover, N.H.: University of New England Press for the Chipstone Foundation, 2014), pp. 2–18.

2. Welland, personal communication, November 16, 2015.

3. *Norfolk Chronicle*, March 23, 1782.

4. *The Universal British Directory of Trade, Commerce and Manufacture*, 5 vols. (London: British Directory Office, 1793), 3:338.

5. *Norfolk Courant*, June, 16, 1744.

6. Llewellynn Jewitt, *The History of Ceramic Art in Great Britain*, 2 vols. (New York: Scribner, Welford, and Armstrong, 1878), 1:464–66.

7. Welland, personal communication, November 16, 2015.

8. John May and Jennifer May, *Commemorative Pottery, 1780–1900: A Guide for Collectors* (London: Heinemann, 1972), p. 90.

9. Oxley Grabham, *Yorkshire Potteries, Pots and Potters* (1916; reprint, Wakefield: S. R. Publishers, 1971), p. 59.

10. Charles Mayo, *A Compendious View of Universal History, from the Year 1753 to the Treaty of Amiens in 1802*, 4 vols. (Bath, Eng.: Printed for G. and J. Robinson, 1804), 4:1.

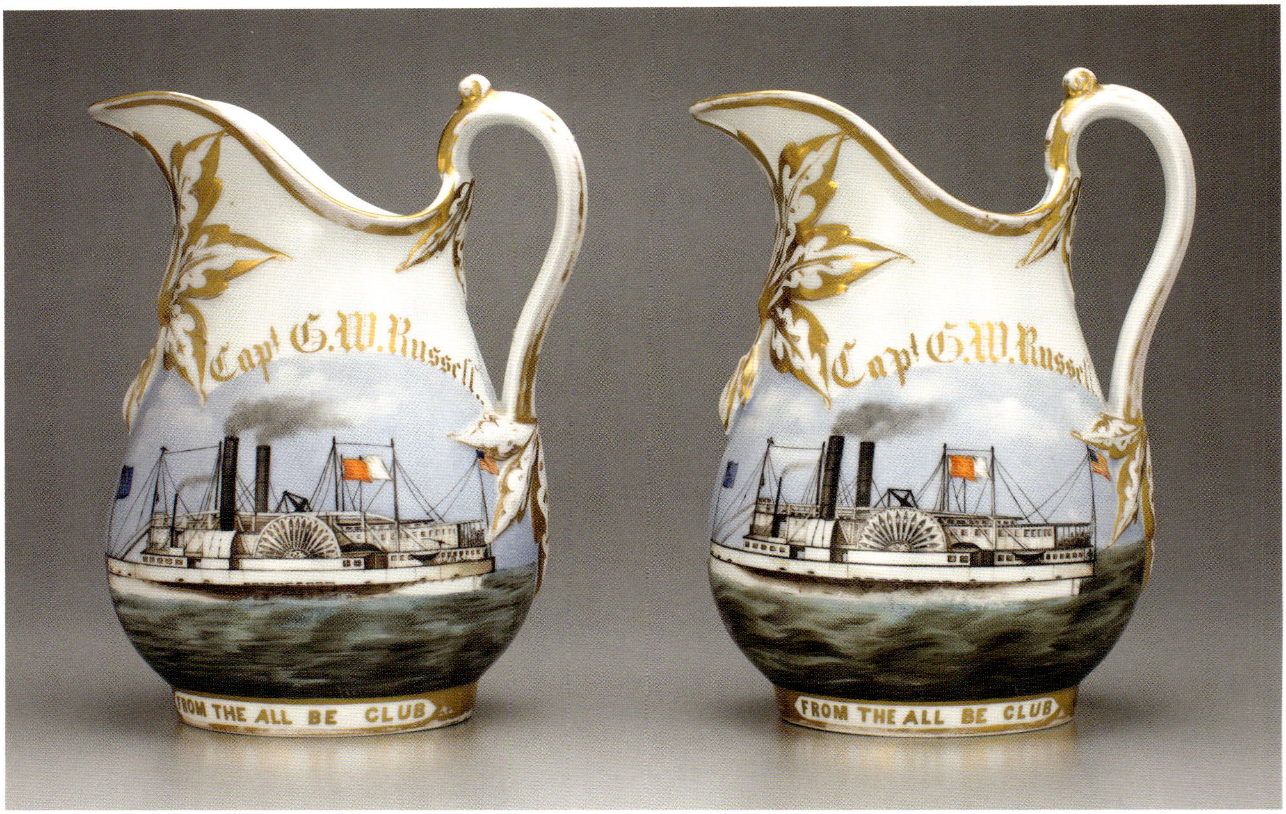
Capt G.W. Russell
FROM THE ALL BE CLUB
Capt G.W. Russell
FROM THE ALL BE CLUB

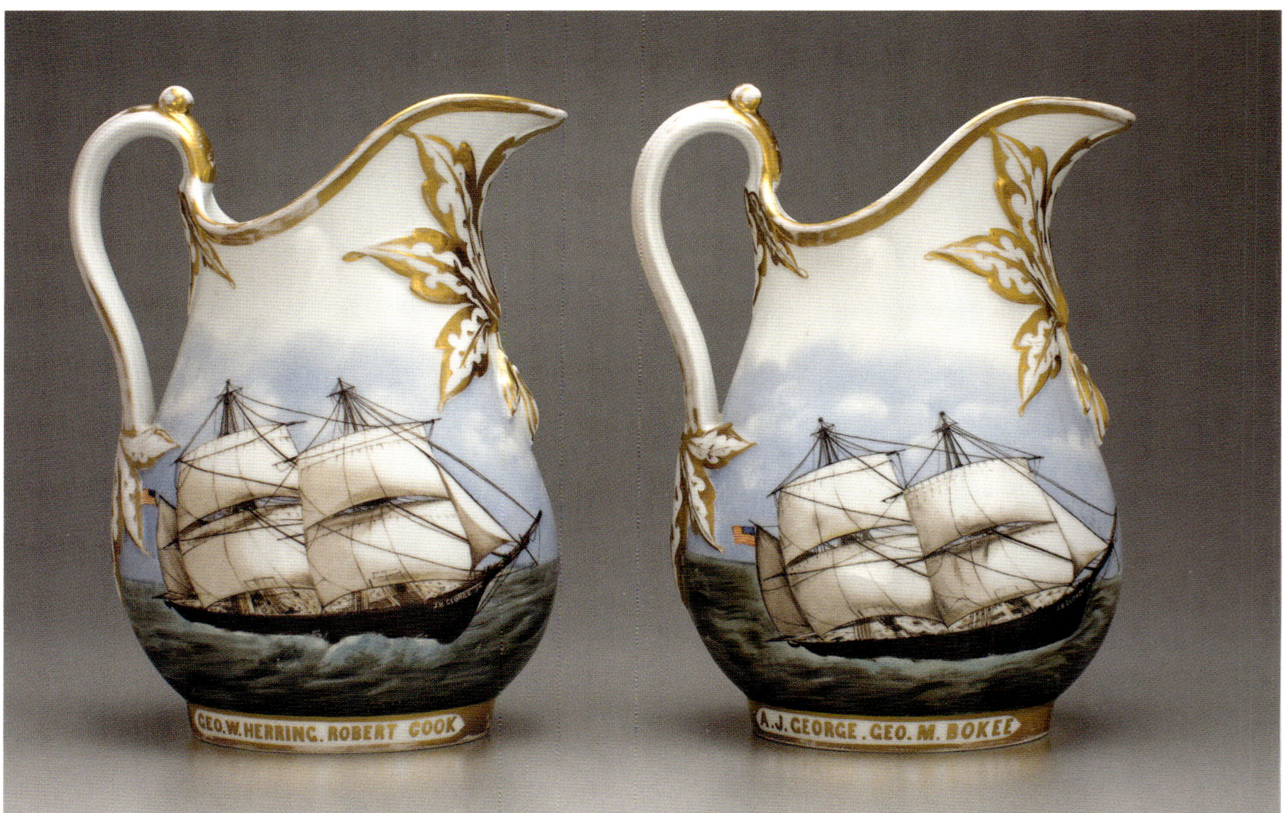
GEO.W. HERRING. ROBERT COOK
A.J. GEORGE. GEO. M. BOKEE

Robert Hunter and Oliver Mueller-Heubach

The Captain George W. Russell Presentation Pitchers

Figure 1 Presentation pitchers, attributed to Limoges, France, 1855–1859. Porcelain, enamels, and gilt. H. 11½". Gilt inscriptions: on steamer *Louisiana* "Capt G.W. Russell"; around the foot rim: "FROM THE ALL BE CLUB" (Private collection; unless otherwise noted, all photos by Robert Hunter.) The commission of the pitchers dates to 1859. This view shows the enameled decoration.

Figure 2 The reverse of the pitchers illustrated in fig. 1, with the enameled decoration of the brig *James B. George*. Around the foot rim is the gilt inscription: (*left*) "GEO. W HERRING. ROBERT COOK"; (*right*) "A. J. GEORGE. GEO. M. BOKEE"

▼ YESTERDAY MORNING *a committee of the members of the Allbe Club, consisting of Major George W. Herring, Col. Robert Cook, Adjt. A.J. George and George M. Bokee, visited the residence of Capt. George W. Russell, and presented him with a pair of beautiful and costly china pitchers. The vessels were about one foot high, and bear on one side a painting of the steamer Louisiana, which Capt. R. commands, and immediately over it the name of the recipient. The opposite side bears a painting of the brig James B. George under full Canvas. Around the bottoms are placed the name of the club and the parties who presented it. The presentation was made by Major Herring and responded to by Capt. R. After the presentation the company partook of a fine collation, and all seemed highly pleased with the occasion.*[1]

For students of ceramic history, this notice, which appeared in the *Baltimore Sun* on December 8, 1859, might have served as the only record of the "pair of beautiful and costly china pitchers" mentioned within. The preservation of ceramic objects is always highly variable; the probability for the survival of the fragile medium is infinitesimal. In spite of the odds, these attractive porcelain pitchers, which recently surfaced from a tag sale in eastern Virginia, have remained in remarkably good condition (figs. 1, 2). While the particulars of their descent from the hands of their original owner—Captain George W. Russell—are yet to be fully understood, the anonymously decorated pitchers stand as significant examples of early china painting in the American context.

The newspaper article clearly tells us that Captain Russell was commander of the *Louisiana*. His connection to the sailing vessel *James B. George*, however, is not stated, but research indicates he was an owner or co-owner of that vessel. Although the inscription on the pitchers gives the name of the brig as "J.B. GEORGE J^{R}," official records consistently give the name without the suffix.

The members of the "ALL BE CLUB," as inscribed on the pitchers, were all prominent participants in Baltimore's china, glass, and crockery trade: Major George W. Herring, Colonel Robert Cook, Adjutant A. J. George, and George M. Bokee. Their connection to Captain Russell is circumstantial. They may have been among his many admirers and friends or, more likely, had direct business dealings with him relating to shipping their wares on the brig *James B. George*, which called on ports from New York City to New Orleans. Regardless of their relationship with Russell, the All Be Club members are part of the neglected story of the expansive mercantile traffic

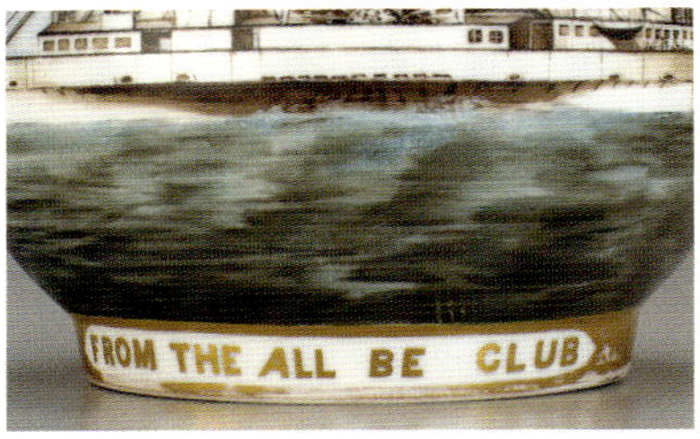

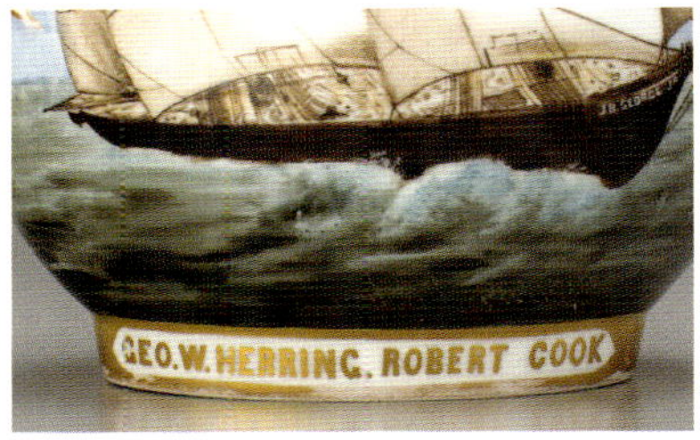

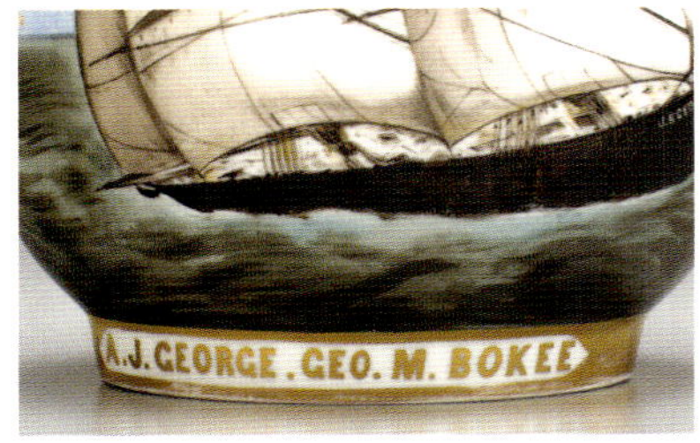

Figure 3 Detail showing the inscription on the foot rim of the pitchers illustrated in fig. 1: "FROM THE ALL BE CLUB" in gilt block letters.

Figure 4 Detail of the inscription on the foot rim of the pitcher illustrated on the left in fig. 2.

Figure 5 Detail of the inscription on the foot rim of the pitcher illustrated on the right in fig. 2.

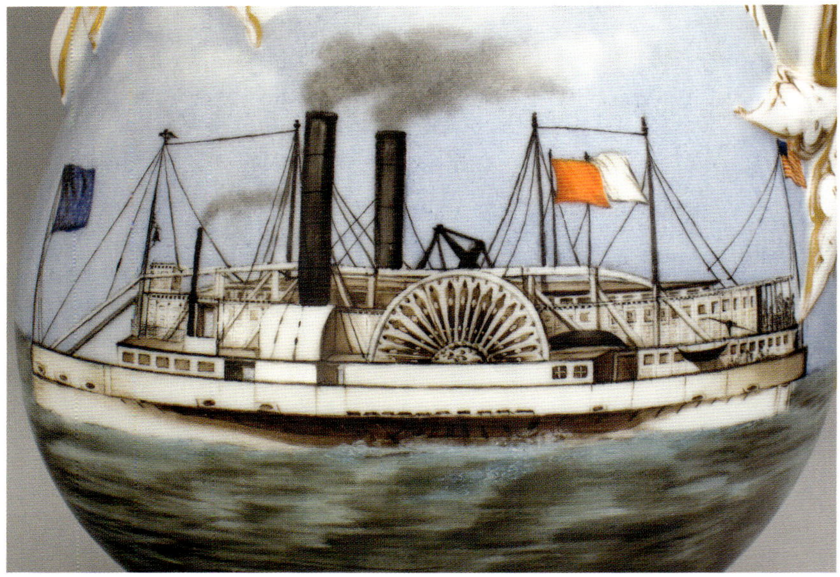

Figure 6 Detail of the steamship *Louisiana* on one of the pitchers illustrated in fig. 1.

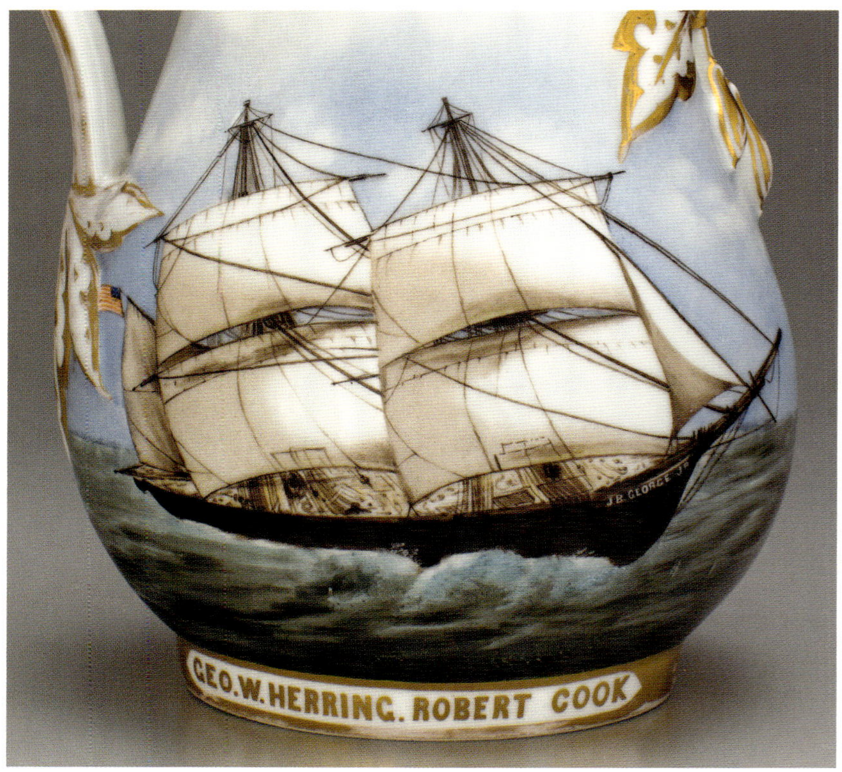

Figure 7 Detail of the brig *James B. George* on one of the pitchers illustrated in fig. 1.

by American china and glass dealers in the mid-nineteenth century with hubs in Baltimore, Boston, New York, Philadelphia, and other major cities.[2]

This article examines the contextual evidence leading to the presentation of these pitchers to Captain Russell. Particular emphasis is given to his tenure as captain of the steamship *Louisiana* and owner of the *James B. George*. In addition, an effort is made to understand Russell's relationship with the members of the All Be Club. In general, historical information was hard to come by, the best evidence being provided by newspapers entries, such as the eyewitness statement of the presentation of the pitchers in 1859. These rich accounts are reproduced relatively intact in the discussion that follows.

Description and Historical Context of the Presentation Pitchers

Measuring 11½" in height, the molded porcelain pitchers are of European manufacture, most likely from a factory in Limoges, France.[3] They would have been glazed and fired but left in the white when they arrived at an American decorator shop, ready to be enameled or gilded. After they were decorated, a low-temperature firing would harden the applied enamels and gilt. Each pitcher has marks scratched in the bottom that look like "21" or "Z1". These marks probably represent a potter's tally marks rather than a particular manufacturer. Molded fig leaves provide decorative flourishes under the spout and on the handle terminals, in keeping with the Empire style of the 1850s.

Both pitchers include the wording "FROM THE ALL BE CLUB" in gilt block lettering (fig. 3). One of the pitchers has "GEO. W. HERRING. ROBERT COOK" around the foot (fig. 4). The other has "A. J. GEORGE. GEO. M. BOKEE" (fig. 5). Overall, the enamel painting is remarkably intact. Some of the gilding has been rubbed from larges areas of the handles and under the spout, evidence of substantial use at some point.

Contemporary images of the *Louisiana* would have been available for a decorator to capture its distinctive architectural details (fig. 6). No picture of the *James B. George* has been discovered, but the brig class of ship was a fairly standard and popular sailing vessel with two square-rigged masts (fig. 7). The lettering on the bow of the sailing vessel is scratched through the black enameling on the hull (fig. 8).

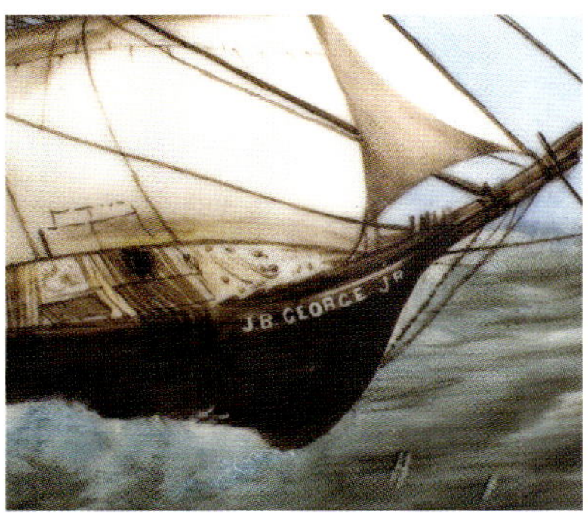

Figure 8 Detail of the inscription on the bow of the brig. Although the inscription reads "J. B. GEORGE J^R^," no evidence for this exact name exists, the closest being *James B. George*, of which in 1859 Captain George W. Russell was a co-owner.

The rendering of the ships is carefully drawn and shaded with black enamels; the white of the porcelain creates form and contrast. The sky is a light blue enamel with soft white clouds painted in. The waves of the sea are executed with hurriedly applied streaks of green, aqua, and black and is the least successful of the otherwise realistic decoration. Bright colors of red, orange, and blue are added to flying flags.

No evidence is seen for an inscription identifying the *Louisiana*, and no other marks or signatures are presented within the paintings or on the bottom of the pitchers. The application of the gilt lettering and highlighting of moldings on the spout and handle terminals would have been the final step in the decorating process (figs. 9, 10).

It can be a bit daunting to confront an unsigned work of art. However, since most of the pottery and porcelain made before the twentieth century

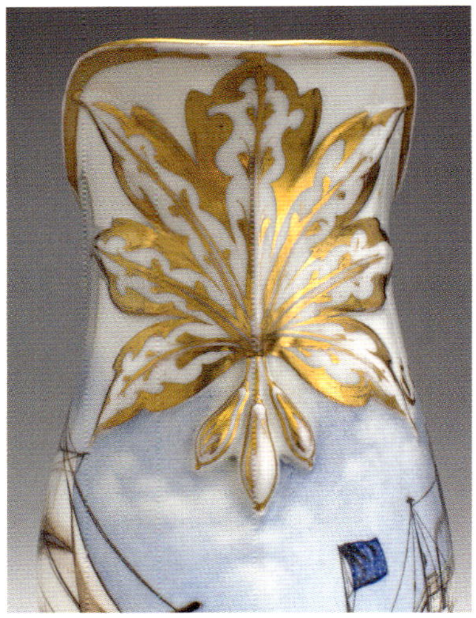

Figure 9 Detail of the molding below the spout of the pitchers illustrated in fig. 1. The molding takes the form of a fig leaf with fruit highlighted in gilt.

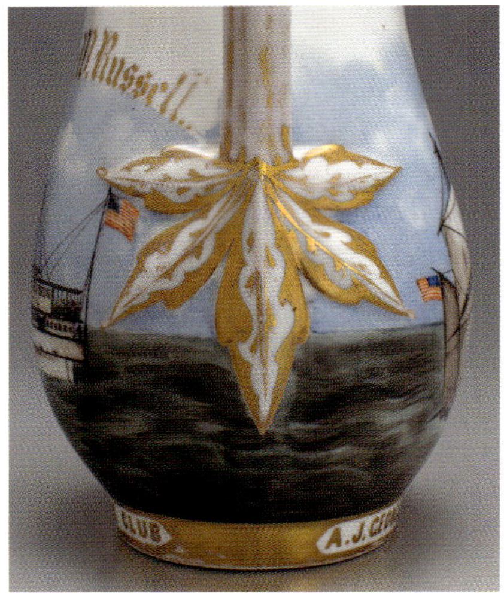

Figure 10 Detail of the molding of the handle terminal of the pitchers illustrated in fig. 1. The molding takes the form of a fig leaf.

was never marked or signed, it is a common practice for ceramics scholars to establish authorship through the process of attribution. Beyond analyzing the style and materials of the paintings on the pitchers, we can narrow down the possibilities of identity through their historical context.

The decorating of French porcelain blanks was the practice of William Ellis Tucker in Philadelphia in 1826 until Tucker and his partners were able to produce their own porcelain.[4] By the mid-nineteenth century, the importation of European blanks was a common practice by American enamelers and decorating shops. Special orders of elaborately decorated but costly French porcelains could be made through any number of New York City shops. But the same New York shops also decorated imported blanks themselves, often taking custom commissions.[5] The firm of E. V. Haighwout, for example, was among the largest supplier of custom-decorated porcelains in the country. Its work included commissions for two White House services, and it also produced prodigious quantities of custom hotel wares and presentation china.[6] Haviland and Company was another New York firm that imported tremendous amounts of china from its Limoges factories—both decorated and blanks for resale to other decorators.[7]

The Baltimore members of the All Be Club in 1859 were all prominent china, crockery, and glass merchants who made their living by importing pottery and porcelain from British and European sources (for the wholesale and retail trades) and had dealings with numerous American cities. They certainly knew from where and from whom to commission such custom-designed and painted porcelain objects. One connection was suggested by several 1857 newspapers accounts that place George Herring in proximity to the New York partnership of William Underhill and Haviland & Company. Both Herring and Mr. Underhill gave successive speeches at that year's Convention of Earthenware Dealers' meeting in Philadelphia.[8]

Established in 1859, the partnership Underhill, Haviland & Co. was a venture with members of the Haviland family who had been importing French porcelain blanks into their New York shops since the early 1840s.[9] The Haviland Company is best known for its factory in Limoges, established in 1853, where it manufactured and decorated a huge variety of patterns for the American market. Prior to that partnership, New York City directories show that William Underhill operated his own china business from about 1848 until 1859.[10]

In preparation for this article, the photographs of the Russell pitchers were shown to several American ceramics historians familiar with mid-nineteenth-century china painting. Several responders suggested that the decoration on the pitchers might be the work of Rudolph Lux of New Orleans. Lux was a German emigrant who had established a New Orleans china shop by 1857. He advertised his wares and promoted his skills using both German and French blanks for his canvas.[11]

The definitive catalog of Lux's oeuvre has yet to be written. A recent summary by Anthony Peluso reviews Lux's prolific but brief career citing nearly forty known examples of his work.[12] Lux is especially well known for his portraits of Civil War generals and soldiers; he diplomatically accepted commissions from both the Northern and Southern sides of the cause. Because most of these portraits appear on porcelain tea cups and saucers, the resulting decorations are fairly diminutive. Peluso's article also lists nearly two dozen objects decorated between 1862 and 1868 with images of steamboats.[13] Without exception the steamships are ones that traveled the Mississippi and its tributaries from New Orleans to the upper reaches of the Midwest.

Lux's depictions are quite detailed, with accurate rendering of the ships' architectural features. Virtually all published examples of Lux's ship portraits have his signature placed somewhere in the composition. A comparison of the paintings on the unsigned Russell pitchers with signed examples of Lux's steamboats makes an attribution to Lux generally unconvincing

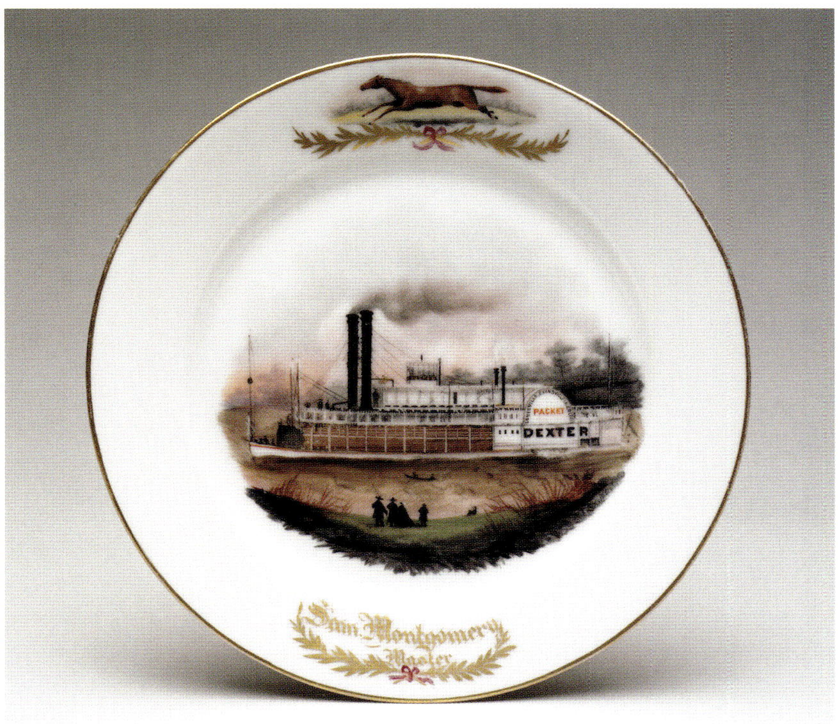

Figure 11 Plate, France or Germany, decorated by Rudolph Lux, New Orleans, Louisiana, 1868. Porcelain, enamels, and gilt. D. 8⁷⁄₁₆". Inscribed: "Sam Montgomery / Master" (Courtesy, The Bayou Bend Collection, museum purchase funded by the Houston Junior Woman's Club.) The steamer is named after the "Celebrated Horse" Dexter, immortalized in an 1867 Currier & Ives print, *The King of the World*.

(fig. 11). The renderings on the pitchers, while highly detailed, lack his delicacy of line and refinement. And although the enameling is of the same general quality and palette of Lux's work, the scale is so much larger than what generally appears on his much smaller tea wares paintings.

The Louisiana *and the Brig* James B. George

The *Louisiana* was built by the Baltimore firm of Cooper & Butler in 1854 at the cost of $234,197. It is the type of steam-powered vessel known as a "side-wheeler," with two paddle wheels, one on either side of the hull amidship. It measured more than 266 feet in length, making it the largest wood-hull steamer of its day. The *Louisiana* was considered the "last word" in steamboat construction, the finest of its type ever to ply the Chesapeake's waters.[14]

Because of the ship's great length, a distinctive timber superstructure was required keep the hull from sagging at the bow and stern. Known as "hog frame" construction, stout trusses ran the length of the ship. This frame is clearly visible in an 1850s lithograph of the *Louisiana* (fig. 12) and a slightly later drawing made during the Civil War (fig. 13). Likewise, the same highly prominent frame, featured on the renderings on the porcelain pitchers, confirms the identity of the *Louisiana* in spite of the lack of an inscribed name (see fig. 6).

The *Louisiana* was owned by the Baltimore Steam Packet Company. The term *packet* originally referenced the mail packets that were transported on a regular basis as part of a contract with the government. The term was applied to any passenger ship operating between designated cities on a regular schedule. In 1858 an advertisement announced that the *Louisiana* would leave Baltimore daily for Norfolk at 5 p.m. except on Sundays; the fare to Norfolk was five dollars. The *Louisiana* remained in service until November 14, 1874, when it collided with another steamship and sank.

The *James B. George*, built in Baltimore in 1856 and listed as 242 tons, was a brig, a classification that was applied to a medium-size sailing vessel intended for cargo. Based on newspaper mentions of its time in various ports, it appears that it would carry anything from passengers to guano.[15] Later newspaper advertisements and registrations show that the *George* remained active in the early 1860s.[16]

The *George* is found in the New-York Marine Register of 1858 showing one Captain Atwell as master.[17] More important, however, is the appearance of the name Russell under the column "Owner or Consignees." This provides the pivotal association with the brig *George* featured on the presentation pitchers.

The *James B. George* appears the following year in *American Lloyd's Register of American and Foreign Shipping, 1859* with a Captain Frazier listed as master. This time, under the column "Managing Owners or Consignees," the entry "Russell & O" implies that Russell was the principal co-owner with one or more partners, but the nature of the business relationships and the merchants involved are unclear.

A clue to these relationships might lie in the name of the vessel. The *Woods' Baltimore Directory, for 1856–57* lists James B. George Sr., designated

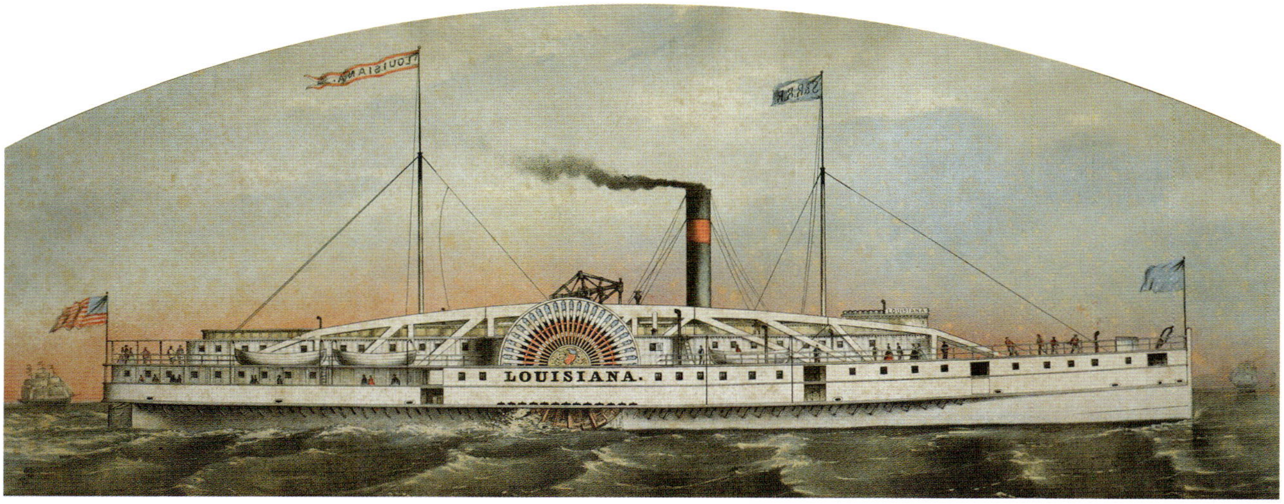

Figure 12 *Louisiana*, unidentified artist, 1850s. Lithograph. 7½" x 19". (Courtesy, The Mariners' Museum.)

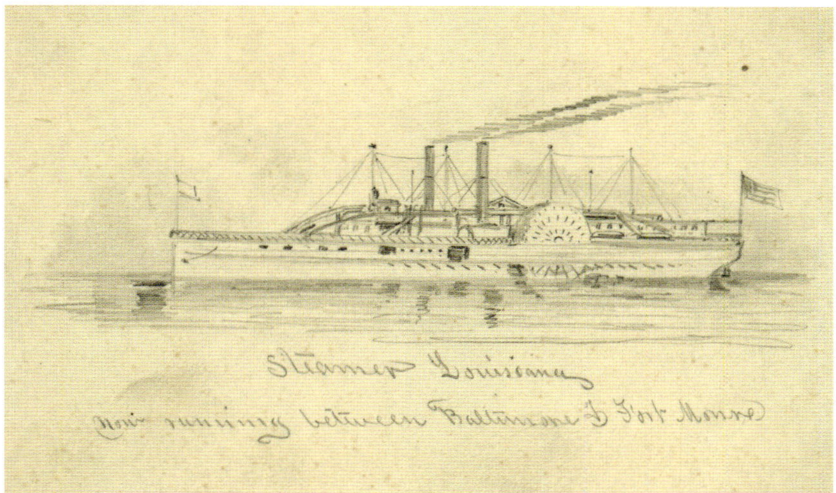

Figure 13 Alfred Rudolph Waud (1828–1891), *Steamer Louisiana now running between Baltimore & Fort Monroe*, 1861–1865. Pencil drawing on paper. 3⅜" x 5⅝". (Morgan collection of Civil War drawings, Library of Congress Prints and Photographs Division, Washington, D.C.; available online at https://www.loc.gov/resource/ppmsca.21373/)

as a property agent and collector, and James B. George Jr., classified as a mail agent.[18] Both activities are closely tied to the shipping business, particularly the aforementioned connection with mail-packet contracts entered into by the steamship companies.

Captain George W. Russell

George W. Russell was born in 1812 and spent his adult career as a professional steamship captain on the Chesapeake Bay.[19] While his biographical information has been only partially revealed, it is clear that he was a celebrity among the fraternity of steamship professionals. By 1839 he was a captain in the newly inaugurated Baltimore Steam Packet Company, whose fleet of boats made regularly scheduled runs between Baltimore and Norfolk in addition to other local excursions.[20] By 1842 Russell was in command of the steamship *Herald*, and the company kept a brisk business in the commute between the major ports of the Chesapeake Bay.[21] In 1846 he was in command of the steamer *Georgia*.[22]

In addition to commuting between cities on the Chesapeake, the Baltimore Steam Packet Company's ship captains were permitted to charter steamers from the company for pleasure excursions offered to the public. Perhaps this is one reason Russell gained such favor with his following of friends and admirers over the years. One of these pleasure excursion offerings was advertised in the *Baltimore Sun* on August 28, 1848, as follows:

> GREAT ATTRACTION. PLEASURE EXCURSION TO ANNAPOLIS, OLD POINT, RIP RAPS, NORFOLK, PORTSMOUTH, NAVY-YARD, &c. &c.
>
> Captain GEORGE W. RUSSELL respectfully informs his friends and the public, that he has chartered from the Baltimore Steam Packet Company the splendid and swift steamer GEORGIA, well known to the public as one of the most commodious, comfortable and seaworthy boats out of Baltimore, and will make an EXCURSION to Norfolk, Old Point Comfort, Rip Raps, Portsmouth, Navy-Yard and big Ship, leaving Baltimore on SATURDAY MORNING, the 9th of September, at 8 o'clock, touching at Annapolis, giving persons an opportunity of enjoying the beautiful scenery of the noble Chesapeake by daylight. Particulars in a future advertisement.[23]

Captain Russell's status in the packet line continued to grow. In 1852 he was given command of the newly built *North Carolina*, a "model boat in every respect," measuring 235 feet in length. By several accounts, Russell was much admired by his friends and colleagues. This private and public admiration is perhaps best recorded by the presentation album quilt made for him by friends and family in 1852 (fig. 14).[24] One of the quilted squares includes an appliqué of a steamboat with smoke billowing from the stacks and a representation of Russell himself with spyglass in hand (fig. 15). Other imagery on the quilt includes symbols that show that Russell was a member of the Independent Order of Odd Fellows (IOOF), a fraternal order of Odd Fellows founded in 1819 in Baltimore, Maryland. The George surname is tantalizingly included among the names of the presenters included on the quilt, suggesting a possible familial connection to James Sr. and James Jr., the future namesakes of the brig *James B. George*.

Writing in 1853, William S. Forrest noted in his *Historical and Descriptive Sketches of Norfolk and Vicinity* that travel by steam packet between Baltimore and Norfolk took a remarkably and comfortably short twenty-four hours. Forrest made special mention of the competence of the captains who commanded these vessels and specifically referenced Russell, who had recently been given command of the *North Carolina*, the newest ship in the line. In what might have been a foreshadowing of events, Forrest commented on the safeness of the Baltimore Steam Packet Company's voyages, which were relatively accident-free:

> We may remark here, that this line is so well and ably conducted, that accidents seldom or never happen. The boats are very superior, kept in the finest order, and are in charge of officers of long experience, and well-tried skill and judgment. The *North Carolina*, a very large and splendid new boat, has recently [been] placed upon this line; and, under the able management of Captain Russell, presents unusual attractions to the traveller, who desires to go north by the pleasant and delightful bay route.[25]

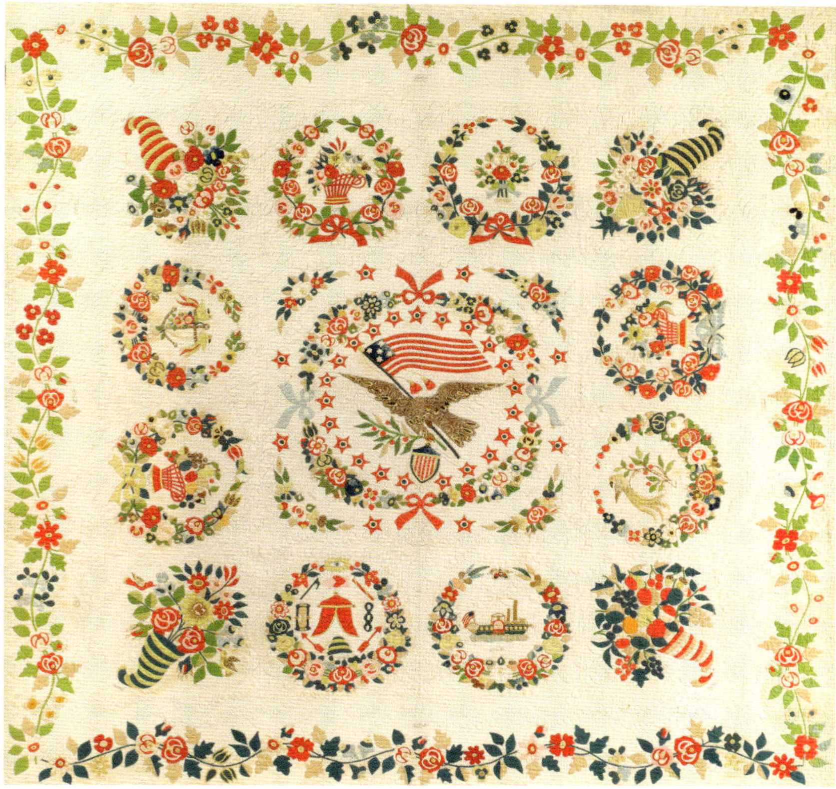

Figure 14 Album quilt, Baltimore, Maryland, 1852. Cotton. 89¾" x 88¼". (Courtesy, The Baltimore Museum of Art, BMA 71.26.1.) Presented by the friends of Captain George W. Russell.

Figure 15 Detail of the album quilt illustrated in fig. 14.

Russell's time on the *North Carolina* did not last long. In 1854 he was given the helm of the *Louisiana*. On November 2, the *Washington Sentinel* reported: "THE NEW STEAMER LOUISIANA,/Captain George W. Russell, is now nearly ready to take her place on the line. She will probably make a trial trip next week to Norfolk—perhaps during the sitting of the Internal Improvement Convention. We would suggest that as a most suitable time."[26] Her maiden voyage was on November 9, 1854, and went from Baltimore to Old Point Comfort and the Virginia Capes. Upon his arrival in the Virginia ports, friends from Norfolk and Portsmouth presented him with a "magnificent silver speaking trumpet," a device to call out orders above the din of the steamship's engines and wheels.[27]

A notice about Russell appears in the *Baltimore Sun* in 1855 commending his action in assisting two distressed ships encountered on his route in the Chesapeake Bay:

> *The Steamer Louisiana*—This fine steamer, under the command of Captain George W. Russell, reached this port on Saturday morning with a freight of over 1,400 barrels of fruit and vegetables, the latter including potatoes, beans, peas, cymlings, cucumbers, cabbage, &c., added to which were 143 gallons of oysters in cans, packed in ice. There were 85 cabin passengers and 23 forward. Captain Russell reports the ship Susan E. Howell, Captain E. W. Rattle, off Poplar Island, leaking badly; brought the captain up to the city. The ship is from the Chincha Islands, loaded with guano, consigned to Barada & Brather, 97 days. Also, a brigantine off the Wolf Trap, bound up, in distress. The steamer gave every assistance to both vessels, and all may be considered safe.[28]

Although Russell was highly competent and a well-respected steamship captain, his profession was nonetheless filled with inherent difficulties, and accidents did happen. Such an accident was to befall the *Louisiana* with Captain Russell in command on the evening of February 20, 1858.

On a "bright moonlight night," the schooner *William K Perrin* was carrying her cargo of oysters down the Chesapeake Bay when she collided with the *Louisiana*. Little or no damage occurred to the much larger steamship, but the *Perrin* was a sailing vessel of just 47 tons; she foundered and was lost. Conflicting claims of right-of-way according to the prevailing rules of seamanship were adjudicated in a well-publicized case heard by the Circuit Court of the United States for the District of Maryland. The courts adjudicated Captain Russell and the owners of the *Louisiana* and held the owners of the *William K Perrin* responsible for $1,700 in damages, and its master, Charles Odgen, for an additional $173.

The circuit court's ruling initially kept Russell's safety record intact; however, the owners of the *Perrin* appealed the case to the U.S. Supreme Court. In lengthy arguments, the circuit court's decision was reversed and Russell and the owners of the *Louisiana* were found at fault. The transcript of the case summarily pronounced the Supreme Court justices' opinion: "[T]he collision in this case, and destruction of the schooner *Perrin*, was caused wholly by the negligence and inattention to their duties of the officers who navigated the *Louisiana*, and that the steamboat should be condemned to pay the whole damage incurred by the said collision."[29]

How the unfavorable decision affected Russell's career and his relationship with his business associates, passengers, and friends is undocumented. It may have been forgiven as an unavoidable accident, a routine occurrence in the hazards of navigation on the Chesapeake Bay.

What is known is that the pitchers were presented to Captain Russell at virtually the same time the Supreme Court judgment was released, in December 1859. Had the All Be Club guessed wrong at the outcome of the case? Or were the pitchers intended as a show of solidarity and support for Russell? Or perhaps the pitchers were meant to signify a new business venture among partners. But the timing of the release of the court's ruling and the presentation seems an unlikely coincidence.

The All Be Club

As recorded on the pitchers, the members of the so-called All Be Club were George W. Herring (1820–1897), Colonel Robert Cook (1824–1912), Adjutant Andrew J. George (b. 1828), and George M. Bokee (1831–1905). These four men appear in all the Baltimore city business directories from 1856 to 1859 as glass and china dealers.[30] Earlier advertisements also helped establish the prominence of these individuals as china merchants. In the February 11 edition of the *American Commercial and Daily Advertiser*, Robert Cook gave notice of a newly established partnership:

> CO-PARTNERSHIP NOTICE.
>
> WILLIAM SHIRLEY has this day associated with him ROBERT COOK in the WHOLESALE AND RETAIL CHINA GLASS AND QUEENS-WARE

business, which will be continued at the old stand, 11 and 13 South Calvert street, under the name of SHIRLEY & COOK.
Baltimore, February 5th, 1853[31]

On August 29 the same year George W. Herring advertised his importing business in Baltimore: "CHINA, GLASS AND QUEENSWARE. / GEORGE W, HERRING & CO., Importers, No. 7 S. / Charles st."[32] And on December 23, 1857, the *Baltimore Sun* published the often mentioned George M. Bokee's china store's numerous suggestions for Christmas gifts:

> CHINA ORNAMENTS FOR CHRISTMAS PRESENTS. A large variety, consisting of BOXES, BOUQUET HOLDERS, COFFEES, / COLOGNES, FIGURES, INKSTANDS, MUGS, / TEA AND TETE-A-TETE SETS. / CIGAR STANDS, VASES, / WATCH HOLDERS, &c. &c., / Now displayed at / BOKEE'S CHINA STORE. / No. 41 NORTH HOWARD STREET, / And offered for sale at less than / AUCTION PRICES, / To suit the times / GEORGE M. BOKEE. / No. 41 North Howard street. / between Lexington and Fayette sts.[33]

A notice in the *Baltimore Sun*, October 7, 1859, is the only other documentation for this group of men:

> *Annual Meeting of the Crockery Trade.*—The annual meeting of the crockery trade was held last night at Allen's, in Light street, when, after the transaction of the ordinary business, the following officers were elected for the year ensuing: Geo. W. Herring, president; Joseph R. Marston, vice-president; Geo. M. Bokee, secretary; and Col Robert Cook, treasurer. The report of the treasurer showed a balance of $119 in the treasury after paying all the expenses of the past year.—Those matters disposed of, the association sat down to a magnificent dinner, comprising all the luxuries of the season, and to which ample justice was done.
>
> The president made a statement of the rise and progress of the association, and of the "All-be Club," which has existed for several years, composed entirely of crockery dealers, and of which one of the most worthy members has always held the post of president.
>
> For several hours there was a good time in the interchange of sentiment, and the best feeling prevailed. There were representatives from the trades in Philadelphia and Wheeling, and the enjoyment of the occasion was full. Before the noon of night the company retired, having spent an evening in rational pleasure.[34]

Apparently, the club constituents were a small, ad hoc subset of Baltimore's crockery dealers, although at this meeting other crockery dealers from Philadelphia and Wheeling, Virginia, were part of the evening's revelry. Of the four members of the club, A. J. George is the least represented in the historical documents. *Wood's Baltimore City Directory* for 1858–59 lists an Andrew J. George as associated with R. Edwards, Jr. & Co.[35] The same directory reveals that Richard Edwards Jr. & Co. were glassware merchants in the city.[36]

George W. Herring appears to have been the most important and influential of the group. Not only was he president of the All Be Club in 1859, but earlier accounts of the Convention of Earthenware Dealers also reveal his considerable standing at a national level. On June 25, 1857, more than 150 representatives from "New York, Baltimore, Boston, Richmond, Pittsburg, Cincinnati, St. Louis, Wheeling, Chicago, and other cities" gathered in Philadelphia at the La Pierre House.[37] The event was reported on in many papers including those in Baltimore, New York, and Philadelphia.[38] In every article, Herring was prominently mentioned. The most extensive and fascinating account of the conference was published in the *Baltimore Sun* on June 27, 1857. This excerpt includes the lengthy and somewhat hyperbolic speech given by Herring with William Underhill co-joining:

> *The Convention of Earthenware Dealers*
> The social conference of the Earthenware Dealers' Board of Trade, of Philadelphia, was commenced in that city on Thursday morning, at the Assembly Building. In the course of the morning a number of invited guests arrived from other cities, especially from New York, Boston and Baltimore. There was no business transacted in the morning. At noon the members partook of a collation, and at 2 o'clock the conference met in private. In the evening the company sat down to a sumptuous banquet at the La Pierre House, to which they were welcomed by the president, Mr. Hacker, in the following appropriate remarks:
>
> Gentlemen—The present scene, at least to me, is one of unfeigned pleasure. It is my duty as president of the "Earthenware Dealers," Board of Trade of the city of Philadelphia—a position not assigned to me because of any talent of speech-making—to stand up and bid you welcome, thrice welcome, to the hospitalities of this social occasion. And I, for myself and my colleagues, now, with open arms and hearts big with the sincerest sentiments of brotherly friendship, do hail you as the guests of our association, and welcome you to the city of brotherly love with a most true and fraternal greeting. We have called our assembling together in this city a *social conference* of the importers and jobbers of earthenware, china and glassware. Our objects are more of a social and business character. And yet we feel that much good will result from our business meeting of to-day and at future sessions.
>
> It is no common sight that we witness this evening. Scarcely in the annals of this country, and certainly not in any other, has there been assembled of one branch of trade; on no occasion such as this, so extended a representation. We have here parties representing fifteen States and twenty-four cities of this Union; and we mean to show to the world that the time has come, when, in the fair and active emulation of commendable business enterprise, there need be no want of social and kindly regard—no jealous rivalry that will mar the harmonies of life; and we shall endeavor to obliterate that *ugly* adage, "*that* two of a *trade* can *never* agree." This . . . saying [has] long been repudiated as false to *our* ears[.]
>
> The earthenware trade of the United States, although limited in amount when compared with other departments of trade and commerce, is yet of vast importance to the general interests of the country. It gives the reward of labor to some thousands in the potteries of Staffordshire, England. Its bulk is so great in comparison with the value of the article that it gives employment to large numbers of the laboring classes in this country, in the departments of packing, storing, draying, &c., and it is of vital importance to the shipping interests of the world; for the groundwork of almost every ship chartered in Liverpool for this country, and for other distant places, is crates of earthenware and china[.]
>
> The number of packages of earthenware shipped from Liverpool to the United States for the past six years, averages about 100,000 crates per annum; the entire shipments from Liverpool to all parts of the world average about 170,000 per annum; the United States, therefore, receive more than one-half of all that is exported.
>
> The bulk of 170,000 crates is equal to 212,500 tons measurement, and would load 212 ships of 1,000 each; being four ships per week for all the year. You can see at a glance how important is the manufacture of this

article to the shipping interest. And then, gentlemen, the vast amount of freight that it gives to our railroads and canals in this country is equally important, for the revenue from it is very heavy, although the value is insignificant when compared to many other articles that are sold and forwarded to all parts of our continent; yet the freight is paid on bulk weight. The average freight from England to the United States is about 5 per cent on its cost.

The manufacture of earthenware can be traced as far back as the building of the tower of Babel—and it is not now confined to England—it is made in some form in every country on the face of the globe; the Chinese, the Japanese and the French, are now famed for the magnificence of these articles of porcelain; and indeed the French China manufacture is becoming rapidly a great source of revenue to this government, and is now a staple article of use in all parts of the United States.

In his response to a toast in honor of the Baltimore trade, Herring said:

Ever since the formation of our government, the office of president has been looked to as a position to satisfy the highest aspirations of the greatest and most ambitious of our countrymen; indeed, to be called president at any time is honorable, but to be called President of the United States, or president of the crockery trade of a city, I think excels all other worldly attractions, and should gratify the ambitious desires of any man; but position brings with it its necessary burdens, and often overtasks its recipients. This is just the position I find myself in to-night; but as most of the presidents have said, I feel my incompetency for the task imposed upon me, and throw myself upon your indulgence. The Baltimore trade, Mr. President, can but feel highly gratified upon this occasion. Scarcely six months since they conceived and carried out in an humble way the idea of calling together around the festive board the members of the earthenware trade of this city, for the purpose of drawing together the bond of friendship, and to do away with the selfish idea which for many years has been prevalent, that competition and rivalry would always prevent merchants engaged in the same business from ever being upon terms of personal intimacy.—The seed we sowed in weakness, you reap in power; and it is my earnest wish that it may grow up to be a mighty tree, under whose branches the entire earthenware trade of our country may seek shelter. I say again, Mr. President, the earthenware trade of Baltimore feel gratified at the sight which meets their gaze to-night. We feel no jealousy because you have excelled us in the beauty and extent of your arrangements. Would Robert Fulton, were he living at this day, gaze with less admiration on a Collias steamer, because it was an improvement on his own humble efforts; would Benjamin Franklin look coldly on Morse, because he had taken his discovery and applied it to purposes of extended good; would Demaratius, father of Tarquinias Pilaeus, who is said to have instructed the Etruscans and Romans in the art of pottery, had he been living [in] 1730, have cursed that year because it gave birth to Josiah Wedgwood; or would Wedgwood, who patented, in 1763, a style of ware called queensware, feel mortified at the improvement in that article of Edwards, Furunel, Alcock and others of the present day? No, sir, the contrary would have been the effect on all of them. And so it is with us of Baltimore. We not only feel proud of the successful effort you have made to excel, but our wish is that others may excel you, and that the motto throughout our country among earthen dealers shall be Excelsior! Well, sir, we are here with other gentlemen of the trade by invitation, and we throw ourselves upon your mercy; and I am fearful that before we get through we will have to exclaim, "save us from our friends!" The Baltimore trade came here, sir, to have a good time, and as we are only passengers, we place ourselves entirely under your control; and although we expect to travel fast, yet the indications are that we will not be able to

> keep our seats; and I dread a collision either with a lamp post or a watchman; and if we keep the track we will be quite fortunate—or do not collapse a flue, as there seems to be no taking in water. In conclusion, Mr. President, I beg leave to return the Philadelphia trade the thanks of the Baltimore trade for their kind invitation to participate in this national convention of earthenware dealers, and hope it may redound to the mutual benefit of all present; and although our numbers are small, and means limited, yet we shall be proud to extend to any of our fellow tradesmen an invitation and welcome to the city of monuments.-

Mr. W. Underhill, of New York, responded at some length to a toast, and among other things said—

> I think I may say, judging from the activity manifested since we have been assembled around this hospitable board, that the earthenware trade, as represented here is in a very healthy state, and when I see surrounding me the majestic soup tureens (figuratively speaking, Mr. President) the capacious covered dishes, the well filled casseroles, the substantial bakers, the useful as well as ornamental sauce tureens and boats, together with the compotiers, baskets and other elegant dessert pieces of which this set is composed, I cannot but congratulate myself upon the perfect entirety which we present, and can fervently ejaculate—"May we never break!"[39]

Conclusions

The point of embarkation for this article began with the discovery of the elaborately decorated pitchers presented to Captain George W. Russell by members of the All Be Club of Baltimore. The most critical discovery was the Baltimore newspaper account recounting the exact time and day that the pitchers were presented to Russell. From that account, we learn that Russell was the illustrious captain of the *Louisiana*, the largest and grandest steamboat of the Baltimore Steam Packet Company. We also found that Russell owned or had ownership in the brig *James B. George*, also depicted on the presentation pitchers.

Subsequent research discovered the identities of the members of the All Be Club recorded on the pitchers as the leading glass and china merchants in the thriving Baltimore trade. A side excursion into the china and glass business revealed the account of the Convention of Earthenware Dealers held in Philadelphia in 1857, which was attended by more than 150 merchants, from many major American cities. The potential for investigating this remarkable event, reflecting the extent and importance of the American ceramic trade network is particularly enticing for future researchers. The identity of the decorator and/or decorating shops, however, remains unknown. In general, the style and quality of the presentation pitchers are in keeping with the output from the china decorating workshops operating in New York City in the mid-nineteenth century. While it is tantalizing to associate the ship portraiture with the work of Rudolph Lux, such a link remains unsubstantiated.

As with many studies of material culture, the stories often begin with an object. While much has been learned on this trial voyage, there is more research to be done. If these pitchers remain anonymous, their aesthetic and intrinsic value alone would still enthrall the historical ceramic community. But it is the continuing stories that are just beginning to be coaxed from these mute objects that will ensure their place among other illustrious examples of mid-nineteenth-century American ceramic art. Where the journey will lead, we cannot know, but it will be a destination that Captain Russell and his friends in the All Be Club would never have imagined on the morning of December 7, 1859.

ACKNOWLEDGMENTS This article benefited from the help and assistance of many people who offered advice, opinions, and supplementary research. Of particular note is Marshall Goodman, who called our attention to these wonderful pitchers. Ellen Denker, Nonnie Frelinghuysen, Anthony Peluso, and Rex Stark provided valuable commentary. Lydia Blackmore, decorative arts curator for The Historic New Orleans Collection, provided access to the collection's research on Rudolph Lux. Joan Quinn provided additional research in the Baltimore records. Robert Doares was extremely generous in sharing his expertise on Haviland china. Angelika Kuettner was instrumental in obtaining images and permissions for this article. We especially thank Mary Gladue for helping to sort out the disparate story lines and bringing cohesion to the final article.

1. *Baltimore Sun*, December 8, 1859, vol. 46, issue 18, p. 1.

2. George L. Miller and Amy C. Earls, "War and Pots: The Impact of Economics and Politics on Ceramic Consumption Patterns," in *Ceramics in America*, edited by Robert Hunter (Hanover, N.H.: University Press of New England for the Chipstone Foundation, 2008), pp. 67–108.

3. Personal communication with Robert Doares, May 11, 2016. Barbara Woods and Robert Doares, *Old Limoges: Haviland Porcelain Design and Decor, 1845–1865* (Atglen, Pa.: Schiffer Pub., 2005).

4. Ellen Denker and Bert Denker, *The Warner Collector's Guide to North American Pottery and Porcelain* (New York: Warner Books, 1982), p. 157.

5. Woods and Doares, *Old Limoges*, pp. 23–26.

6. Alice Cooney Frelinghuysen, "Empire City Entrepreneurs: Ceramic and Glass in New York City," in *Art and the Empire City: New York, 1825–1861* (New York: Metropolitan Museum of Art), pp. 327–54, esp. p. 330.

7. Woods and Doares, *Old Limoges*, p. 25.

8. "The Earthenware Trade of the United States," *Evening Post* (New York), June 27, 1857, p. 2.

9. Woods and Doares, *Old Limoges*, pp. 22–23.

10. *Trow's New York City Directory* for 1859: "Underhill, Haviland & Co. china, 22 Vesey Street." I am indebted to Robert Doares for providing this information.

11. *Encyclopaedia of New Orleans Artists, 1718–1918*, edited by John A. Mahé and Rosanne McCaffrey (New Orleans, La.: Historic New Orleans Collection, 1987).

12. A. J. Peluso Jr., "There Was No One Like Rudolph T. Lux, Ever," *Maine Antique Digest*, August 17, 2015, available online at www.maineantiquedigest.com/stories/there-was-no-one-like-rudolph-t-lux-ever/5224.

13. Ibid.

14. Alexander Crosby Brown, "The Old Bay Line of the Chesapeake: A Sketch of a Hundred Years of Steamboat Operation," *The William and Mary Quarterly* 18, no. 4 (October 1938): 389–405.

15. See, e.g., the following reference, which was published May 20, 1861: "NEW-YORK . . . SUNDAY, May 19. / BELOW — Brig. James B. George, from Rio Janeiro" (online at www.nytimes.com/1861/05/20/news/marine-intelligence-arrived-sailed-miscellaneous spoken-c-foreign-ports.html).

16. *American Lloyd's Register of American and Foreign Shipping*, 1861, available online at http://library.mysticseaport.org/initiative/SPSearch.cfm?ID=1891.

17. *New-York Marine Register* (1858), p. 163 (online at http://library.mysticseaport.org/initiative/SPSearch.cfm?ID=18868).

18. *Woods' Baltimore Directory, for 1856–57*, vol. 544, p. 125 (online courtesy of Maryland State Archives, http://aomol.msa.maryland.gov/000001/000544/html/index.html).

19. Dena S. Katzenberg, *Baltimore Album Quilts*, exh. cat. (Baltimore, Md.: Baltimore Museum of Art, 1981), p. 116.

20. Alexander Crosby Brown, *Steam Packets on the Chesapeake: A History of the Old Bay Line since 1840* (Cambridge, Md.: Cornell Maritime Press, 1961), p. 26.

21. Ibid., p. 30.

22. Ibid., p. 34.

23. *Baltimore Sun*, vol. 28, issue 79, August 28, 1848, p. 3.

24. Katzenberg, *Baltimore Album Quilts*, p. 116.

25. William S. Forrest, *Historical and Descriptive Sketches of Norfolk and Vicinity* (Philadelphia: Lindsay and Blakiston, 1853), p. 155.

26. *Washington Sentinel*, vol. 3, issue 33, November 2, 1854, p. 2.

27. Forrest, *Historical and Descriptive Sketches of Norfolk* p. 39, originally cited in Harry W. Burton, *The History of Norfolk, Virginia: A Review of Important Events and Incidents Which Occurred from 1736–1877* (Norfolk, Va.: Norfolk Virginian Job Print, 1877), p. 19.

28. *Baltimore Sun*, vol. 37, issue 33, June 25, 1855, p. 1.

29. Haney v. Baltimore Steam Packet Company, 64 U.S. 287 (1859), on appeal from the Circuit Court of the United States for the District of Maryland.

30. See, e.g., *Wood's Baltimore City Directory, for 1858–59*.

31. *American Commercial and Daily Advertiser* (Baltimore, Md.), February 11, 1853, p. 3.

32. Ibid., August 29, 1853, p. 4.

33. *Baltimore Sun*, vol. 42, issue 31, December 23, 1857, p. 2

34. Ibid., vol. 65, issue 123, October 7, 1859, p. 1.

35. *Wood's Baltimore City Directory, for 1858–59*, p. 179.

36. Ibid., p. 154.

37. *New York Tribune*, vol. 17, issue 50, June 26, 1857, p. 4.

38. *Evening Post* (New York), vol. 56, June 27, 1857, p. 2.

39. *Baltimore Sun*, June 27, 1857, vol. 41, issue 36, p. 1.

Book Reviews

Brian D. Gallagher. *British Ceramics, 1675–1825: The Mint Museum*, with contributions by Barbara Stone Perry, Letitia Roberts, Diana Edwards, Pat Halfpenny, Maurice Hillis, and Margaret Ferris Zimmerman. Charlotte, N.C.: Mint Museum, in association with D Giles, Ltd, London, 2015. 272 pp.; 230 color illus., notes, bibliography. $79.95.

The Mint Museum of Art in Charlotte, North Carolina, is named for its building's origins as a U.S. mint. Built in 1835 to a design by architect William Strickland, the two-story Greek revival Charlotte Mint branch opened its doors two years later and began producing coins from North Carolina gold. Use as a mint continued after a fire and rebuilding in the 1840s, but with the advent of Civil War, production ceased forever. Slated for demolition in the 1930s, the mint building was acquired from the U.S. Department of the Treasury by a coalition of private citizens and moved a few miles south of downtown Charlotte. In 1936 it was dedicated as the Mint Museum of Art, the first art museum in North Carolina. In 1968 the museum added a wing to house the Delhom Gallery and Institute for Study and Research in Historical Pottery and Porcelain.

In *British Ceramics, 1675–1825: The Mint Museum*, Brian D. Gallagher, the museum's curator of decorative arts, along with a team of contributing scholars, presents two essays and more than two hundred of the finest examples from the museum's collection. The essays discuss the collectors, patrons, and curators who made a public collection possible, focusing on key figures in the creation of the Delhom Gallery: M. Mellanay Delhom (1908–2003) and Daisy Wade Bridges (1932–2015).

Barbara Stone Perry's introductory essay, "In the Beginning: Ceramics at the Mint," details the origins of the collection. Miss Delhom, a resident of Chicago, had, out of personal passion, amassed one of the largest collections of British and Continental ceramics in the country. Bridges worked for Wedgwood in New York City, where she developed a lifelong love of pottery and porcelain. In 1960 Bridges left the firm and moved to Charlotte with her husband, becoming a devoted supporter of the Mint. In 1962 the two women met at the Wedgwood International Seminar in Chicago and quickly developed a relationship based on their common interest. Bridges encouraged Delhom to offer her collection to the museum in Charlotte, and in 1963 Delhom gave the Mint twenty-five examples of English ceramics. In 1965, with the counseling of Bridges, she agreed to sell the balance of the collection—2,080 pieces—to the museum over a fifteen-year period. Delhom subsequently accompanied the collection to the Mint as curator of ceramics. This large acquisition inspired a number of other benefactors to support further accessions, and several of them gave their ceramics to the museum. These gifts added Spanish and Italian maiolica, French faience, Dutch Delftware, Native American wares, and North Carolina pottery, among other wares.

The second essay, "Fit for a Queen," by Letitia Roberts, vividly paints Delhom's life story. The author easily draws in the reader, describing Delhom giving a talk at a ceramics forum in Pennsylvania on a most unpleasant September day in 1973. The tent was airless, and the hot, damp weather invited countless mosquitoes. Apparently Delhom was unfazed and easily delivered her presentation. Roberts, who suffered, observed, "This was a lady to be reckoned with—a lady whose very persona inspired questions that demanded answers." Born in 1908 in Fort Worth, Texas, by the age of twelve Delhom embraced theosophy and began studying oriental history, religion, philosophy, and design. She attended Texas Christian University in Fort Worth and at nineteen moved to Denver to study dance and acrobatics. At thirty she settled in Chicago, and at thirty-two made her first ceramic purchase, a porcelain Song dynasty piece. Thereafter, she joined ceramic organizations, including the Wedgwood International Seminar, American Ceramic Circle, English Ceramic Circle, and Oriental Ceramic Society. In the meantime, Chicago's Shangri-La Restaurant employed her as a bookkeeper and hostess, positions that helped subsidize her growing collection. Delhom collected both British and European ceramics but concentrated on fine eighteenth-century English pottery and porcelain, with creamware a favorite.

Gallagher and contributing scholars Diana Edwards, Pat Halfpenny, Maurice Hillis, and Margaret Ferris Zimmerman provide well-written entries on selected examples of the Mint's pottery and porcelain. The catalog is divided into sections on earthenware, stoneware, and porcelain, each with text explaining clay production in England. In part 3 several porcelain manufactories are discussed. Within each part, entries are arranged chronologically and illustrated in color. The catalog features exquisite objects from the major centers of production: Wedgwood, Chelsea, Worcester, and Staffordshire. A sample to whet readers' appetites includes a William Littler earthenware sweetmeat dish, circa 1765–1770, a 1723 stoneware loving cup, and a pair of Derby soft-paste porcelain vases with landscapes, circa 1760. The essays and catalog are each followed by detailed endnotes, and the bibliography is extensive.

The text is more descriptive than analytical. The engaging essays by Perry and Roberts might plausibly have been combined into one, as their narratives occasionally overlap. However, the tale-telling is fascinating, and it would be a shame to miss Roberts's firsthand recollections of Delhom.

In 1972 Delhom formed the Delhom Service League, and today it remains an active affiliated organization of the Mint, bringing together its members to study ceramics, from ancient to contemporary. The league generously supported the funding of this catalog, which accompanies the ongoing exhibition, "Portals to the Past: British Ceramics, 1675–1825," at the Mint Museum Randolph. Together, they handsomely celebrate the fiftieth anniversary of the museum's acquisition of the Delhom Collection of British and European ceramics.

Remi S. Dyll
Bayou Bend Collection

Stella Beddoe. *A Potted History: Henry Willett's Ceramic Chronicle of Britain*. Woodbridge, Suffolk, Eng.: ACC [Antique Collectors' Club] Art Books, 2015. 352 pp.; 812 color, 6 b/w illus. $89.50.

Henry Willett was a born collector. He assembled significant and extensive collections of natural history, Sussex folk life, old master paintings, and what Arthur Lee, a visiting American collector and author, called "the most interesting ceramic collection in England" (p. 11). Reading *A Potted History: Henry Willett's Ceramic Chronicle of Britain*, we learn about the man and his ceramic collection, and it is obvious the two are inextricably woven together. An enterprising and successful businessman with strong liberal sympathies, Willett practiced religious tolerance, advocated the abolition of the death penalty, opposed enclosure of common land, and campaigned for education for the poor. While this kind of background sometimes does little to help us understand a collector's fascination with ceramics, in Willett's case his interests informed his choices. He believed that his collection had two purposes: first, to document the history of his country on "homely pottery"; and second, to reflect on what might be "found on the mantelpieces of English cottage homes," representations of what their inmates or forefathers "admired, reverenced, trusted in."[1] In short, it is material culture depicting the material culture of everyday life.

Wanting to share his collections with his local community, Willett played a leading role in establishing Brighton Museum. The plan was conceived in 1872, and by September 1873 Brighton Free Library, Museum and Picture Gallery opened with exhibits including Willett's collections. The loan of his ceramic collection seems especially altruistic, as Willett had just lost more than sixty pieces of pottery and porcelain lent to an exhibition at the newly completed Alexandra Palace, London, which was destroyed by fire in June 1873, only sixteen days after it opened. In 1879 a text-only *Introductory Catalogue* was published to accompany the Brighton ceramic display, but as it grew in size and importance, a much larger catalog illustrated with line drawings was produced in 1899 in time to accompany a London showing of the collection.[2]

Written by Willett, the 1899 catalog featured more than 1,700 ceramics of every type spanning three centuries of use in English homes. Organizing such a large and wide-ranging collection would be daunting had it not had such a dominating focus. Dividing the catalog into manageable topics, he wrote, "The classification, while confessedly arbitrary, has been made not so much in reference to the maker, the time and place of manufacture, but with regard to the greater *human interest* which each object presents."[3] The 2015 publication authored by Stella Beddoe also covers over 1,700 objects and honors the spirit of Willett's original organization. Some of the categories have been renamed, others have been merged or separated, but essentially the nineteen sections of the new book credibly represent Willett's twenty-three topics. As in the original, the first subject is "Royalty and Loyalty" and the last is "Domestic Incidents," but in between Beddoe has sensibly combined some topics, creating such practical headings as "Scripture History & Religion" and "Combat Sports & Field Sports."

The book is not fully a catalog nor is it a history textbook, but, appealing to the widest possible market, it combines a contextual discussion with extensively captioned illustrations. Inevitably there are compromises, perhaps the most irksome of these being the lack of information on marks. Admittedly Willett chose not to focus on this aspect of his objects, but as scholarship has evolved it has become standard to state whether attribution to a factory or artist is based on something tangible. That being said, the extended captions deliver all other necessary information, including estimated date, pottery type, dimensions, and description. Accompanying text is designed to expand our understanding of the historical or social importance of an object or group of objects, and here the author has admirably achieved her aim. But this is not just a collection of British history souvenirs, it is also a major ceramic collection, and anyone interested in pottery and porcelain can find a wealth of fascinating pieces, including rare and perhaps unique examples. The ceramics span three centuries—the earliest dating from the mid-seventeenth century, the latest from 1899—and while the collection comprises mainly English wares, pottery and porcelain from Europe and Asia are also present. Every kind of ceramic used in English homes is represented, from the relatively few pieces of slipware to large numbers of delftware, salt-glazed stoneware, creamware, and pearlware. Porcelain examples include soft and hard paste, and bone china and parian also occur. Wares typical of the Victorian period are represented by white earthenware, stone china, and smear-glazed stoneware, large numbers of which are illustrated and discussed. Whether you like printed or painted decoration, figures or vases, tableware or ornamental pieces, you can find examples in the Willett collection.

Two pieces caught my eye in the "England & America" section. The first is a circa 1781 William Greatbatch creamware teapot printed and enameled with a scene of British troops firing on retreating American militia; the other is a 1791 creamware plate with a black overglaze printed profile portrait of William Penn. In almost fifty years of looking at pots, I have never seen these prints before, and although it is likely they were made in some quantity, these are rare survivors.

Having a soft spot for figures, I found plenty to enjoy, especially in the wide range of examples depicting courtship in the "Domestic Incidents" chapter. I particularly liked the romance of embracing lovers in Derby porcelain, courting couples in salt-glaze stoneware and creamware, and the Victorian portraits of romantic bliss; the whole section is a brief encounter that is designed to amuse and inform.

Every time you open the book there is a new surprise: sometimes there are unexpected relationships between objects, occasionally we see reminders of historical events long since forgotten, and there is something to delight everyone. Stella Beddoe knows and loves this collection better than anyone since Willett. Her dedication to the collection shows in the strength of this volume, which should be on the shelf of anyone professing to love ceramics.

Pat Halfpenny

1. Victoria and Albert Museum, *Catalogue of a Collection of Pottery and Porcelain Illustrating Popular British History* (London: Wyman and Sons for H. M. Stationery Office, 1899), n.p., quotations from introduction.

2. Ibid.

3. Ibid.

John L. Scherer, photography by John Yost. *Art for the People: Decorated Stoneware from the Weitsman Collection*. Albany, N.Y.: New York State Museum, 2015. 296 pp.; 340 color illus., bibliography, index. $75.00.

In 1971, as an eleven-year-old growing up in western New York State, my hobbies included reading Nancy Drew mysteries, practicing my oboe, and playing kick the can. In 1980, when Adam J. Weitsman of Oswego, New York, was eleven, a chance discovery of a couple of stoneware beer bottles led him to embark on a mission to collect decorated stoneware. He began to amass an important and valuable collection while still in his teens and eventually donated his collection, valued as high as $10 million, to the New York State Museum in Albany. Adam Weitsman and I may have grown up only nine years and 110 miles apart, but it's safe to say that his adolescence was a bit more industrious than mine.

Weitsman's stoneware collection is the subject of a beautifully produced new catalog, *Art for the People: Decorated Stoneware from the Weitsman Collection*, written by John L. Scherer. As curator of decorative arts at the New York State Museum from 1967 to 2009 and now historian emeritus, Scherer is particularly well qualified to write this book. He helped acquire the collection for the museum, beginning in 1996 with Weitsman's first donation of 100 pieces of stoneware. In an introductory statement, the museum's director, Mark A. Schaming, credits Scherer with having spent the past two decades carefully studying each of the 221 pieces featured in the catalog.[1] Scherer began work on the book in 2006 but delayed finishing because Weitsman, who paid more than $125,000 to produce the catalog, continued donating new pieces to the museum and wanted them included in the publication.[2] It was worth the wait—if nothing else for the fabulous West Troy Pottery 4-gallon jar decorated with a marching elephant that Weitsman obtained for the museum in July 2015, just as the book was being laid out at the publishers.[3]

Art for the People is organized by location of manufacture, with a chapter devoted to each of six areas of stoneware production throughout New York State: Manhattan, Mid-Hudson Valley, Upper Hudson Valley, Central New York, South Central New York, and Western New York. Chapter 7 looks at Vermont potteries in Bennington and St. Johnsbury; the eighth chapter covers selected stoneware manufacturers in Connecticut, Massachusetts, New Jersey, Pennsylvania, and Ontario, Canada. A final chapter examines pieces in the collection whose makers are unknown. Forty-six potteries are listed in the table of contents, and each is the subject of a short historical essay followed by color photographs, descriptions, and interpretations of their products.

Schaming notes that the book's regional organization helps bring the Weitsman collection into focus. In fact, the stonewares found throughout this book share many attributes and would seem to transcend regionalism in a number of ways: they were formed, glazed, and fired in similar fashion; they were mostly meant to be utilitarian containers; they feature incised and cobalt decoration; farm and domestic animals were popular themes; and the development of Spencerian script in the United States about 1850 contributed to more stylized freehand decoration on wares decorated after that time.

But regionalism is important, Scherer argues, not only because makers and decorators often worked at multiple local potteries, but also because some of the finest pieces in the Weitsman collection serve as material evidence of the communities in which they were made. A key strength of *Art for the People* is Scherer's attempt to tease out the ways in which the collection shines light on nineteenth-century life in America. According to the author, stoneware potters and decorators frequently depicted people, places, and events in their towns. Some designs were commissioned, such as the circa 1865 water cooler made by Leander Fenton and Frederick Hancock of St. Johnsbury, Vermont, for the Civil War general Asa Peabody Blunt and his wife, Mary (pp. 248–49), and the 1828 pitcher decorated with the "Charter Oak" of Hartford, Connecticut, made for Jonathan Montague and attributed to local potter Daniel Goodale (pp. 256–59).

Even on production pieces, however, Scherer argues that potters and decorators observed the world around them and drew inspiration from local people, animals, churches, race tracks, forts, boats, and even a cemetery. He employs contexualizing graphics—an advertisement, a cartoon, a drawing of a building, or a lithograph of a fireman—in his exploration of the relationship of the ceramic artist to everyday life in the community. An 1885 Fulper Brothers jug featuring a freehand depiction of a female acrobat was likely inspired by performances at the Flemington, New Jersey, agricultural fair (pp. 266–67), while circuses traveling through town may have played a role in the exotic animals decorating the wares of a number of makers in the collection.

Any shortcomings of *Art for the People* are far outweighed by its strengths. The book, and presumably the collection, do not broadly document decorated stoneware in the United States but, rather, salt-glazed stoneware of the mid-Atlantic and New England; I was surprised that the existence of southern decorated-stoneware traditions was not even acknowledged. In his desire to discover evidence of artistic inspiration, Scherer often has to resort to words that hedge his statements, such as "perhaps" and "probably." While his admitted reliance on the work of William C. Ketchum and other ceramic historians means there is likely little new information about the potters and potteries, the focus on the social context of decoration and the pots themselves make *Art for the People* a valuable addition to any museum's or stoneware collector's library. It is the most exciting survey of American stoneware that I have seen in quite some time. I am the curator responsible for the ceramics and glass collection at the Smithsonian's National Museum of American History, and the experience of poring over this book made me eager to reexamine our stoneware collection to compare decorations and map out new research projects. It also made me very envious of the New York State Museum's fabulous collection.

Bonnie Campbell Lilienfeld
National Museum of American History, Smithsonian Institution

1. Somewhat inexplicably, the ca. 1798 Thomas Commeraw half-gallon jar featured on p. 39 is noted as being in the collection of the Metropolitan Museum of Art in New York City, rather than the New York State Museum, although its catalog number is similar to the rest of the pieces shown in the publication.

2. "Art for the People: Decorated Stoneware from the Weitsman Collection | The New York State Museum," www.nysm.nysed.gov/exhibitions/ongoing/art-people-decorated-stoneware-weitsman-collection (accessed May 31, 2016); and Paul Grondahl, "Scrap Metal Mogul Donates World-Class Art Collection to State Museum," *Times Union*, www.timesunion.com/local/article/Scrap-metal-mogul-donates-world-class-art-5553066.php (accessed May 30, 2016). The money raised by the sale of the book will go to the New York State Museum's acquisition fund, according to "Businessman Adam Weitsman's Stoneware Collection Featured in New York State Museum Book | Lifestyles | Auburnpub.com," http://auburnpub.com/skaneateles/lifestyles/businessman-adam-weitsman-s-stoneware-collection-featured-in-new-york/article_0187f7e2-e043-5a2c-b69e-fbf6df1b91b0.html (accessed May 30, 2016).

3. "Art for the People: Decorated Stoneware from Weitsman Collection," www.antiquesandthearts.com/art-for-the-peopledecorated-stoneware-from-weitsman-collection/ (accessed May 30, 2016).

Susan J. Montgomery. *"The Endless Possibilities": American Arts and Crafts Tiles from the Two Red Roses Foundation*. Palm Harbor, Fla.: Two Red Roses Foundation, 2016. 360 pp.; 329 color, 350 b/w illus., introductory essay, appendix, notes, selected bibliography, index. $75.00.

With her seminal writings about William H. Grueby and the companies he organized, Susan Montgomery has been a major influence in advancing knowledge of our national tile industry, the artistic movements that influenced its development, and the tiles themselves.[1] Her most recent work for the Two Red Roses Foundation (TRRF) expands and continues this advancement with meticulous and insightful scholarship.

"The Endless Possibilities" is concerned with Arts and Crafts tile production from the 1890s to World War II. It is organized by the tile's intended use into seven overlapping and interrelated chapters: "Individual Tiles"; "Modified Tiles" (such as advertising and souvenir tiles); "Panels and Plaques" (discrete compositions); "Borders and Friezes"; "Fireplace Facings and Overmantels"; "Installations" (which is partially a compressed rendition of Montgomery's *The Aloha Boathouse and the Iris Bathroom*); and "Faience and Fountains."[2] Montgomery explains that these divisions "provide a framework for considering the design, production, and marketing of tiles at a particular point in American culture, the transition between the highly mechanized manufacturers of the nineteenth century and post World War II

independent studio artists, when craftsmanship and individuality were paramount" (p. 8). An introductory essay, "Color in Architecture: Tiles and Faience in America," sets the historical scene for the catalog.

In her narrative, Montgomery focuses on the ways in which tile designers and makers balanced the ideals of the Arts and Crafts movement with the need to succeed commercially. This was, in essence, the same contradiction found in the British Arts and Crafts movement, on which the American movement was based. The British Arts and Crafts reformers, reacting against the harshness of nineteenth-century industrialization, wished "to foster spiritual harmony through the work process and to change that very process and its products."[3] Their goal was to end the divisions between art and industry and between art and craft and to unite these into one. In the United States, Arts and Crafts goals "were the revival of handcraftsmanship, the creation of more satisfying working conditions and the promotion of simple, uncluttered houses and interiors achieved through a unification of all art forms."[4] According to Montgomery, in the final analysis Arts and Crafts tile makers wanted "to create beautiful, useful objects that could make a contribution to the built environment" (p. 19).

Montgomery focuses on the manufacture of Arts and Crafts tiles as a collaborative effort between the designers/tile makers and the other artisans who worked for them, such as glazers and mold makers. Usually the designer—George P. Kendrick or Addison LeBoutillier—or the manufacturer—Grueby, Hartford Faience, Rookwood—was given the credit for a tile, but in reality many artisans were needed to produce a work of ceramic art, and a mistake by one of these could cause the piece to fail. Montgomery tries to give as much credit to these individuals—mainly glazers—as possible in the catalog entries, but this effort is limited by the anonymity of most and by the greater credit given to the designer/maker.

Montgomery writes a history of each design in the collection—how or why it was developed, how it was crafted, the clays and glazes used. In the first, and largest, chapter, "Individual Tiles," each type of design—for example, flora, boats/ships, animals, knights—is examined from maker to maker whenever possible. The book's structure and Montgomery's documentation also allow the reader to visualize a design's change from maker to maker, as well as its progression of use from, for example, individual tile to frieze to architectural piece.

An appendix, "Designers and Tile Makers," gives brief background summaries of each designer/tile maker represented, acknowledging that a "comprehensive history of each of these tile manufacturers would require a monograph of its own" (p. 304). The histories of the designs, their designers, and the manufacturers were fascinating, and I found myself constantly downloading and reading sources cited in the text. This volume only whets one's appetite for more comprehensive treatments of each of the makers represented.

Montgomery writes in the preface that the TRRF collection is not comprehensive; it is an annotated catalog reflecting founder Rudy Ciccarello's personal taste and choice (p. 8). This is both a strength and a weakness. Ciccarello has chosen the best items he could find for his collection and museum and has had the advice of exceptional Arts and Crafts experts. Sixteen tile makers/designers are represented—fourteen American and two British companies, Doulton and William Morris.[5] The American companies include some but not all the innovative designers and artists who contributed to the development of an American Arts and Crafts aesthetic.[6]

This shortcoming is not of Montgomery's making. The book is limited by the collection itself, which is top-heavy with tiles from one company, Grueby. Why only these tile makers? Is this a collecting choice, an availability choice, or both? Mr. Ciccarello began his collection by acquiring "rare individual tiles and sets" made by leading designers and manufacturers (p. 6), but a collection of American Arts and Crafts tiles should be as broad and representative of this movement as possible and include lesser-known companies and designers.

This being said, Montgomery's book is a monumental work with lavish photos and is well researched with a wealth of supporting documentation. It is an indispensable reference work about Arts and Crafts tiles and the Arts and Crafts movement in the United States.

Michael Padwee
"Architectural Tiles, Glass and Ornamentation in New York" blog
(http://tilesinnewyork.blogspot.com)

1. Susan J. Montgomery, *The Ceramics of William H. Grueby* (Lambertville, N.J.: Arts and Crafts Quarterly Press, 1993); *Grueby Pottery: A New England Arts and Crafts Venture; The William Curry Collection*, exh. cat. (Hanover, N.H.: Hood Museum of Art, Dartmouth College, 1994); Susan J. Montgomery, *The Aloha Boathouse and the Iris Bathroom: Two Installations by the Grueby Faience and Tile Company* (Tampa, Fla.: Two Red Roses Foundation, 2013), among others.

2. For installations, see Montgomery, *The Aloha Boathouse and the Iris Bathroom*.

3. Elizabeth Cumming and Wendy Kaplan, *The Arts and Crafts Movement* (New York: Thames and Hudson, 1991), p. 8.

4. Ibid., p. 143.

5. American tile makers represented are Art Academy/People's University, Batchelder, CALCO (California Clay Products), California Faience, Grueby Tile and Faience Company, Hartford Faience Company, J. B. Owens Floor and Wall Tiles, Marblehead Potteries, Mosaic Tile Company, Newcomb Pottery, C. Pardee Tile Works, Paul Revere Pottery/Saturday Evening Girls, Rookwood Pottery, and Van Briggle Pottery.

6. American tile makers who made significant contributions to the American Arts and Crafts movement but are not represented include, to name a few, the Mueller Mosaic Tile Company (even though Herman Mueller is represented through the Mosaic Tile Company), Enfield Pottery, Flint Faience and Tile Company, Franklin Tile Company, and American Encaustic Tiling Company.

Edmund de Waal. *The White Road: Journey into an Obsession*. New York: Farrar, Straus and Giroux, 2015. 401 pp.; 45 b/w illus. $27.00.

Sometimes the little pieces in the center of life's large white jigsaw puzzle fit together effortlessly, and for a moment life feels like its own great reward. Such was the case with Edmund de Waal's obsessive journey on "the white road" to porcelain's origins and my fascinated observation from the edge—the easy part of every puzzle. My part began on November 17, 2015, when I attended the author's talk at the Colony Club in New York City. As de Waal spirited us through his porcelain adventures, made all the more gripping by his anecdotal style and often slightly unfocused but revelatory slides, initial attention turned to captivation, and by the end of the talk, all of us were leaning forward, on the edge of our chairs. I, an inveterate lecture attendee, declared it the best lecture I had ever heard on any subject and floated home, clutching my autographed copy of *The White Road*.

In spite of mixed reviews from friends, many of whom had praised de Waal's first book, the best seller *The Hare with Amber Eyes* (2010), which I had not read, I found this second literary effort both engaging and enlightening. It is written in a highly personal style: his use of words, while economical, is expressive and pictorial; his descriptions are remarkably visual; and he traverses the line between autobiography and biography (the lives of the participants in the porcelain story) with an interesting fluidity.

The author begins his narrative with a quotation from Herman Melville's *Moby-Dick*: "What is this thing of whiteness?" Edmund de Waal is a man who works in white. The clay of porcelain is white, both before and after it is fired, but he needs to know more about this whiteness than just its chemistry and technicalities. He wants to comprehend the entire history of true hard-paste porcelain: its discovery and creation, from the mining and mixing of the clays (chiefly kaolin for the "bones" and plasticity, and petuntse for the "flesh," the hardness and translucency) to the shaping of the vessels and the firing of the kilns. And he needs to know how, often with little more than vague clues largely gleaned from the early-eighteenth-century letters of Père François Xavier d'Entrecolles (1664–1741), a French Jesuit missionary in China, the knowledge of porcelain making traveled from Jingdezhen to potters in and near Paris, and to Plymouth, England, while in Dresden the secret of the porcelain recipe, the "Arcanum," was being discovered through scientific experimentation and luck.

The cities of Jingdezhen, Dresden, and Plymouth (with detours to other locations that provide various significant links along the way) are the destinations on de Waal's journey to trace the origins of porcelain: his "white road" to the "three places where porcelain was invented, or reinvented, [his] three white hills in China and Germany and England" (p. 3). He writes, "I need to get to these places, need to see how porcelain looks under different skies, how white changes with the weather. Other things in the world are white, but for me, porcelain comes first. This journey is a paying of dues to those that have gone before" (p. 3). "White is also my story. From my very first pot. I was five" (p. 10). This is a deeply personal journey.

The book begins: "I'm in China." In the opening paragraph the author establishes the history, mystery, and eternal fire of Jingdezhen, "the city 'like one furnace with many vent holes of flame' . . . It is the city of secrets, a millennium of skills, fifty generations of digging and cleaning and mixing white earth, making and knowing porcelain, full of workshops, potters, glazers, decorators, merchants, hustlers and spies" (p. 1). In China de Waal has found his first white hill, Mount Kao-ling (High Ridge).

He proceeds to Dresden via "a dogleg through Versailles and the court of

Louis XIV. I'm tracking the Jesuits and this is where they are, this is where ideas and images of China come into focus. And because this is where porcelain is talked of, I need to listen" (p. 113). Here de Waal finds Ehrenfried Walther von Tschirnhaus (1651–1708), the twenty-four-year-old German mathematician, philosopher, and physicist hired by French minister of finance Jean-Baptiste Colbert (1619–1683) to be his son's tutor. The author follows Tschirnhaus to Dresden and the Goldhaus (a community of scientists), where he experiments with porcelain making. In Dresden they both meet the porcelain-obsessed elector of Saxony, Augustus II "the Strong" (1670–1733), and the deceitful braggart Johann Friedrich Böttger (1682–1719), the apothecary's assistant, who falsely claims the ability to make gold. In 1708 he and Tschirnhaus discover the "Arcanum" and fire the first piece of porcelain in the Albrechtsburg Castle in Meissen. When Tschirnhaus dies soon after and Böttger snatches the credit, de Waal feels the loss of a friend for whom white is light, and light is porcelain. Having found his second white hill in Dresden, he notes, "So I finish Saxony and write *Good Work*" (p. 209).

On to Plymouth and the birth of English hard-paste porcelain through the acute observations, experiments, and hard work of a Quaker apothecary, William Cookworthy (1705–1780). William is the only member of the ceramic troupe whom de Waal consistently refers to familiarly, by his first name, as if the two are old friends, sharing qualities of inquiry and mindfulness. The reader senses a real empathy in the agonizing story of Cookworthy's eventual discovery of the right china clays in Tregonning Hill in Cornwall (recognized through descriptions of the material in the published letters of Père d'Entrecolles) and his trials with the creation and production of a new type of porcelain in Plymouth. The royal warrant for Cookworthy's patent was published in 1768, the year of William's first successful firing. de Waal has reached his third white hill.

The manufactory in Plymouth was not a commercial triumph, and it was transferred within two years to Bristol, where it was managed with only slightly less misfortune by another Quaker, Richard Champion (1743–1791), a man constantly challenged by the indefatigable and powerful Josiah Wedgwood (1730–1795). The struggle between Champion and Wedgwood involved the 1775 extension of Cookworthy's patent for hard-paste porcelain (which he had assigned to Champion for a fee and a percentage), but in the background lurked *unaker*, the mysterious white "Cherokee clay" from Ayoree Mountain in North Carolina, which by the 1740s had been introduced to the Bow factory producing soft-paste porcelain in London, and by the mid-1760s was also known to both Champion and Wedgwood. Economically, however, it was the accessible Cornish clays, not *unaker*, which attracted Wedgwood, and he made a deal to dig them for pottery, not porcelain. The Bristol factory was doomed, and in 1784 Champion and his family escaped the humiliation of bankruptcy and emigrated to South Carolina.

De Waal and his young son take a road trip to America to find Champion's grave (ironically on a seam of kaolin) but also to search for the *unaker* that the Cherokees used "in complex ways": for house insulation and decoration ("imagine the luminescence of a white space, the faintest glimmer of mica. A porcelain room" [p. 317]), for pots, and "for pipes, not just because it is a fine clay that burns cleanly, but because white is a central ritual colour. It symbolises peace" (p. 317). This becomes his fourth white hill.

While preparing his installation at the Victoria and Albert Museum in London, which he calls "Signs & Wonders," de Waal decides he needs to delve into the ceramics of revolutionary Russia and the Bauhaus in Weimar, Germany. "It is the last part of my journey, revolution, *Porcelain 1919*" (p. 332). This is a world of sharp edges, of assemblage over skilled craft, of repetition over imagination, and by the 1930s it is a world of metal, concrete, glass, and imminently of clean, white porcelain. de Waal Googles the name of a glass artist, Wilhelm Wagenfeld (1900–1990), to determine whether he ever made porcelain, and up pops the name Allach—an unrecognized factory in Dachau, near Munich. This blank whiteness draws the author to Dachau. The factory was founded in the town of Allach in 1934, taken over by Heinrich Himmler (1900–1945), and moved in 1940 to larger quarters within the Dachau concentration camp, where it produced primarily white figures of extraordinary virtuosity until its protective shut-down before the liberation of Dachau by U.S. troops in 1945. Among the forty-five monochromatic views of key people, historical events and documents, ceramic factory interiors, places, and porcelain objects—well chosen and appropriately placed, although their illustrative value is limited by their generally small reproduction size on nonglossy paper—is a photo of Hitler and Himmler admiring the Allach figures that the latter gave to the former for his birthday in 1944. Allach is the final stop on de Waal's white road, but understandably, it is not a fifth white hill. It is a tragic reminder of the fragility of life, and of porcelain. It is a moment of darkness in this journey of white obsession.

In the author's final reckoning, he ruminates on the dues paid by porcelain makers who preceded him: "This white porcelain . . . costs. Porcelain consumes . . . the wood on the hills, it silts the rivers and clogs the harbors, enters the deltas of your lungs. . . . I remember my years in the workshop, sweeping. And if it costs me, that is one thing. But it is the cost to others. . . . This, I think, is what I've been trying to trace, the glimpse of white rising and then sinking below the waves again. The wind catching and eddying white dust, settling and resettling" (p. 387).

Sitting at his wheel he makes 2,455 small porcelain vessels "glazed in whites. I use all the accomplished, attempted, consolatory, melancholy, minatory, lambent whites from my journey" (p. 389). These pots are for his 2010 exhibition in New York, where he is asked at its opening, "*How is it possible to make white things?*" (emphasis in original), to which he replies that "making porcelain is a way of starting again, finding your way, a route and a detour to yourself" (p. 390).

In finding himself, de Waal has traversed an enormous amount of history, much of it told anecdotally, woven with his own experiences as a potter, with personal comments, and with philosophies, and by the alternation of past and present he avoids didacticism. Through his historic acquaintances along the journey, the story becomes a multiple biography as much as an autobiography, and as such it is perhaps best enjoyed by those with some knowledge of, or at least interest in, porcelain and its protagonists.

The White Road isn't just an obsession. It isn't just a journey to the white hills of kaolin and to the origins of true porcelain. It isn't just an answer to Melville's question, "What is this thing of whiteness?" It is a journey of self-discovery and of the ultimate revelation that the white road is the road home.

Letitia Roberts

Amy Kurlander, with essays by Joey Brackner and Michael K. Brown. *Texas Clay: Nineteenth-Century Stoneware Pottery from the Bayou Bend Collection*. Houston: Museum of Fine Arts, Houston, 2015. 99 pages; 262 color, 11 b/w illus. $26.95 (softcover).

"You may go to hell and I will go to Texas."[1] These famous words of iconic American pioneer Davy Crockett ring softer but no less true in a groundbreaking study of the state's nineteenth-century stoneware manufactories. *Texas Clay: Nineteenth-Century Stoneware Pottery from the Bayou Bend Collection* is the catalog for an exhibit of the same name at the Bayou Bend Collection and Gardens of the Museum of Fine Arts, Houston, September 5–November 1, 2015. The focus is on pieces in the Bayou Bend Collection, at least 159 of which were donated by Houston collector William J. Hill in 2012. Two Bayou Bend curators—Amy Kurlander, exhibit curator and project fellow for the William J. Hill Texas Artisans and Artists Archive at the museum, and former curator the late Michael Brown—and Joey Brackner, director of the Alabama Center for Material Culture, provide thoroughly researched essays on a subject that has been understudied and often overlooked.

Amy Kurlander's "An Introduction to Stoneware Pottery in Nineteenth-Century Texas" offers an excellent background on the art of stoneware production and discusses its emergence in the state of Texas in the 1840s. Ceramics enthusiasts will be impressed with the depth of Kurlander's research and presentation, down to a map of Texas potteries and their natural clay sources within the state (p. 16). Kurlander draws connections between the Edgefield, South Carolina, pottery style and Texas stoneware as she highlights two of Bayou Bend's most outstanding American ceramic jars. Both, made circa 1850, display a distinctive ovoid form with four applied lug handles at the shoulder, a trait typically associated with the work of Edgefield potters. One (fig. 26), produced by Thomas Chandler in Edgefield, South Carolina, is covered in a greenish alkaline glaze with iron slip decoration and includes the seldom-seen incised signature "Chandler Maker" under one handle.[2] The other (fig. 24), covered in a reddish-brown alkaline glaze, bears the impressed signature of Marion County, Texas, entrepreneur Jefferson S. Nash, who, Kurlander notes, was born in Georgia and had familial ties to Edgefield potters (p. 17). This relationship between Edgefield and other pottery complexes farther west is a theme further emphasized in essays by Brackner and Brown.

Kurlander also discusses an associate of Nash, the somewhat mysterious African American potter Milligan Frazier (ca. 1848–after 1910). The Louisiana-trained Frazier produced stoneware near the Nash site during a later period and is well known among collectors of Texas stoneware for vessels displaying "thick, separated runs of dark brown glaze" (p. 19).[3]

Kurlander highlights a lesser-known ceramic folk artist, John L. Stone, who made a temperance jug in Oletha circa 1870–1872 (p. 22). Stone produced this masterwork in the style of the infamous Kirkpatrick brothers of Anna, Illinois, by whom he was previously employed. Kurlander includes a striking full-page image of his piece, which she describes as "a ground for encircling rattlesnakes, creeping spiders, crawling lizards, and a doomed, sinking man" (pp. 22–23). The jug (fig. 31), with its various applied animal and human figures, is a significant acquisition that stands out among the many utilitarian pieces in the collection. In 2015 it was purchased and graciously donated by Bayou Bend docent Leslie Bucher and Museum of Fine Arts, Houston trustee Brad Bucher, who funded the exhibition and catalog.

Brackner, in his essay "The Shaping of Texas Pottery: Shards along the 'Thousand-Mile' Journey from the Carolinas to Texas," takes an ambitious approach to the origins of Texas stoneware production, tracking the migration of eastern potting families to Randolph and Chambers counties in Alabama, and onward to Texas. The author, citing late stoneware authority Georgeanna Greer, aptly refers to Alabama's booming pottery complex in these counties as a "Crossroads in Clay" (p. 33), where potters of different origins commingled and shared ideas. Much of Brackner's essay discusses the "Rock Mills" school of potting, named for a pottery-dense region in Randolph County where a "mix of Edgefield-trained potters alongside those from North Carolina and Georgia would create a style of pottery that would in turn influence other communities" (p. 34).[4] Mississippi, Louisiana, and, ultimately, Texas are also discussed. Rife with original census research, photographs of the potters themselves, and images of shards from various kiln sites, Brackner's essay is truly authoritative.

Brown's "The Wilson Potters: An African-American Enterprise in Nineteenth-Century Texas" takes an in-depth look at the Guadalupe Pottery of John M. Wilson Jr. (1808–1881) and the Capote stoneware establishment managed by his ex-slave Hyram Wilson (1836–1884). Brown's analysis of the history of the Wilson potteries is extensive and includes images of the Wilson potters (pp. 49, 59), twentieth-century photographs of the Capote site (pp. 60–61), and a two-page photograph of Hyram Wilson products, all of which bear the stamp "H. WILSON & CO." (pp. 62–63). His commentary on this master-slave relationship is compelling. He is quick to note, just as Kurlander and Brackner have, that Texas stoneware craft was ever changing, often driven by its potters' training or markets in the regions where they formerly worked. He expounds on Kurlander's discussion of Jefferson S. Nash and the Edgefield influence revealed in pieces bearing his mark. He credits Ohio-trained potter Isaac Suttles with bringing salt-glaze techniques to the shop of John Wilson (p. 59). Other innovations were possibly homegrown, as he thoughtfully points out "significant differences" between ware produced by the Wilson potters at the Guadalupe Pottery and later pieces made by the same men under H. Wilson and Company (pp. 64–65).

Pictures are worth a thousand words, especially in the realm of decorative arts, where the nuances of a craft are difficult to convey through the written word alone. One of the many strengths of this publication is its abundant use of high-quality color photographs (most by Thomas R. DuBrock) of the stonewares. Several full-page images, most featuring the museum's best pieces, allow the reader to see the subtleties of the objects' alkaline- and salt-glazed surfaces. Additionally, photos of each pot depict the sheen of the glaze without reflective hot spots that mask color and form. Particularly appealing to students of Texas stoneware is the photographic survey, by county, of the state's stoneware products. This grouping of more than 180 color photographs of pieces in the Bayou Bend Collection clearly shows the stylistic differences and similarities between the work of virtually every nineteenth-century Texas stoneware manufacturer. The format of this section, which displays six to eight images per page, allows for easy comparison between regions and potters.

I have often wondered why Texas stoneware has received little fanfare over the years. In my work at an auction house, I have handled a number of Texas-made pieces exhibiting a high level of sophistication in their craftsmanship. A few, I have felt, were comparable in quality to the work of the great nineteenth-century stoneware manufactories of the eastern and midwestern United States. Why, then, has interest been so limited or, at the very least, so regional? The answer is unclear. However, I hope that the rapidly growing enthusiasm for southern decorative arts and the inclusion of areas farther west, coupled with publications and exhibits of the quality of *Texas Clay*, will increase awareness of the state's stoneware craft. From their studies of the work of figures like Nash, Stone, Frazier, and the Wilsons, the authors insist that Texas stoneware is not just a subject of American history or material culture. It is a matter of art.

Mark Zipp
Crocker Farm

1. This quotation is a popular paraphrase from Crockett's actual quotation, printed in *Niles' Register*, April 9, 1836, p. 99, available online at http://martinparmer.tripod.com/id11.html (accessed July 18, 2016).

2. Numerous examples of stoneware bearing impressed maker's marks by Thomas Chandler are known. The fact that the signature on this jar was carved into the clay by hand is highly unusual. The piece was donated to Bayou Bend by legendary collector and philanthropist Ima Hogg in 1974.

3. A discussion I recently had with southern stoneware authority Dr. Arthur Goldberg brings into question the attribution of a jar to Milligan Frazier, illustrated in *Texas Clay* (p. 94). The jar includes two incised slash marks at the shoulder, often referred to as "slave marks." According to Dr. Goldberg, these markings, coupled with the jar's form and glaze, indicate an Edgefield origin. He suggests that an elemental clay analysis be performed to create a body of work concretely attributed to Frazier.

4. For a more comprehensive study of nineteenth- and early-twentieth-century stoneware produced throughout the state of Alabama, see Joey Brackner, *Alabama Folk Pottery* (Tuscaloosa: University of Alabama Press, 2006).

Index

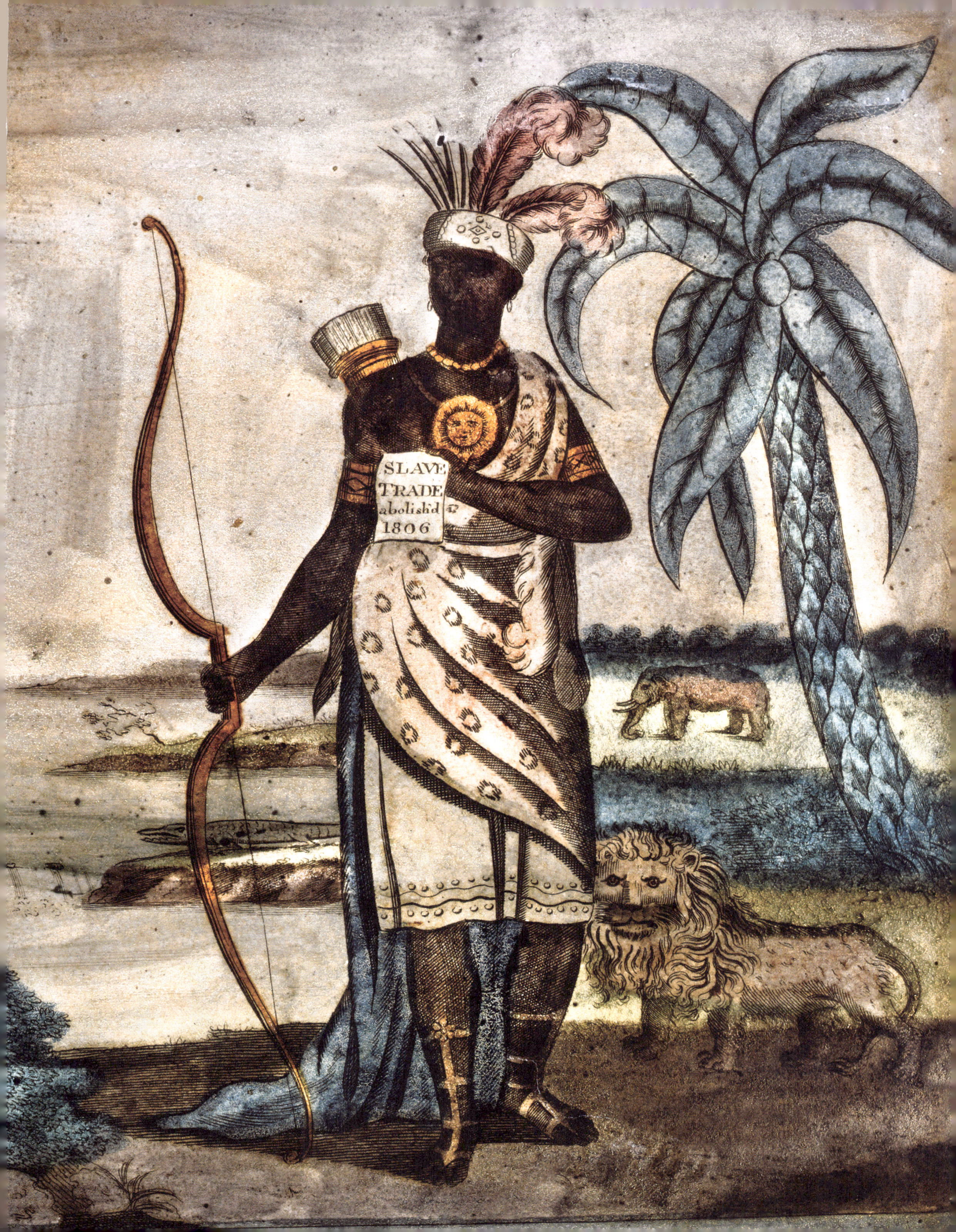
SLAVE
TRADE
abolish'd
1806

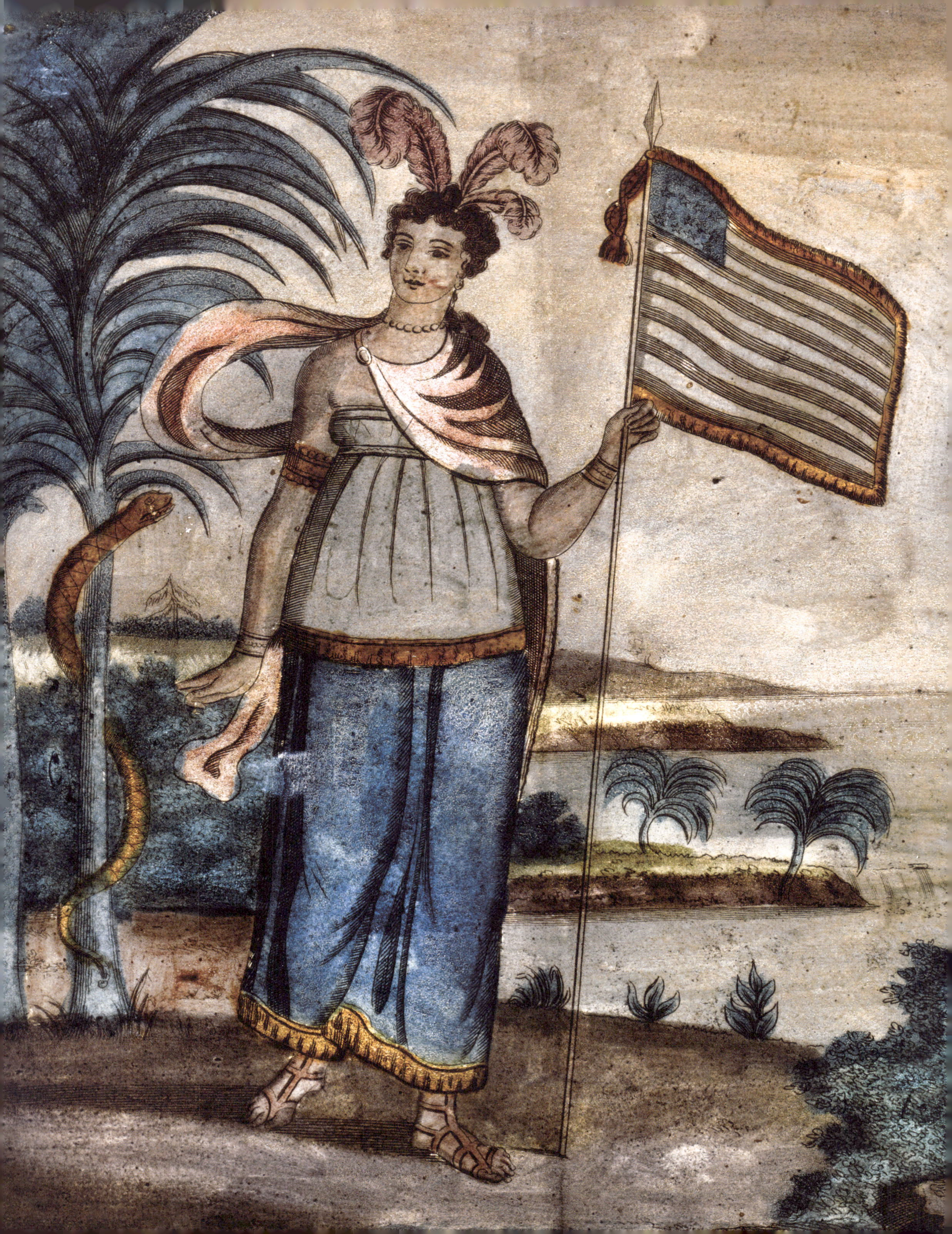